How Music Works

Rolf Bader

How Music Works

A Physical Culture Theory

 Springer

Rolf Bader
Institute of Systematic Musicology
University of Hamburg
Hamburg, Germany

ISBN 978-3-030-67157-0   ISBN 978-3-030-67155-6   (eBook)
https://doi.org/10.1007/978-3-030-67155-6

This Springer imprint is published by the registered company Springer Nature Switzerland AG
The registered company address is: Gewerbestrasse 11, 6330 Cham, Switzerland

Introduction

This book is about how music, and therefore, how art and culture work: as a self-organizing system. Such systems maintain a low level of entropy or a high level of order. All living systems, men, animals, or plants are self-organizing. They are maintaining their lives through self-organizing processes. So do music, art, and culture. The book tries to show this in the field of music on a physical basis, in musical acoustics, music psychology and brain research, and music ethnology, the music from all over the world.

The Physical Culture Theory therefore is:

1. **Culture is a system of physical human-made objects**, like musical instruments, **as well as human brains and bodies** interacting to extend life by **building self-organizing systems aiming for maintaining life** as a state far from random. The spatial and temporal fields of culture are then an interplay of order with intermediate chaotic states.
2. **Conscious content**, which is perceptual phenomena, qualia, notions, or feelings, **are electromagnetic spatiotemporal fields in the brain**. Cultural notions and understandings, sounds and timbre, vision, memories, self-consciousness, and all kinds of conscious content, like electromagnetic fields, are, therefore, only one element in culture, which consists of both, man-made objects and human—as well as animal—brains and bodies.
3. **Culture can be modeled purely in the physical domain by spatiotemporal fields of energy bursts, impulses**, on different time scales, interacting with the subsystems, objects, and brains, nonlinearly to arrive at self-organizing states of order and chaos. These impulses can be wave packages in musical instruments, spikes and spike bursts in the brain, or bits and data packages in internet information streams.

Three major findings can be derived for music and arts from such a Physical Culture Theory:

- **Music is food**. Due to the self-organization in musical instruments, the sounds they make have a low entropy, or a high level of order. When listening, this low entropy is lowering the entropy of our brains on a physical basis. Just like food feeding us through its energy with low entropy, music is feeding us in the same way.
- **Music, art, and culture are Human Rights**. Human Rights are therefore derived from man's characteristics. Musical instruments have been built by humans. The instruments work such that they have a low entropy, interacting with us in a way of feeding us. Therefore, we have built musical instruments to enlarge our bodies and brains. This holds for arts and culture in general. Therefore, music, art, and culture determine us as humans, and therefore, they are a Human Right.
- **Music, art, and culture are ethical. Entropic imperative**. As all life is slowing down the increase of entropy on earth by an endless complexity of structures, to maintain life, one needs to build such complex structures. Music, art, and culture are such self-organizing systems. Therefore, they help to maintain life. So we can formulate an entropic imperative: **'Act such that your doing builds structures which slow down the increase of entropy!'** This entropic imperative also holds for environmental, political, or social issues.

Therefore, this book introduces a Physical Culture Theory exemplified with music, showing how musical instruments work, how their sound is processed in self-organizing brain structures, how conscious content appears in humans, and how this is dealt with in the music around the world.

Energy and entropy are closely aligned. A steam engine has a tank filled with heated water. So there is a lot of energy in this water. Still, to use the heat for driving the engine, there needs to be cold air outside the tank. Only because of the difference in pressure and heat inside and outside the tank, this pressure and heat wants to get out of the tank. If outside the tank, the same pressure and heat would be present, the steam in the tank would not move out and could not be used for doing work, for driving the engine. Therefore, no matter how much energy there is, the energy needs some kind of order, low entropy, to perform work. With the steam engine, order means that not the same temperature and pressure exist all over the place, but that there is a region of high pressure, the tank, and a region of low pressure, outside the tank.

The sun provides the earth with heat, which gives life to everything. Still, on Mars, the sun shines too without any life up there. The reason Mars is dead is that Mars has nearly no atmosphere, no plants, animals, or any other self-organizing being with a very complex molecular structure to keep entropy low. There is an endless cascade of processes, structures, and performances of living beings to lower the entropy as much as possible. Physics tells us that entropy always increases. Disorder always wins. Still, life on earth slows down this increase of entropy by beings that take the energy and use it for maintaining their own lives, self-organize in a way to maintain their very complex structures. Energy does not arise from nowhere or from nothing and does not vanish into nowhere or into nothing. It is only changing its nature, from heat to force, from speed to height, etc. Still, within this process, the energy is distributed more and more equally over a spatial field, it lowers its order, entropy increases. Of course, in the end, all energy is distributed equally. Entropy has won. Still, life on earth slows down that increase of entropy tremendously, with its endless diversity of plants, animals, humans—and culture. On Mars, the sun's energy heats the Mars surface during the daytime, which cools down at night nearly entirely again. Therefore, entropy has maximized within one day. On earth, trees take over the sun's energy, grow wood and leaves, storing energy in complex structures, maintaining low entropy, sometimes for millions of years as oil or gas.

Music is energy with low entropy. Humans are living by self-organizing themselves to a system of low entropy. Music, therefore, generally speaking, is food. It helps to maintain our lives. Of course, it is a different kind of food compared to vegetables or meat. Music is mental and soul food, still, on a physical basis. We can decide which food we choose and feed us with the physical energy of low entropy to recover, rethink, feel good or bad. With music, we can also choose how to use music to some extent, how to use the low entropy we feed ourselves with. So stating that music is food is not an analogy, it is a physical reality.

Musical instruments are built like living beings, as self-organized systems. Only because of the high complexity and nonlinearity of tone production, a very simple and stable output is produced, a harmonic sound with simple mathematical frequency relations of 1:2:3:4... Musical instruments are culture. They have been invented and built by us. We rebuilt these instruments fitting our physiology, our brains, and interestingly using very similar physical principles as keeping us alive. Music, therefore, is an extension of our lives. Musical instruments have many properties like those of friends or relatives. Some musicians might love their guitars more than some of their companions, maybe because they act like living beings in many respects (of course not in all). So enlarging ourselves with music, may it be by playing music or by listening to it, is part of our very nature, of extending our self-organizing principle within the world. This is culture. Following the Age of Enlightenment reasoning, that Human Rights are no arbitrary values free to choose, but are derived from determining humans, which need to eat, live, work—and make music, arts, and culture, clearly, music and arts are a Human Right.

Having determined how music, art, and culture works, we can formulate an ethnic principle, just like Kant did with his categorical imperative. When we want to maintain life, ourselves, plants, animals, and nature, we need to build structures that can slow down the increase of entropy. This might sometimes include destroying old systems and building new ones. Still, we can claim an ethnic principle of physical reasons, to act such that your doing builds structures that slow down the increase of entropy. Music is one of them, in a way, meeting this principle.

The Physical Culture Theory (PCT) proposed in this book is deterministic, still not simply mechanistic or materialistic. Nonlinear, self-organizing systems are that complex that a simple relation between single elements is very seldom possible. It needs to consider the system as a whole. It is also not pure materialistic as it also covers our conscious content, the qualia in philosophical terms.

Therefore, the hope is also to bring hard and soft sciences closer together again. In terms of musicology, the Pythagoreans found everything to be number, where music was an essential part of it. They vowed to the Tetraktys, a mathematical relation of 6:8:9:12, from which all musical intervals can be derived. Still, the so-called smiths' legend needs to be wrong. Pythagoras one day passed a smith and heard that the relations between the hammer length were just like the relations of the musical pitches these hammers produced when hitting an ambo. This needs to be a legend, as the pitches a hammer produces are not relative to their length but to their volumes. So although experiments were performed in ancient times, not all experiments seem actually to have been done. Still, this does not mean that the relations are not there. They are only more complex. This relation between music and the rest of the world was maintained in Renaissance times, e.g., with Keplers Harmonia Mundi, where the Quadrivium included music, geometry, arithmetics, and astronomy, while the Trivium contained dialectics, rhetorics, and grammar, making it the seven free arts.

The split between hard and soft sciences might be found, starting with Descartes, claiming the mind/body duality. The slip continued during the development of modern hard sciences, especially physics in the nineteenth century. The core of this duality is the qualia that we have a conscious content of aesthetic nature, the color red, a musical timbre, a smell. Still, in modern brain sciences, we are pretty sure to connect physiology, which in the end, is physics, with conscious content. Starting from the end of the nineteenth century, the field of psychophysics, which for music is psychoacoustics, is combining physics with the idea Franz Brentano called 'inner measurements', of listening tests in music psychology. Still, to bridge this gap again fundamentally, we need to explain consciousness and conscious content out of physics. As what we measure when performing EEG, fMRI, or other brain scanning techniques is electromagnetic fields, and as electromagnetism covers all three fundamental forces in physics, electromagnetism, strong and weak forces, known as the standard model (the forth force of the standard model is gravity, still astronauts also have consciousness, therefore gravity does not qualify as origin of consciousness), therefore, it is straightforward to expect electromagnetic fields changing over time and space to be conscious content. This might sound esoteric to some. Still, the opposite is true. Looking for another source of consciousness rather than that we know and can explain what we perceive and measure seems more esoteric, at least to me. Also, claiming a reason for consciousness itself, next to a conscious content, would only make sense if there could be consciousness without content. Still, even a consciousness of emptiness has such a content, the emptiness. Therefore, the conscious content we have, hearing sounds, melodies, and rhythms, thinking, and feeling all seem to be special spatiotemporal electromagnetic fields.

Such a fusion of hard and soft sciences gives endless possibilities for understanding culture using objective measures rather than subjective, heuristic ideas, feelings, or world views. This does not mean that subjectivity is something wrong, still, it is not scientific. The book proposes an Impulse Pattern Formulation (IPF) as a mathematical method to understand musical instruments and brain activity in the realm of music. As this method is computationally fast, arbitrary scalable to very small activity as well as to very global events, and as it is able to cover billions of active, nonlinear subjects, it might be one way to come close to explaining the highly complex interactivity of culture in physiology, nurture with nature. The Impulse Pattern Formulation is a Physical Modeling (PM) method. It models the physical system in an iterative way, time after time, to arrive at a prognosis of a future in case certain actions have been taken or events have happened. As it is a physical model of self-organizing, nonlinear nature, it is able to detect long-range developments, as well as sudden phase changes. It is also able to predict events that never have occurred in the past. This is what makes physical modeling techniques superior to those of artificial intelligence, which mainly learn from the past, and therefore, are not able to predict events that have never happened before. Also, AI is often predicting long-range developments much better than sudden changes. Still, of course, AI will also play a role.

These methods, PM and AI, can serve as objective measurements to detect racism, populism, or tell invented from real history. They are ways to politically enable crowd-sourced decisions using internet platforms of open-source algorithms to analyze culture and suggest future cultural politics and actions based on objective measurements. Still, this is only a first approach, and therefore this book is also a starting point of a scientific program which I would guess will be outlined over the next decades to come.

Artificial intelligence is a term basically used in two ways. Immanuel Kant, in his Critics of Pure Reason, suggested a two-stage process of perception. First, in transcendental aesthetics, the sensory input is collected and sorted using four types of machinery categories: quantity, quality, relation, and modality. These are processes, machines. Therefore, categorical perception is not putting things into boxes, it is a way for arranging the endless sensory input in a meaningful way, to hear a pitch from a complex time series, to detect a musical instrument playing. This transcendental aesthetics Kant calls *Verstand*, may be translated as understanding. This is what is often meant by intelligence in the Anglo-American literature. Still, Kant then proposed a second stage of perception (the a-perception), which is *Vernunft*, reason, spontaneity, human freedom. This *Vernunft*, reason, is our ability to intellectually freely deal with what the categorical perception gave us as organized data. Reason needs to be guided by ethics, according to Kant in his Critics of Practical Reason, through the categorical imperative, 'Act such, that your doing could always be used as a foundation of a public law.' In German, often, his reason is taken as *Intelligenz*, which is often also similarly used on the European continent.

We must not mix these two, understanding and reason. Artificial intelligence that serves people shall perform the first task and leave the judgment of what to do with the sorted data to the people. Of course, it can also perform the second task, be creative, reasonable, free, guided by ethnic principles. Still, the output of such an AI should not be taken as the only choice. It is not 'alternativeless', a rhetoric term often heard in recent years. Although there might be many actions that are not reasonable, this does not mean that there is one and only one reasonable action. AI needs to serve the people. And by the way, everybody programming AI knows how far we are away from a computer taking over our lives. It is always another human that might use a computer to take over our lives. Still, this human could also determine our thoughts and actions, not using a computer. In the former Eastern German state, people were watched by simple microphones and by evaporating letters.

Musical instruments are no human beings. They do not have consciousness. Still, they are living to some extend, as for us, they often act like intelligent, sensitive beings. This animistic view was strongly suggested by Ernst Cassirer in his philosophy of symbolic forms. All living beings around us, trees, rivers, fields, animals interact with us. They change in time, rivers flow or fall dry, fields grow through the year. We can help them and might be threatened by them. This is not restricted to ancient times. Many talk about their computers or cell phones, just like active persons in their daily life. People find computers interact with us, get sick, need to be repaired, do what we want them to do, etc. Although we all know that computers do not have consciousness, we need to deal with them as changing objects, or even subjects, just like we need to deal with our friends and relatives, with dogs and cats, living beings we assume they are conscious. What makes a guitar or a saxophone different from a stone or from water is that, when played, it exhibits a self-organizing behavior. It produces harmonic overtone structures we only know from animals or humans, very seldom from 'dead' things. This book will go deep into such systems to show that we do not deal with analogies, but complex physics, meeting our ears' and brains' demands.

Therefore, what the PCT takes culture to be is more than ideas and thoughts. It is also more than a society of brains. It finds musical instruments, LPs, CDs, concert halls, recording studios, architecture, machines, books, computers, all to be part of the culture. Like us, they are physical, and therefore, together, we are an extremely complex system, where here and there conscious content pops up. We are nature, physics, self-organizing systems, which decays entropy increase in billions of ways, what we call life, and what makes life so fascinating and worth living.

Sometimes people find that knowing how music works destroy the magic in music. After decades of doing research in systematic musicology and as a professional musician, I find that the opposite is true. The original magic when listening to a new musical piece, and the pleasure when hearing songs I always admired, remains the same. Still, knowing how this music came into place, how the instruments work that produce it, and how it is processed in my brain is magic too. Sometimes it is breathtaking to understand how music works and to see how all the features, physics, psychology, and culture interact. This second layer of magic is what makes the pleasure even more intense.

Indeed, science itself is a trip. Arriving at new findings of scientific problems, new principles, or new models is a very special experience. Like Plato said: leaving a dark cave and suddenly seeing the light. One might really get addicted to such states, and often it is hard to get back to everyday life, which seems quite dull then.

Math is such a case. Music always had much to do with math. Many people find it boring and hard to understand. Still, working through a differential equation and really understanding how they work and how all parameters of an equation are able to model millions of things working in real life is magic. It is a new kind of understanding, one which is so precise that often one would never have thought this to be. There is much work until one reaches these states. Still, it is rewarding and such a wonderful state-of-mind.

Such people are called Renaissance people, interested in all aspects of life, and try to find connections between them. In music it is self-explaining, as it is produced by musical instruments governed by physical laws, it is perceived and made by humans governed by neural networks and psychology, it is performed by ethnic groups all over the world, it has a political, social, theoretical, commercial, or philosophical impact. So understanding how music works is only possible by understanding all these aspects. Systematic Musicology, as an academic discipline, therefore has taken Musical Acoustics, Music Psychology, and Music Ethnology as their main parts, none for itself, all interwoven.

Although today, many scientific fields are getting specialized more and more, Systematic Musicology cannot work that way. Restricting oneself to a single problem makes us blind in this field. MP3 is such an example. In the 90th, people started downloading music from the internet, which was pretty slow then. Therefore, a compression algorithm was needed. Many suggestions were around, but the one who made it combined signal processing and psychoacoustics. When a sound is loud, the ear is deaf for a small sound portion right behind this sound. But when we do not hear a short time period anyway, why then store this time period. Leaving out some parts of a musical piece leads to a smaller file size, a compression. So a psychoacoustic finding improved a computational method.

It is necessary to do just this, deal with all aspects of music as much as needed to understand the basic principle of how music works.

Dealing with music by understanding how it works means understanding a lot about the world as a whole. The same physical laws which govern musical instruments govern all other things. Music perception and emotional reaction, composition as a creative process, or making music is done by the human brain, and therefore, needs to follow the rules of neural networks the brain consists of, just as all things we think, feel, perceive, or do need to go through. The rituals of music performance are in close relation to other customs, political thoughts, religious or spiritual attitudes of a culture, being part of the global culture in a global music industry, web content, and music production cycles.

Therefore, understanding music means understanding a great deal of the world. But it is more than this. It understands how these diverse aspects, physics, psychology, culture, or politics interact. In music, all these aspects are always there, right at the same time. Understanding music, therefore, is also understanding the interactions between these fields, ending in a global or holistic world view, which will hold for many other aspects of life and nature too.

Furthermore, this understanding is present in an aesthetic moment of hearing and making music. All of these aspects are heard and perceived. So understanding music does not only lead to an abstract piece of mind. It also leads to an instantaneous event of understanding in an aesthetic way.

Still, this aesthetic understanding is only really present when we know about music, when we know about the physical processes underlying it, the neural networks which process the acoustic input in our ears and nervous system, the culture and history of musical styles, instruments, composition techniques, and usage of music.

Otherwise, music is just some sound, some rhythm, some melody. This might be nice and beautiful. Still, it misses a lot! And even understanding a rhythm, being able to tap a foot, is quite some effort for the brain. Being able to sing along a melody is not trivial. This becomes clear when we look at the effort musical software needs to go through to model this perception. Until now, there is no software that is able to perform as good as a normal listener does. So we have some kind of understanding of music already anyway, we have learned it by listening to music all our lives. And without this understanding, music would not make much fun. It would just be some sounds changing in time. How would be dance without a beat or a rhythm? In most cases, we would not dance at all.

So we already have an understanding of music, which is instantaneously there. Listening to music is understanding music.

This book suggests a basic mechanism of how music works: self-organization or synchronization. This is not evident from the start, it is also not part of most textbooks about music, may they be physical, psychological, or ethnological. Still, going deep into the production of music in musical instruments, fascinating things appear. One of them is that without self-organization, music would not be music, all sounds produced by instruments and the singing voice would be rough and harsh. Only a self-organizing process makes them sound harmonic.

Interestingly, self-organization and synchronization are also basic principles in the brain. Neural networks, the only thing our brain consists of, are all self-organizing. They consist of simple units, neurons connected in a very complex and nonlinear way interacting all the time. Each neuron is quite simple. Still, the output of many is highly complex. It is not random. It is ordered and structured. This structure and order are the results of self-organization the brain automatically performs.

This is what we want to discuss in this book. The question is the fundamental question of Systematic Musicology, the discipline which tries to figure out fundamental laws, systems, and interaction principles music consists of. It has a history starting from Pythagoras, trying to relate numbers, strings, and cosmology to music, a world view still present in Renaissance times with people like Kepler or Descartes. In modern times, around 1900, this discipline was established anew when it became possible to record music around the world on wax cylinders. It is a field still very active and growing all over the world.

In Part I, the book gives a brief introduction to some aspects of acoustics, psychology, and ethnology of music. This is not exhaustive at all, still featuring important findings relevant to the following discussion. The development of the three disciplines, Musical Acoustics, Music Psychology, and Music Ethnology, will be discussed briefly. To understand concepts and ideas around today, one needs to know about some historical developments. Of course, this is not a comprehensive overview, but it tries to cover the most important aspects and stresses those we need to understand how music works.

The main reasoning of the book is found in Part II, systematically introducing the acoustical, psychological, physiological, or philosophical reasoning of how music works. We find musical instruments to be self-organized systems. This aligns with how the human brain works, and we will see the same physical laws holding for both.

Part III is viewing these findings in terms of culture, ethnology, or philosophical ideas. It summarizes and discusses what the future might bring and what to do next.

Sometimes our discussion will be quite technical. Still, this is only done when it is really necessary to get to the point of our story. So sometimes a reader might need to work a bit, the same way a scientist needs to work on new findings and ideas. Still, I hope you will find the same pleasure working yourself through the subject as I do!

Contents

Contents

Part I

Some Fundamentals of Musical Acoustics

Friedrich Chladni tried to make the invisible visible.[1] Sound can be heard but not seen. Guitar or violin bodies are moving when the instruments are played. Still, we cannot see these movements. When touching the surface, we can feel a movement, yet, we do not know how these surfaces are vibrating and how they are radiating these vibrations into the air around the instrument. Also, the sound in air is not visible. How does the vibration of the guitar come to our ears? We do not see the air vibrating, although there is influence over a distance.

Science in the 19th century was very much aware of forces acting on distances. Electrical and magnetic energy which act upon us all the time but which we cannot see, or lunatic forces driving the tides were undeniable, however not understood.[2]

Although religious or occult dogmas were assuming angles, ancestors, and other spirits to act invisibly, science was to tell the real forces from mistaken influences, physics from imagination. In Renaissance times, it was assumed that, having a personal item of someone, it is possible to influence this person instantaneously over hundreds of miles.[3] So if one had a towel with some blood on it of a wound a person once suffered, by spreading salt on this towel, the wound would start bleeding again. And such an instantaneous remote action was also found with electricity. When coming close to an electrically charged device, our hair stands up without touching the device. How do all these influences act through a *fluidum* or aether? It was only Michael Faraday (1791–1867) who, during mid 19th century, introduced the idea of *force-lines* on electricity and magnetics, which are between two interacting bodies in space. This is what we call electrical and magnetic fields.[4] This replaced the model of a remote action, which was taken as the standard model in physics until then, as promoted by Weber. Faraday showed that an electrical or magnetic interaction is the state of a field within both bodies the influencing and the influenced were laying. It became clear that energy is the state of a field built by multiple bodies, and interactions are field states. Energy is a field notion.

So science not only tried to tell true effects from false ones, it also tried to put them all into one system of reasoning and make the outcome of the system understand as

© The Author(s), under exclusive license to Springer Nature Switzerland AG 2021
R. Bader, *How Music Works*, https://doi.org/10.1007/978-3-030-67155-6_1

an interaction of all the system parts. But which system was that of vibrating bodies? For Chladni, this was not only a scientific question as he was also building musical instruments and made a living from traveling, giving lectures and demonstrations, publishing scientific books, and making music with his self-built musical instruments [141]. There he discovered so-called longitudinal waves and build several instruments from it, the *clavicylinder* [62, 122] and the *euphon*. He kept the secret of the mechanisms, and even when Napoleon wanted to know, he refused, pointing to the fact that he was forced to make a living from it. Napoleon accepted and seemed to have enjoyed Chladnis performance on the instrument in a private concert the famous mathematician Lagrange and Poisson had arranged to persuade Napoleon on funding the translation of the book *The Acoustics* of Chladni into French. Napoleon agreed after Chladni had shown him his famous Chladni-figures, making the invisible visible. These famous Chladni figures are easily obtained by putting some sand or grain on a plate and bowing this plate at one side with a violin bow. The vibrations of the plate make the grain jump around the plate and settle at places where the plate is not moving. So after a while, the grain is ordered along lines on the plate, which means that all other parts of the plate are moving.[5]

This effect is very well known to us, but in those days it was very surprising that a regular pattern could be seen on the plate and that this pattern changed when the sound produced by the bow was changing its pitch, the higher the pitch, the more complex the figure. This looked systematic indeed, but how to understand such a system?

Strings are simpler to understand than plates. At least one might think so. The vibration of strings is more obvious because the movement of a string is visible to some amount. We see it moving, still, the movement is so fast that we cannot say how it is moving. Brook Taylor (1685–1721) was the first to discover the wave equation for a string using the Newton laws of forces and therefore relying solely on the principle of forces balancing each other. Again forces are not visible but a pretty abstract thing. We can see a string displaced from the resting position, we can see it move with a velocity, but force = mass times acceleration and acceleration is not visible.

Also, it refuses to act according to the law of contingency. This simple law means that if a body wants to move from one point to another, it has to do so continuously, moving through the space between these two points. If it would suddenly be at the other point, it would be magic. The problem of contingency was much discussed by Descartes (1596–1650), another major contributor to the physics of mechanics. He had a dream when he was a young musketeer during the Thirty years' war stationed in Neuburg at river Donau over the winter of 1619. He dreamed that a book covering all knowledge of the world would approach him. He tried to find the page of a question he had and found this page torn off the book. Then, suddenly, the book had disappeared, and after a while, again suddenly, it was there again. From this dream he concluded that what tells sleeping from being awake is that in the domain of waking things act with contingency, while in dreams, this is not the case.

Coming back to mechanical principles, this holds for the places things are, and for the velocities. If a velocity changes, it needs to do so continuously. Still, acceleration can jump suddenly from one value to another. This happens when a body is suddenly hitting another body. Of course, the displacement of the bodies and their velocities change fast but contingent. Still, the acceleration changes from zero to a certain amount instantaneously. So, using force, which means acceleration, as the basis for mechanics was quite some abstraction and meant to accept that the principles of mechanics can only be understood taking the invisible as their basis.

This genius idea that mechanics can be understood as interacting forces is what makes mechanics possible and what made Tayler able to derive the wave equation. The first solution to it was suggested by Daniel Bernoulli (1700–1782), who discovered that each complex vibration can be build by adding together much more simple movements, those of sinusoidal movements, of sine waves. We know this today as the Fourier theorem and Fourier (1768–1830) showed that adding infinite numbers of sinusoidals, with different strengths or amplitudes, is the solution to any function in a very general sense. For strings, d'Alambert (1717–1783) suggested the correct mathematical solution of two waves of the same shape, one traveling from left to right and one from right to left, overlapping. But there are endless possible shapes. A guitar might look different from a violin string, the sound of a drum different from a saxophone. The movement of a violin string, which looks like a sawtooth, was only made visible by Hermann von Helmholtz in this famous book *Sensation of Tone as a Physiological Basis for the Theory of Music* [128] in 1863 by using a stroboscope and film. But in the 18th century, the only way to make these movements understand was a mathematical one.

So at Chladnis days, the math of a string and its solution was known, still, the math for a plate was not known, and therefore Lagrange and other famous French mathematicians like Laplace, Lacroix, or Malus decided to arrange a competition on the subject of a mathematical description of the vibration of a plate. Such competitions were common these ways and in line with the political movement of democracy, that the best shall win, ignoring the social status. Sophie Germain (1776–1831) was the only one contributing to this competition, but her suggestion did not satisfy the committee at first. Only after arranging the competition three times—with three times only one contributor—the suggestion of Germain was accepted and still is today, with several differentiations of course.[6]

So why was it so difficult to find a mathematical solution. The answer is very close to the Faraday idea of energy being a field in space. With strings and plates, one needs to consider the solution as an instantaneous interaction of all parts of the plate or string with all other parts. Only when assuming that everything is connected with everything, the correct solution can be found. This is a different picture than if a phenomenon can be explained by only taking local interactions into consideration. A car is moving on a street no matter if other cars are moving on it too, or if it is alone there. It might need to slow down sometimes, but still, at first, it is moving on its own. This is not the case with strings or plates (or electric or magnetic fields) and therefore needs a more complex mathematical description, another form of equation.

Such an equation suiting vibrations of bodies is called a *differential equation* in mathematics and is fundamental to the understanding of our world in many ways. Still, it was only developed by Leibniz (1646–1716) and Newton (1643–1727) independently one from another around the beginning of the 17th century, where today we use the notions of Leibniz. It was developed when trying to square the circle. This is impossible, but still, we can try to come close to it. The problem is to calculate the area below a cured line and was first discussed by Archimedes (287–212 BC) in antique Greece. He suggested to use two curves—or line segments to be precise— one bigger and one smaller than the actual curve. Leibniz made a new suggestion to use only one line, a compromise between the line too large and that too small, and therefore was able to integrate the area below the line within one step. This was groundbreaking, as then the integration could be done with one equation. From this equation, all the mathematical rules for integrating and differentiating a curve can be derived, and only then the system of integration and differentiation could lay the foundation of most science in our modern world.

This compromise between two integrations, the one too large and the one too small, made it necessary to approach the real solution for a point on the circle by only referring to it from the outside. This referring-to does not have any clear boundaries but may be taken from a very near area around, or it may be taken from very far apart. In any case, this referring means to understand what is happening at one point by what is happening at other points. As there is no spatial limit for such an influence, one can only understand what is happening at one point by understanding what is happening everywhere else. This means an instantaneous interaction, and it means that there can only be one solution of the whole body and not just single solutions of parts of the body. Such a body then has a field of forces, displacements, velocities, or accelerations. So the figures Chladni showed did not only make the invisible visible. They also showed that what in the mathematical treatment of the centuries before was suggested, the mathematical idea of instantaneous interaction is real. And indeed, the suggestion Germain gave in the contest was such a differential equation, similar to that of the string, but considerably more complex.

The better scientists understood what vibration was, the better they realized that there were new opportunities for sound production. The secret Chladni did not want to discover to Napoleon were longitudinal waves he had discovered and used in his instruments. These are waves vibrating in the plane of a plate or a string. These waves can hardly be heard under normal conditions, as they can only drive the air outside the plate or string at their rims. But a plate or string has only a very small rim, and therefore, these waves are not loud. Chladni, therefore, coupled the rim of a rod to other objects which took over these vibrations and were able to radiate them into space. This was simply sound, too. Still, the longitudinal waves are much faster than other wave types and therefore are pitched much higher, leading to a spherical, ambient sound. We know such sounds from synthesizers nowadays, at Chladnis time, these sounds were very seldom and unheard to most people who meant to hear the angels sing or to hear the ancestors whisper. Many such instruments were built in the follow, and people very much associated these with spirits and voices from the beyond, which we will have a closer look at in the next chapter. Although these voices were simply sound and nothing more, for the reasoning we try to follow in this book, it will be important to keep this in mind.

Play It Loud!

Western art music during Baroque times was played at royal courts or in churches. The audience, especially in the case of Baroque theater music, was small. The turn of the French Revolution towards democracy brought along the wish of a wider audience to attend art concerts, and larger concert halls were built. Still, the ensembles performing at court were quite small and, therefore, low in volume, much too low for a large concert hall. Therefore several adjustments needed to be done to fill the halls with so many people. An obvious one was the enlargement of the ensemble, which found its peak in the 8th symphony of Gustav Mahler (1860–1911), also called the 'symphony of thousand', where indeed at the premier in 1910 over 1000 musicians and singers performed.

Around 1800 several other adjustments were made. One was to increase the tuning. In Renaissance and Baroque times, the tuning of instruments was not fixed and differed in about a tritone around the contemporary standard tuning [123]. Starting from the 1970th, there is a music scene that attempts to play old music, that of the Renaissance and Baroque, the way it was played way back then. They often prefer to pitch the standard tone $a^1 = 415\,Hz$. Still, this is only convention, tuning the a to g#. The standard tone of $a^1 = 440\,Hz$ was only defined by the International Federation of the National Standardizing Associations in 1939 (ISO 16). Still, in terms of art music, there is evidence that the pitch did raise around 1800 from 370 Hz to about 440–450 Hz [137]. The advantage of such an increase is an increase in loudness of the instruments. This does not mean that the physical energy radiated from the instrument is considerably larger. Still, the energy in the higher frequency range around 1–4 kHz is increased a bit. The human ear is most sensitive to frequencies around 3 kHz mainly because of the middle ear channel, the channel we place in-ear headphones to. This channel is a tube with a resonance frequency around 3 kHz, amplifying energy of such frequencies. So by increasing the string tension, the instrument sounds louder.

Still, this has a trade-off, as an increased tension means a serious stability problem. The instrument might get cracks or even break down completely with too much string tension. The situation even became worse as in those days, the violin neck became longer and got its tilted shape. Then the bridge needed to be higher to make it possible for the player to play all four strings. Violins were needed, which were able to stand such tension. Therefore, the bass-bar was enlarged. This bass-bar is a short piece of wood glued below the lowest bass string, the G-string. It is not meant to make the bass sound louder, it is there for stability. This meant opening the violin, removing the old bass-bar, and putting another one at the old position. Also, in terms of statics, a curved violin top plate is a good idea. Still, it must not be curved too much, but with a smooth in- and decrease of the curvature. The violins, gambas, or other stringed instruments in Baroque times had a very strong curving of the top plate. They were known for their sweet and lovely sound. Still, the instruments at these days with a quite flat curvature were Stradivari or Guarneri violins. Therefore, they were considered the best instruments, and other builders started building such instruments too. So Stradivari violins were not chosen because of their sound but of their superior stability. Still, also these instruments were altered, and nowadays, there is barely a Stradivari violin in its original form. Indeed, today we find a similar tendency to associate quality not with sound but with other aspects of musical instruments. The high prices for Stradivari or Guarneri violin are also not caused by a superior sound but solely by the name 'Stradivarius'. This is also true for vintage jazz guitars or other such gear.

To find a clever construction of violins in terms of sound, we need to come back to the ear, especially the inner ear later. Instrument builders have often adapted instruments according to specifications of the ear, and with the violin one of these adjustments is the bridge. It has a resonance frequency of around 2,5 kHz, very close to the 3 kHz where the ear is most sensitive. As the energy of the violin strings needs to travel through the bridge to arrive at the body where only there it can be radiated, the 2,5 kHz resonance of the bridge is boosting this frequency region. Therefore the violin is sounding louder than it would with a bridge having another resonance frequency, maybe around 1 kHz, or over 4 kHz. So although the energy of the radiated sound is not very much more or very much less than with another bridge, with the standard violin bridge the violin sounds much louder. This trick is also known in today's radio broadcasting, where increasing the energy of song at around 3 kHz will subjectively make the radio channel sound louder compared to other radio broadcasters.

Around 1800 the first systematic investigations in Musical Acoustics were done, mainly by Felix Savart (1791–1841). He was a physicist, doing investigating in many fields, and is also known for the law in magnetostatics, the Biot-Savart law. Savart did many experiments with violins, building a trapezoid-shaped one and experimenting with the soundpost, a small wooden stick clammed between the top and the backplate, around the place where the right food of the violin bridge is located. All violinists know that if this soundpost is falling down, the sound of the violin breaks down too, and is much lower in volume[7] Savart was the first to find that the reason is the rocking motion of the bridge. This rocking motion, to the left and to the right, is fundamental to the violin string-bow interaction. We will discuss this interaction later, as it is a wonderful example of the self-organizing nature of musical instruments. It results in a sawtooth motion, the one Helmholtz had first visualized around 1863. During the days of Savart, this was only partly known, still, it was clear that the bridge is rocking, and, therefore, the to plate is rocking too. So if the left part of the violin top plate is going down, the right one is going up and vice versa. This repeats very fast, so when playing a note a^1 440 Hz, this back and forth is happening 440 times each second. Still, then the air above the top plate is pushed up at one side and torn down at the other side. This also means that in the middle of the plate both, the pushing-up and the tearing-down are canceling out each other, and therefore, there is no movement at all.

Canceling out is no good idea when the aim is to be loud. It would be much better if both sides were going up at the same time, and therefore the whole energy of the instrument would be radiated into space. But this is not the case, and therefore, the instrument sounds quite low. Now the idea of the soundpost is simply to make one side of the instrument very stiff, so that it is not moving very much anymore. Then the other side can more or less freely move and radiate its energy into the air. And indeed, when inserting the soundpost, the instrument is sounding much louder.[8]

Still, of course, this is not the ideal situation as half of the energy is wasted, as the soundpost is permitting nearly half of the vibrations. An alternative is the so-called 'hole-violin'. Here one foot of the bridge is not standing on the top plate but is prolonged through a hole in the top plate down to the back plate. Indeed, such a violin is not only sounding louder, it also sounds much more free and lively, as the body is able to vibrate much freer. Still, tradition wants the violin without a hole in the top plate, and therefore, the sound is less interesting than would be possible.

Loudness is still a problem with violins, as classical musicians are not always satisfied with amplification. Modern violin builders, therefore, experiment with hybrid top plates where carbon or other material is used. As this material is much stronger, it can be carved much thinner than wood and is, therefore, more flexible. The more flexible a plate, the more air it may move above it, the louder is the instrument. These violins are still covered with a thin wooden layer to make them still look like normal violins.

The run for loudness, the classical loudness-war[9] starting around 1800 is a pity nowadays, as obviously, sound quality must suffer when concentrating on loudness.

This is not a problem with Pop or Rock music or contemporary Bluegrass, Country, Western, Independent, Metal, or Grunge or all that stuff around. Here musicians simply amplify and modify their instruments electrically and electronically. This allows for a much more sophisticated tone, as these instruments need not to be loud at first. The development of the electric guitar is such a case. The first electrified guitar for sale was the Gibson ES-150 introduced in 1936.[10] It was electric Spanish (ES), which means it was not a Hawaiian lap steel guitar, which was much more popular in Northern America these days. Additionally, it costs 150 $ way back then, explaining its name, which was typical for Gibson guitars. So the Super 400 sold for 400 $ and other models were named accordingly.

Now the ES-150 was more or less a Western guitar based on the L-Series the Gibson company sold before, especially the L-5 or L-7, with an additional pick-up for amplification. It was sold with an electrical amplifier, as these days one could not go into a music shop and buy a guitar amp, there were no. The guitar would only sell if it still would serve as a Western guitar alone, as players were shy to buy a pure electrified instrument. The L-5 and L-7 sound character, for our ears, sound straight, loud, and right to the point, still not with a very differentiated sound. Nowadays, we call their sound 'vintage' and pay much money for the old instruments, a bit like with Stradivarius violins, still by far not as much.

Times progressed, and in 1949 Gibson brought the ES-175 to the market, again costing 175 $ in the shop. This is one of the models the company produced over a very large time span. Its sound is the typical Modern Jazz sound played by Joe Pass or, in its ES-275 version, Wes Montgomery. This guitar is still a hollow-body model, but with top and back plates considerably thicker. As thick bodies vibrate with much less amplitude than thin ones, the radiated sound from such bodies is much lower, and therefore it is not very loud. Still, then the instrument can be built with much more interesting sound character and variability, fitting the needs of Jazz musicians in the 60th or 70th, to vary from sweet soft sounds to bright ringing tones or to hard and heavy attacks. The low sound radiation is not a problem, as the pickup electrifies the sound, and the guitar amplifier will produce any loudness necessary. The main change with this guitar is that when the top plate is thicker, the guitar has more sustain. This idea was introduced most radically in the solid-body guitars introduced only later by Less Paul and others. Solid-body guitars, like the Les Paul named after his inventor, or the Stratocaster guitar invented by Leo Fender, reduce the sound level even more, as the guitar body is made of a solid piece of wood. The Gibson Les Paul guitar, introduced in 1952, was the first commercially available.

Les Paul was a pioneer in many aspects doing the first multi-track recordings by overdubbing and doing pitch-shifting. Such manipulations were done at about the same time only in studios of electronic music, like at the French radio by Pierre Schaeffer (1910–1995) in his Musique concréte, the studio of Electronic Music in Cologne founded by Meyer-Eppler (1913–1960) at the Phonetic Institute at the University of Bonn, where people like Karlheinz Stockhausen (1928–2007) or Iannis Xenakis (1922–2001) worked, or in Ottawa, where Hugh Le Caine (1914–1977) developed his new electronic instruments. These were all institutes in the academia and some experimental radio programs, following and expanding the Western art music tradition around and after WW II, and were therefore only recognized within the academia, while the recordings of Les Paul sold millions of copies. Indeed, his music was much more mainstream, strongly relating to the Django Reinhard guitar tradition and that of American Standard Jazz, and he was considered a real virtuoso in his times. The new developments were happening because it was then possible to do so from a technical standpoint. In the 1940th in the electronic studio in Cologne, a monochord, a melochord, noise generator, filters, and a four-track recordings device were the main tools to produce music.

Les Paul reported that he was attaching a string to a railroad log and said he could pluck it, go to lunch, and when returning, the string was still vibrating. Although this might be too optimistic, what he built was a device that is transferring the string energy, its vibration, very weakly to the body the string is attached to. This is because a thick piece of wood is not vibrating very much on its own, and therefore a driving vibration, like a string, will not very much succeed in making this body vibrate. So the energy transfer from the string to the body is very weak. But this means that the energy stays in the string. In other words, the body is resisting vibrations from outside. Physically, the relation between the force of the string acting on the wooden piece and the velocity the wooden piece is vibrating is called resistance or impedance. Generally, the tinner a guitar body, the less resistance it has, the faster the energy of a string goes into it, the faster the string loses its energy, the shorter the tone. And vice versa, a thick wooden body produces a sustained string vibration, a sustained tone.

It is unlikely that Les Paul himself invented this. In the South of the US in the 80th of the 20th century, black Americans still remembered that it was usual for poor musicians to nail a string to a wooden bar, maybe at the veranda of a house, and play music there (with low loudness of course). A nice demonstration can be found in the 'American Patchwork' series of Alan Lomax, an American ethnomusicologist [182].

Alan Lomax was the son of an anthropologist John A. Lomax, who was interested in the American people. During his fieldwork, we discovered many Blues musicians like Leadbelly or Muddy Waters, who he visited at this house on a plantation in the American South, and was also supporting Woody Guthrie, the founder of the singer/songwriter scene in the 1940/50th.

The loudness war of mechanical musical instruments was always raging, especially in outdoor music, which needs to be loud. Everyone experiencing a *gamelan* orchestra in Bali [247] or Java (which is a doubling, as 'gamelan' already means orchestra), playing in a temple with no roof, is struck by how loud this ensemble really is. The instruments are made of bronze, which can be struck very hard (without braking) and sound very long, this time because of low internal damping of the instrument. Percussion instruments decay their sound mostly because of the loss of energy when the sound is radiating into the surrounding air [222]. Of course, this energy is gone, and therefore the instrument decreases in volume. Still, a wooden plate of a xylophone and a metal plate of a *gamelan* instrument sound different in timbre and loudness. The metal plate will sound longer. This is because metal has much less internal damping than wood. The reasons for this difference are not perfectly understood yet, and research needs to go down even in the quantum mechanical domain to find answers. We will come back to this later in more detail.

A typical, very loud European folk instrument for at least the last more than two thousand years is the bagpipe. In ancient times, the Phrygian people who lived in the south of today's Turkey were considered the best musicians around. The phrase 'Playing after a Phrygian' was similar to a hopeless try, especially during the Olympic games, where music was part of in these days. The most famous Phrygian musician, [198], even had a statue in Rome where one could see him playing the bagpipe. Contemporary bagpipe music in the south of Turkey, but also all over Europe, from Scotland to Spain, from Russia to Crete, is still alive. Bagpipes also exist in Africa and India for a long time. Pieter Breugel the Elder (1525?–1569), a Dutch Renaissance painter, in many of his paintings like *Peasant Dance* or *Wedding Feast* shows them in oil.[11] The instrument was often associated with trance, tarantellas, or the devil. This might be due to its loudness, or due to people dancing to it. As we will see, the connection between acoustics and psychology is much more complex.

Bagpipes most often are wind instruments working with two reeds, double-reed instruments. The reeds are most often made of bamboo but can also be made of plastic. Nepalese *sahanai* or Singhalese, southern Sri Lanka *horonai* double-reed instruments even have layers of leather as 'reeds'. And they are much louder than single-reed instruments, like clarinets or saxophones. The reason for this is quite complicated, and we will discuss the airflow of reed instruments below as a crucial aspect of how music works. Still, the strong loudness of double-reed instruments again makes them a perfect candidate for playing music outside. The Turkey *zurna*, also a double-reed instrument, was therefore used in the Ottoman Janissary marching military band *Mehter* or *Mehteran* most likely formed around the times of Murad (1362–1368) [105]. The instruments were different kinds of *zurnas*, double-reed instruments, drums, trumpets, and cymbals. They did perform daily, sometimes even three times a day. Due to the influence of the Ottoman Empire over the centuries, due to war but mainly due to trading, this ensemble was introduced mainly to countries in the east, known as drum-and-shawm bands. The Indian and Nepalese wedding ensemble *pancha baja* [250] use the *sahanei*, most likely a derivative of the *zurnei* of the Central Asian Janissary ensemble. In Buddhist temples in Sri Lanka, the *horonai*[12] is played, again double-reeded with drums, where the name might be derived from the Turkish *ney*, a flute. In Myanmar, the main ensemble for performing puppet and dance stage shows, the *hsain wain* [19][13] is a mixture of drums, gongs, and a *hne*, double-reed again. In Nothern Myanmar, in Kachin state, the marching band for *wunpawng* rituals consists of *dumba* and *sum-pyi*, double-reed and flute, performing with drums.[14] In Thailand, the orchestra performing during Thai kick-boxing shows consists of double-reed instruments and drums. In other words, the thing seems to work, it is loud.

The Mysterious Sound of Friction

But of course, there is much more to instrument building than loudness. Another aim is to make instruments talk. One example is the use of friction to play instruments. Today we use synthesizers to create ambient sounds and all kinds of magical timbres, sounding like from outa space. In the 18th and 19th century, there were no electronics, and to achieve mysterious sounds many friction instruments were invented. Everyone knows the sound of wine glass rubbed by the finger around its rim. The process of such rubbing is not generally different from the violin bow-string interaction, which we will have a closer look at later. The sound is soft and special, as it is hard to produce such a sound with other instruments or the human voice. Still, when playing a melody with such glasses, one needs many of them to get all the pitches a melody consists of. In the 18th and 19th century, many attempts have been made to get such a sound in a more simple way.

One of the first proposing such a musical instrument was Benjamin Franklin, known for his inventions of the lightning rod and the definition of the plus and minus of electrical current. In 1755 he invented the glass harmonica, or *armonica*, which it was sometimes called for reasons of public relations. It consists of many glass bowls placed next to each other on a bar which goes through the rotating center of each bowl. Each glass bowl has its own frequency when rubbed, and when placing adjacent pitched bowls next to each other, one has a musical scale. Now the whole system, the bowls attached to a bar, is rotating using a mechanism of a sawing machine of the old days, that is with the power of the feet. The lower side of the bars goes through a water box, and therefore the bowls are always wet. Then one only needs to touch the rotating bowls to produce a sound. As the bowls are close next to each other, with two hands one can play the instrument, using the ten fingers for different pitches.

This instrument became very popular in America and Europe, and it was the young Mozart, next to others, who composed pieces for this instrument, finding their first performance in the garden of Anton Mesmer, who was very much improvising on the instrument himself. We will get closer to Mesmer in the next chapter when having a look at the development of modern music therapy, as friction instruments and music psychology was closely related then.

Another friction instrument was introduced by Chladni, as mentioned above already. The *clavizylinder* consists of metal bars that are robbed. This produces a longitudinal wave that cannot be heard. So the end of the bars is attached to a plate that is radiating the sound. The instrument produced a sound similar to that of a glass harmonica. Chladni made a living from performing and writing, and therefore he performed the *clavizylinder* all over Europe.

Keyboard instruments, like harpsichords or hammer pianos, were standard instruments of that time, and therefore the idea to have a friction instrument with a keyboard arose. One was the terpodion invented in Berlin by Johann David Buschmann (1775–1852) in 1805, which was build and distributed by his son Friedrich Buschmann (1805–1864) [122]. Its name was taken from one of the seven muses of ancient Greece, the terpsichore, which is known to play a stringed instrument. An alternative name of the terpodion was the uranion, another muse. There a wooden cylinder laying horizontal in the instrument box is rotating, again using the mechanism of a sawing machine. The instrument looks like a regular keyboard instrument at first, and when the player presses a key, a bar covered with leather is pressed against the rotating cylinder. The tone is again produced by a friction process, where the bar is sticking and slipping along the cylinder, producing a pitched sound—well, most of the time.

And not all of the time, as due to the very complex interaction in such sticking-slipping interactions very different sounds may be produced on this instrument too. Some low notes may sound like a drilling machine when played with strong key pressure, they sound like a regular pitched sound with medium pressure, and may become noise with low key pressure. Middle pitched keys may sound like normal tones, still not glass-like, but much harsher and stronger, they can easily squeak or whistle. High pitched keys may sound normal, but when changing the key pressure, they might sound very voice like, often like a cry or a moaning. So the instrument shows a wide range of sounds also found in human or animal voices or with machines. All this is caused by the complex mechanisms of the friction process (Fig. 1).

The terpodion could be played in different volumes according to the strength the keys were pressed down. And of course, different from a piano, the tone continued as long as the key was pressed. The decay of piano tones has often been found a weak point, and alternative pianos were built, like the sostenente piano [34], which continues the sound with a mechanism similar to that of the violin where a continuous bow is driving the string as long as the key is pressed. Such mechanisms were also built in music automata at the end of the 19th century, which mimics whole bands and ensembles. Still, this was an exception, and the sound of these pianos reminded of those of violins and, therefore, did not have a distinguishable sound on their own anymore. A continuous tone is also known from organs, but again also organ keys only have one volume. They are either played or silent. Therefore having those features combined, a continuous tone and different volumes together with a distinguished sound of the instrument was a crucial point for inventing the terpodion.

Fig. 1 The terpodion is a keyboard instrument where sound is produced by friction. A wooden cylinder, in the top of the picture, is rotating. When pressing a key, one of the hammers is pressed against the cylinder, causing a friction process and producing a sound. The pictured terpodion is one of the very few remaining. The author thanks the Viandrina in Frankfurt a.d. Oder to have granted access to the instrument

The voice-like character of the instrument was very much felt, which in a Romantic age where people believed in the presence of the spirits of dead ancestors was an obvious association. For the reasoning of how music works we follow in this book, it is an interesting case to associate a physical tone production so closely with elements found in voices and with machines, and we will come to this in much detail in the later chapters.

Until the mid 19th century, there was only a loose feeling that Musical Acoustics and Music Psychology were related, and the mathematical understanding, as well as experimental verifications, were only beginning. The experimental side was, most of the time, again trying to make the invisible visible. It was Thomas Young (1773–1829) who tried to make sound in tubes visible [141]. Therefore he was using smoke in tubes, which can be seen.[15] Today we find considerable differences between sound in air and the flow of air, describing both with very different differential equations. This is a crucial point for the following discussion about musical instruments and our conscious perception of them.

Young was the first to state that "...undoubtedly they cross, without it disturbing each other's progress; but this can be no otherwise affected than by each particle's partaking of both motions..." [270], p. 130. He, therefore, discovered superposition, a basic principle of linear acoustics. Superposition means that two waves are not interchanging energy between each other and therefore do not couple in any way. When they are moving through space at the same time, they are simply additive and often form quite complex wave shapes, but when traveling further, they perfectly split again and look the same as before.[16]

Indeed, Young was right in proposing this principle for sound. Still, he was not perfectly right when it comes to air flow where he first examined it. Air flow, like water, is behaving in a much more complex, nonlinear way. Therefore we would expect two waves crossing each other to interact in such a way that afterward they look considerably different. Still, this is not the case. The first to make this understand was John Scott Russell (1808–1882), a Scot, who, in 1843, examined at the Union Canal near Edinburgh that a boat suddenly stopping produced a "...solitary elevation, a rounded, smooth and well-defined heap of water..." [224] before it. This wave was traveling with high speed, not changing its shape. Russell was following the wave on his horseback for about two miles and was convinced that we had made an important examination. These waves are called solitons today and are the fundamental basis of modern telecommunication using optical solitons. Still, it was only Joseph Valentin Boussinesq (1842–1929) who was able to give the first mathematical explanation of solitons in 1871, which was reconsidered by Diederik Korteweg (1848–1941) and Gustav de Vries (1866–1934) in 1895 and now is called the Korteweg-de Vries equation [160].

This equation is very complex, still one can show mathematically that one of its solutions is a soliton wave. When saying mathematically, it means on a piece of paper with much math on it. Still, the Korteweg-de Vries equation, like many others, is so complex that it is no longer possible to solve it on such a piece of paper, and we then only can use a computer, which solves the equation for us step by step. When doing this with the Koreteweg-de Vries equation for two waves crossing each other, it appears what Young had observed, that after wave crossing, they split up again, having their original shape back. Still, the reason for this behavior is very complex and part of the theory of surface waves. The superposition principle holding for acoustic sound in air looks the same at first sight, but when examining the reason, in this case, it appears to be much more simple than that of solitons.

Linear Math and Nonlinear Nature

So air normally behaves linearly, and water moves nonlinearly. Linear and nonlinear are terms that we need further on a lot. The difference is not difficult to understand, and there is a mathematical as well as a physical and a psychological interpretation that is interchangeable. The mathematical formulation is most simple. Assume two parameters, say a force acting on a frame drum and the displacement of the drum head caused by this force. When increasing the force on the membrane, the displacement increases too. Generally, when increasing one parameter by a certain amount, the other parameters will change too. If a certain increase of a first parameter leads to a certain increase of a second one, and this relation is true no matter how large the first or second parameter is, then the relation is linear. So when we double the force we hit a membrane with, the displacement of the membrane doubles too. Or when a string is vibrating with a certain strength, it vibrates with a certain amplitude. When driving this string with twice that strength, it vibrates with twice the amplitude. When it is driven by a force three times that of the first force, it is vibrating with three times that amplitude, and so on. This is a linear process.

This is a very simple law, but unfortunately, in reality, such relations are very seldom. Most often, this law does not hold, and the processes are nonlinear. When a real string, like a guitar or bass string, is driven with a very strong force, the amplitude is not increasing linearly anymore but increases much less. All guitarists or bassists know this, when applying much tension to a string, it will no longer displace further. And when applying too much tension to the string, it simply breaks, a very strong nonlinearity. In reality, in the physical world, mathematical 'pure' linear relations often do only hold for a very small range of a parameter, and beyond this range the systems becomes nonlinear. So when displacing the string striking the membrane only very slightly, the other parameter, the string or membrane displacement, are indeed in a linear relation to the causes of their displacements. Still, linear laws are mathematically much easier to handle compared to nonlinear cases. Therefore a researcher will try to simplify a problem to a linear model at first and hope that he is not too far off from reality.

Nonlinearities may appear in many possible ways. With a string, the relation between force and displacement above a certain force is about cubic, a power of three. Then, when wanting to displace the string to the double of its present displacement, it is no longer enough to double the applied force like $2^1 = 2$, one would need to increase the force by $2^3 = 8$. Still, this cubic relation is only one possible nonlinear relation, the relation might also be that of a square or a logarithmic one, or of any wired relation. So although there is only one linear case, there are endless possibilities of nonlinear cases, which makes the world on nonlinearities much more complex.

In terms of psychology, the relation between a physical parameter, like sound pressure level, and the psychological parameter, like loudness, is nearly always non-linear, too, very often it is logarithmic. To perceive a sound double as loud as another sound, much more physical sound pressure level is needed than just doubling it. The base of the logarithm is ten in this case. So if the pressure of a fist sound wave is p_1 and that of another is p_2 the perceived loudness L, measured in decibel (or dB in short), is

$$L = 20 \times \log_{10} \frac{p_2}{p_1} .\tag{1}$$

This is a famous measure, and most people have heard that 120 dB is the threshold for pain and damages of the ear. (Indeed, this is only a very rough estimate, and we will differentiate this in the following chapters). Many astonishing aspects come along with this measure. 120 dB corresponds roughly to a physical pressure of 1 Pascal (or simply Pa). On the other side, the minimum sound pressure level we hear around 3 kHz is about $p_0 = 5 \times 10^{-5}$ Pa. This corresponds to a displacement of the inner ear basilar membrane of about 10 Ångström, which is 10^{-9} m or 1 nm, which again is about the size of ten hydrogen atoms. If we would hear even better, we would hear the quantum noise. This would be pointless, as the quantum noise is always there, and we would hear a constant noise.

So the ear covers a tremendous range of possible sound pressures. To make this wide range not too confusing for us, the ear compresses it into a small range using a nonlinear function, the logarithm of ten. Therefore the ear is most efficient when it comes to present to our consciousness what is happening outside. Indeed, efficiency is a basic principle of many aspects of nature, and of music as we will see further on.

It is important to note that the mathematical equation above is relating physics and psychology, and is therefore in the domain of psycho-physics, or psycho-acoustics in this case. Psychophysics has been introduced most prominent by Gustav Theodor Fechner (1801–1887) and Ernst Heinrich Weber (1795–1878) again in the 19th century, as an attempt to map physical parameters to a content of consciousness, loudness, pitch, timbre, but also seeing, touching, smelling, etc. This has been applied very successfully, still with several drawbacks, especially when it comes to measuring the content of consciousness, the perception. Loudness is a good example, as the phenomenon is much more differentiated than the formula above. This is much under debate today in terms of possible damages of the ears because of loud music, but also in terms of environmental noise and the tremendous health damages caused by it in modern societies. There is a reason why one can frighten people by loud sounds used in wars [104], and why noise can also be a political strategy to control people [8]. Still, Noise is also a musical genre, where the destruction is used to make changes and explore new sonic words [124, 130].

Hermann von Helmholtz

The first who related acoustics, physiology, psychology, and tonal systems in modern times was Hermann von Helmholtz (1821–1894). He was working on many topics, discovering basic principles like the physical law of energy conservation, inventing the eye mirror, discovering fundamental mathematical equations, was working on electrodynamics, or on music. Because of the wide range of fields he made basic findings, he may be considered as one of the few genii of all times, maybe on the same level as Aristoteles or Leonardo.

Helmholtz, in 1863, was publishing a book *On the Sensations of Tone as a Physiological Basis for the Theory of Music [Die Lehre von den Tonempfindungen als physiologische Grundlage für die Theorie der Musik]* [128]. The title is program, and with many experiments and much mathematical background, he tried to show that the Western tonal system of the major and minor scale has its origin in Acoustics, Physiology, and Psychology. At first, he found experimentally what has been estimated by Georg Simon Ohm (1789–1854) before, that a musical sound, and all sounds, consists of single frequencies, sinusoidals. He, therefore, developed a set of resonators called Helmholtz resonators. These were bottles of glass with a neck, through which the air is breathing in and out. This system, in principle, is like attaching a heavy body to a spring and let the body move up and down, driven by the spring. The air in the bottle is similar to the spring, which is compressed and depressed by the air in the neck of the bottle, which serves as the moving mass. Such a system is also working in musical instruments, the sound holes of guitars or the f-holes of violins are working that way in combination with the air inside the instrument. The sound is considerably loud, which makes it usable for musical instruments. When a sound traveling through air is meeting such a bottle, the bottle is resonating strongly, amplifying the frequency of the bottle, making it much louder than before.

Helmholtz was producing sounds with several generators, most often with a siren. Still, he was also constructing the first electric synthesizer with several frequency generators added together, a system nowadays called additive synthesis. So he also was the first to construct a synthesizer in the modern sense.

What Ohm had only proposed, Helmholtz could show, that sounds consist of frequencies many single frequencies, the principle of superposition of Young. Harmonic sounds, like guitar, violin, or organ tones, the human voice, and so on, consist of many frequencies which are harmonically related, i.e., the frequencies of the sound are integer multiples of a fundamental sound frequency. So if a sound has a fundamental frequency 100 Hz, the overtones or partials will 200, 300, 400 Hz, and so on, all with different strengths. When changing this strength, the timbre of the sound changes, and therefore different instruments sound different, a violin sounds different from a guitar, a flute from an oboe. A synthesizer may have all possible sounds, as one can choose the strength of the overtones at will.

Still, there was one part of the sound Helmholtz, with his experimental possibilities, was not able to investigate, the beginning of a tone, its attack, or initial transient. He estimated that there might be different phenomena going on, but was not going into details. Only some years later, Carl Stumpf, a psychologist and co-founder of modern Systematic Musicology, found that when the initial transient is cut off, it is very hard to tell a violin from a saxophone or a guitar from a piano. There are several important issues with such transients, and we will consider them later in more detail.

Helmholtz was going beyond acoustics and considered the human ear to analyze the sound in terms of its frequencies, separating them. His idea was that the ear consists of several strings that would resonate when a sound came to the ear which meets the frequency of that string. This is not perfectly true, and it was only Georg von Békésy (1899–1972) who understood the mechanism several years later. There are no resonating strings in the ear, still, the ear is indeed decomposing the sound into its single frequencies. It does so by a complex mechanism where the sound is traveling along a membrane, the basilar membrane. This membrane has a changing tension along its way. The mathematical equation is, therefore, one of a normal wave but with an inhomogeneous tension. The resulting wave on the membrane then looks very different from a normal wave. Its amplitude changes from low to a maximum at a certain place on the membrane, only then to fall off strongly after this point. The point of maximum amplitude is different for each frequency, therefore, this system is able to sort the frequencies of the incoming sound.

So although Helmholtz did assume a wrong mechanism behind this frequency decomposition, he was right in terms of the composition itself. Only then he could argue how tonal systems like the major or the minor scale developed, and why listeners prefer them over other scales. He was considering two tones played at the same time, say on a piano. If one tone is c and another is played simultaneously an octave higher, both will have their harmonic overtone series on their own. But now both series are sounding at the same time and therefore might disturb each other. Disturbance in this sense means that they might be so close to each other that a beating occurs. A beating is the amplitude of a tone rise and drops several times per second. So when two frequencies of 100 Hz 105 Hz are sounding simultaneous, the beating will 5 Hz, so each second, the amplitude will raise and drop five times. This is a nice effect well known from science fiction films from the 50th or 60th, where these effects were done right this way.[17]

So how often two frequencies are beating per second depends on the distance the frequencies have. With slow beatings, one can follow them perceptually. If they are getting faster, above 16 or 20 times per second, one can no longer follow them and therefore perceive a sound that is often called rough. This roughness increases with frequency distance, and Helmholtz assumed a maximum roughness when the frequency difference 33 Hz, so, e.g., 100 133 Hz were sounding at the same time. If the distance increases even more, the roughness decreases again because the frequencies stop disturbing each other.

When two piano tones are played simultaneously, each of them may have up to 50 partials or overtones, and so there will be 50 times 50 possible combinations. Still, most of them will be more 33 Hz apart. But those within this region will beat and therefore produce roughness. The more frequencies disturb each other, the larger the roughness. Therefore Helmholtz was trying all possible intervals from unisono, c and c, to the octave c and c' with all tones in between, i.e. all 12 keys of a piano (and even many smaller intervals between two semitones). For all cases, he was calculating the roughness and plotting this in a curve. Interestingly this curve has positions of a minimum roughness right at the intervals we use in Western music, the fifth, fourth, major and minor third, major and minor sixth, and the seconds and seventh. Still, in this list, the first had the smallest roughness while the last, the seconds, and seventh show considerable roughness. So Helmholtz followed that the tonal system we use is caused by the minimum of roughness.

In other words, he had a definition of harmonicity, the absence of roughness, found by combining acoustics and psychology.

The First Synthesizers

As mentioned above, Helmholtz was the first to build an electric synthesizer in the modern sense around 1860, still only for the purpose of research. He used tuning forks of different frequencies in harmonic relations as loudspeakers (loudspeakers were only invented about 60 years later). These forks were driven by an electrical circuit. The fork with the lowest frequency was closing and opening the electrical circuit when it moved towards or away from an electrical contact. Closing the contact made electricity flow through electromagnetic coils, which then produced a magnetic field caused by the flowing electric current through them. This magnetic field made the tuning forks move. As this process repeated with the fundamental frequency of all forks, they sounded continuous. The fork sounds were amplified by tubes in front of the forks having respective frequencies. Still, the purpose of this experiment for Helmholtz was not to make music, but to determine if the phase relations between the single frequencies of the whole sound were audible. The phases of frequencies are their relative starting points. By changing the distance between the forks moving ends and the ends of the tubes opposite to them, the phases of the amplified sound could be changed. In this continuous sounding synthesizer tone, he found that changing the relative phases did not change the timbre of the sound. It is amazing that Helmholtz was constructing such a sublime synthesizer where the phases can freely be chosen at his times. Such a feature was only to be found in synthesizers later with electronic music appearing over a century later.

From then until synthesizers became commercially available from around 1970 on with the Buchla or the Moog synthesizers, there were very many synthesizers being invented. The earlier synthesizers were used for entertaining people via telephone as single units, for use in then-contemporary art music, as well as for speech synthesis. History is full of technically and musically fascinating inventions and designs [223]. Although there are too many to mention, we still need to discuss a few examples which are interesting for the reasoning to follow.

One was the telharmonium of Thadeus Cahill developed in the 1880th.[18] In these days of Thomas Edison or Graham Bell, inventors of new technology also needed to be businessmen, salesmen, and distributors, in other words, everything from the first idea to the distribution. Cahill was to explore the new possibilities of the telephone invented before by Bell. At that time, nobody understood what the telephone could be used for, as messages were distributed via Morse code fast and reliable. Indeed, nearly nobody had a telephone, but everyone could be reached via a postal address, so telegrams would find their addresser. Cahill, therefore, had the idea of using the telephone to listen to music. So one could pick up the phone, and the operator was asking about the kind of music one was preferring to hear, from light classical music to folk or now also to electronic music produced by the telharmonium of Cahill.

The instrument was powered by a steam engine that was driving huge metal cylinders, some about two meters in height. Each cylinder had several wheels with a changing amount of ribs or teeth. When the cylinder was rotating, the different wheels were producing different pitches according to the amount of ribs passing a magnet during each second. Therefore, in the electrical coil around the magnet, a changing current was induced with a frequency of the amount of ribs per second passing. The shape of the ribs or teeth was determining the timbre of the tones, where sinusoidal-like sounds were produced. Adding sinusoidals resulted in different timbre, mimicking musical instruments like wind instruments or violins, just like Helmholtz had discovered.[19] This principle of rotating wheels and electric coils as sound generators was later also used in the Hammond organ, built since 1935.

The telharonium is an interesting example of an early synthesizer, as it was starting to use the telephone only emerging in these days. It was therefore linking electronic music and sounds to the phone, which was also motivating another electric instrument or a sound manipulation device to be more precise, the vocoder.

The vocoder was developed for the telephone at first for reasons of compression or data rate reduction, an early MP3, so to say. The capability of the first cables placed in the Atlantic ocean, connecting Europe to the USA, was very restricted. Therefore a data reduction was needed to be able to transmit more telephone channels. The vocoder was therefore invented to reduced speech to its very basics, its initial transient, and its steady-state. The transients are the plosives like k, t, or p or the frictives like f or sh, the steady-state sound are the vowels like a, e, i, o, u. The latter consists of harmonic overtone spectra, where a fundamental or lowest frequency f_0 exists together with other frequencies which are in simple mathematical ratios of $1 : 2 : 3 : ...$ to that fundamental frequency. So if a speaker pronounces an a with a fundamental frequency of $f_0 = 100 Hz$ this a will have additional frequencies 200, 300, 400 Hz, and so on. Now all these frequencies might be stronger or weaker, have different amplitudes. The mixture of these amplitudes make the sound color of such vowels and tell an a from a i or o. Furthermore, all amplitudes of the spectra of all frequencies normally change in time considerably. So 200 Hz frequency might start with a strong amplitude and decrease within some hundred milliseconds to a low amplitude while, say, 400 Hz frequency is going the other way, starting low and becoming strong during this time. Indeed, all frequencies are constantly changing their amplitude, and one needs to take this into consideration when designing a speech synthesizer.

The vocoder does take this into consideration, and in those days, used ten of the frequencies and recognized one amplitude value of each frequency every 20 ms. This makes 50 amplitude values per second multiplied by ten frequencies, which is a total of 500 numbers representing a vowel lasting one second. Compared to the continuous signal of real speech, this is a tremendous simplification—and indeed also sounds like this. Still, the speech is recognized easily even with such a simplification, and therefore the vocoder can be used to transmit speech. It reduced the amount of data effectively and therefore could be used as a data reduction tool. Today nearly all data transmission via the internet or cell phones is done using data reduction. This reduction may be lossless when the output is the same as the input, or it may be lossy, so that data is lost, just like with the vocoder, still keeping the important information. All data transfer, therefore, need to be coded at the sender and decoded at the receiver. With the vocoder, the coding is described above. The method or algorithm produces a code, 500 numbers per second, which are transmitted via a channel, wire, or wireless, and need to be put together again at the receiver side. In terms of the vocoder, the receiver takes the amplitude information of each frequency and synthesizes the speech again by adding the ten frequencies together, each with its amplitude at the respective time-point. This technique is called additive synthesis, as it simply adds frequencies with their amplitudes together. So the decoder with the vocoder is an additive synthesize type. As a sender also want to receive information back, both sending and receiving might be present at the same time, and therefore a coder and a decoder need to be implemented. Such a system is a coder/decoder, or a *codec* for short, and present in most sound cards, computers, or cell phones. So we might consider the vocoder as a very early form of such a codec.

The vocoder was invented for data reduction but never used for this purpose, as the cables through the Atlantic ocean were becoming better and faster than the development of the vocoder. Still, the vocoder has a fascinating history which includes the development of one of the first voice synthesizers, a secret military communication channel during World War II and made a great career in Pop and electronic music, in the music of Kraftwerk (Trans Europa Express), Laurie Anderson (Oh Superman), Herbie Hancock and many many others [253].

Following this development, an automatic speech synthesizer, the voder, as built between the 1920th and 1930th on the principle of the vocoder.[20] It also was adding frequencies with changing amplitudes to synthesize vowels. The plosives and frictives were produced by using a random noise generator. This machine was much like a musical instrument, where operators needed to practice for about a year on a daily basis to be able to 'speak with it'. Still, then it could not only speak but also sing. This is because the fundamental frequency of the synthesized vowels could be varied, and when the operator was using frequencies fitting to a melody, the machine could sing a song. The machine was not sounding as natural as modern synthesized speech, which one can enjoy when waiting in a call center loop or at a train station. Still, the speech was recognizable and an astonishing achievement.

Voice from Impulses

Now to get even closer to the reasoning this book is about, how music works, we need to take a look at a home-grown speech synthesizer, the electric shaver. The idea is simple, and everyone can perform it during a morning shave. When holding such a shaver to one's neck, the sound of the shaver will enter the throat and sound out of the mouth. Now when moving the mouth, just like when speaking, without actually making a sound, the sound of the shaver will be shaped by the mouth producing vowels like *a*, *i*, etc. So the shaver takes over the work of the human larynx, the vocal folds, a mechanism we will discuss in a later chapter in much detail.

Still, the sound of the shaver needs to be periodic and not just noise, otherwise one would hear some noise shaping, but it would not sound like a vowel. This means that the vocal folds produce a periodic spectrum, an impulse train, of many impulses per second according to the fundamental frequency. A shaver might rotate 20 times per second, which works ok but is not the best choice. To make the point more clear let us assume a shaper 100 Hz fundamental. So each time it moves the neck, it gives a small impulse to it. Indeed, such an impulse train is right what the vocal folds produce. Figure 2 shows the opening and closing of the human vocal tract while singing. This movement is very impulse-like. It is not a complex waveform. The vocal-fold vibration results in a sound-wave that is also impulse-like, as we will discuss below. Only when this impulse travels up the vocal tract through the singer's mouth, it is

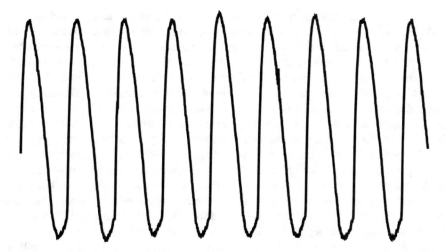

Fig. 2 Glottography of a singing voice, showing the regular opening and closing of the vocal folds while singing. Like the movement of the vocal folds, the waveforms produced at the folds are again very impulse-like. These impulses are further shaped up in the vocal tract to form vowels and consonants that might have much more complex wave-shapes. Still, the basic waveform produced at the vocal folds is impulse-like, just like those in wind or stringed instruments

shaped to form vowels and consonants, more complex waveforms. For itself, the impulse-waveforms sound quite sharp and bright. Still, it is reshaped through the vocal tract, the mouth, tongue, lips, and teeth, to result in the vowels we use in speech and music.

Such an impulse train is a musical tone with a definite pitch—and not a noise-like sound. This is important for understanding how music works, and therefore we will need to consider the relation between such a sound wave and the frequencies it consists of, its spectrum. A waveform, like that of an impulse train, is called a signal, and the science dealing with signals is called signal processing. There are two basic domains in signal processing, the time-domain and the frequency-domain. The time-domain is simply the waveform itself. The frequency-domain displays the waveform in terms of frequencies. A basic and important feature of these two domains is that they perfectly correspond one to another. So if we have a waveform in the time-domain, we can transfer this into the frequency-domain and get a spectrum, all frequencies with their amplitudes and phases the waveform consists of. Vice versa this spectrum can be transformed back into the time-domain where we get a waveform, which is perfectly the waveform we started with. In other words, the transformation from one domain into the other is lossless, and we have perfectly the same information in both domains. Only because of this perfect correspondence it is useful to do such transformations. Thereby the transformation from the waveform into the frequency spectrum is called a Fourier transform, and the transform back from the spectrum into the waveform is an inverse Fourier transform.

The fact that the Fourier transform is able to do a transformation which is loss-less, one which gives the same waveform when going from the time-domain into the frequency-domain and back, is not at all trivial. It is only case because the sinusoidals are completely independent one from another. We can see this when doing a simple calculation with two waveforms. The Fourier transform is basically doing this: it is multiplying two waveforms all along the way and then summing up all multiplica-tions. Suppose the case of two waveforms, a slower one which only has one cycle, and a second one with two cycles where the first has only one. The second waveform is twice as fast as the first one and therefore its frequency is doubled compared to the first frequency. So when the slower waveform has traveled halfway, the second has already run through one complete cycle. Now when multiplying both waveforms at each time-point, both, the slower and the faster waveform, are positive. Therefore the result will be all positive, as a positive number multiplied by another positive number will always end up in a positive number. The same result will be obtained if both waveforms are negative, as negative times negative becomes positive. In the other cases where one waveform is positive and the other negative, the result will be negative.

Now, as the waveforms are perfectly symmetric, the amount of positive and neg-ative portions of the multiplication results are perfectly the same. Therefore when summing over all results, the final result is zero. Mathematically this reads

$$\int_{t=0}^{t=T} \sin(2\pi f_1 t) \sin(2\pi f_2 t) \, dt = 0 \, . \tag{2}$$

Here the two waveforms are both sinusoidals, which is abbreviated by sin. They have frequencies f_1 and f_2 and run over time t = 0 to t = T. The 2π is needed to run through a whole cycle and is not important now, we will deal with this later. Finally, the $\int_{t=0}^{t=T}$ is an abbreviation of *integration* which is the sum we were talking about. As we sum over all points, and theoretically the points are infinitesimally close together, the sum becomes an infinite sum which is written as $\int$. This equation is stating nothing else than when multiplying the two waveforms and add all results together, the result is zero.

We needed to go into these details to get to our point above, to show that two sinusoidal waveforms are totally independent one from another. Basically, the Fourier transform is doing nothing else but multiplying two waveforms and summing up over time. The difference is just that shown in the example above of the two multiplied sinusoidals. When analyzing a real musical sound using the Fourier transform, one waveform is again a sinusoidal with a certain frequency, but the other waveform is the real sound.

Now, all real sounds consist of sinusoidals with some amplitudes. To get from the sound to the spectrum, all we need to do is to perform the calculation above for all frequencies of the sound. So when we ask if a certain frequency is in the sound or not, the Fourier transform will tell us. If the waveform is strongly present in the sound, the result will be a large amplitude, if it is there only weakly, the calculated amplitude will be small. By doing this with all possible frequencies, we will obtain all amplitudes, which is called the spectrum of the sound. The spectrum, therefore, is the result of the Fourier transform, all amplitudes of all possible frequencies in the sound.

So now we can return to the impulse train the vocal folds produces. Such an impulse train can be transferred into a spectrum too, of course. The result in such a case depends on the shape of the impulses. Suppose the impulse would be infinitely short—which is unphysical—then the spectrum would contain all frequencies with the same amplitude. This is a special case that cannot be reached. Still, it is interesting as it means that when adding all possible sinusoidals together all having the same amplitudes, they would sum up to an infinitely large and short impulse at one time-point and would sum up perfectly to zero at all other times.

This behavior is useful in many cases, even if one is not able to construct such an impulse and has brighter impulses in practice. Such impulses are used to understand and reconstruct concert halls or other acoustically interesting rooms. Everyone knows that when clapping hands in a room, there will be a reverberation afterward. The clapping is such an impulse that is used to test the room. The room answers with a response, the reverberation, which is called the impulse response, and characterizes the room completely. Sound engineers can record the reverberation and use it as an artificial reverberation in recording studios to make a musical performance sound like if they would have been played in right this room.

Unexpected Harmonicity

Still, this behavior of an impulse, as discussed above, describes one single impulse, one single clap of hands. As we have seen above, the vocal folds produce such an impulse many times each second. The male voice has a fundamental frequency of 100 Hz when talking normally. Then the vocal folds produce 100 impulses per second. When we Fourier analyze such an impulse train, not all frequency waveforms fit in, as would be the case with only one single impulse. This is because the impulses are repeated, and then only sinusoidals which fit right into the periodicity of the impulses, the time interval between two impulses, will have a strong amplitude. All sinusoidals which do not repeat perfectly the way the impulses do have zero amplitudes. But this also means that the frequencies of these sinusoidals will be a simple mathematical multiple of the first, slowest periodicity (or lowest frequency) the waveform fits in.

So then we have simple harmonic ratios between the frequencies. If the periodicity, the time between two impulses is 10 ms, a 100th of a second, then the lowest frequency present in the sound 100 Hz. The next possible frequency will fit twice into the 10 ms and therefore 200 Hz, the third fits three times and 300 Hz, and so on. So the ratio between the first and the second frequency 100 Hz : 200 Hz = 1 : 2, that 200 Hz : 300 Hz = 2 : 3 etc. In the end, we have what is called a perfect harmonic relation between the frequencies of 1 : 2 : 3 : ...

When discussing wind instruments, we will see that this is not at all expected and we would expect much more complex relations. So it could be the case that the first frequency 100 Hz is followed by a second frequency slightly higher or lower 200, 202 Hz or more, which strongly depends on many aspects of the instruments. Still, we hope for a perfect and, therefore, harmonic relation of 1 : 2 : 3 ... as only then we hear a normal pitch. If the relation is getting more difficult, we very much deviate from such a pitch, as then roughness and beatings will appear in the sound. Such imperfect series would not be acceptable as we could not play melodies with them.

Now the reason for the deviations expected but not found with musical instruments are quite complex, and we will come back to them in the next chapters. We will find that musical instruments are only showing harmonic spectra because they are self-organizing systems.[21] In other words, they behave in such a simple way because they are very complex within themselves, where many subsystems interact in such a way, that again a simple output is produced. Only through the inspection of music psychology, physiology, and physical principles at work in music, we can understand the nature of these interactions.

Why Are Musical Instruments Such Complex Boxes?

Of course, the interactions are different for each instrument family. Guitars, pianos, dulcimers, or other plucked or struck stringed instruments seem to be more simple. There is a string which is attached to a body, mostly a wooden box, and when plucking a string instrument is sounding with the frequency of the plucked string. Still, things are not that simple, and it is interesting to consider why these instruments are constructed the way they are.

The basic problem is that strings vibrating without such a wooden box can hardly be heard. Sound is moving air, and musical instruments move this air while vibrating and therefore pushing the air in a certain direction back and forth. A string is no different, and when vibrating in one direction, it pushes the air into this direction. Then in front of the string, the air is compressed a bit which means that the air pressure at right this point is increased. Still, behind the string, the air is stretched and decompressed, and therefore the pressure there is a bit less than in the equilibrium case, that is without any vibrations. Now when two pressure regions of high and low pressure exist next to each other, they balance in such a way that both regions will end in a pressure in between both high and low ones. Right this is happening with the string, and so the vibrating string is only moving the air around the string itself with very little radiation of sound towards a listener. This is also called an acoustic short circuit, analog to an electrical short, where much energy flows, still with no effect.

This is not happening with 2-dimensional geometries like with plates or membranes simply because the air would need to move around the whole geometry, which is far too long to take place. Then the air is pushed upwards and produces a pressure right at the top of the plate. This pressure is different from the pressure a bit upwards in the air above the membrane. Such a pressure difference is called a pressure gradient and is forcing the air to move. This process continues and in the end we have a wave starting at the top of the membrane or plate which is traveling into space towards the audience.

Now only because of the short circuit of a string, we need to attach the string to a plate or membrane, to have considerable radiation and, therefore, can hear the string at all. Of course, such a system is more complicated to build, resulting in a new job, that of an instrument builder.

But there is a second problem, that of a weak bass. The lowest frequencies a guitar or violin plate can vibrate in depends on its size. The larger the size, the lower the frequency. Piano soundboards are large and therefore are able to radiate very low frequencies, starting from 50 Hz. Still, a normal classical guitar top plate only radiates from a frequency of 200 Hz. The lowest tone on a guitar, the low open E-string, has a fundamental frequency 83 Hz. This frequency is still radiated from such a plate, a puzzling thing for itself we will consider later, still, it is weak in sound power. Therefore a string on a wooden plate is still not enough, and a mechanism is needed to radiate low frequencies as well to get a low bass sound.

This bass mechanism is often a Helmholtz resonance discussed above, which is air traveling in and out a box through a hole. With guitars, the box is the guitar, and the hole is the soundhole. It works like a mass attached to a spring that is moving up and down. The mass with the Helmholtz resonance is the air around the soundhole, and the spring is the air inside the box. If the air at the soundhole is moving into the box, it is compressing this air, so the air volume inside the box will react and push this soundhole air mass in the other direction. When succeeding, the air inside the box will decompress, but it will overshoot, and in the end, will have a lower pressure than the air outside the guitar. This will lead the air around the soundhole to again change direction into the box, and the process has passed through one cycle.

Now this movement is slow because air is only moving with a speed of about 343 m per second, while vibrations on wood my have a speed of several 1000 m per second. If the box is large enough and the soundhole has the right size, the frequency of this oscillation is considerably lower than the frequency of the top plate 200 Hz. With the guitar, the Helmholtz resonance is tuned to 100 Hz, with Western guitars it is lower, about 70–80 Hz, to make it more 'boomy'.

So basically, this is why guitars are the way they are, we need a plate to make the string audible, and we need a box to have powerful low frequencies.

Who Wins the Game?

Now to come back to the point of musical instruments being self-organized systems, one thing might puzzle us in the description above. We have argued that the string is moving the top plate. But we can also knock on the guitar body, which gives a completely different sound, that of a knock on wood. But then we are not driving the string, but the body and could then expect the guitar to sound with the body frequencies. Indeed, it does at first, we hear the knocking, but then we will hear the strings sound, low but clearly there. As discussed above, we can only hear the strings because the body is radiating them. But this means that the strings have taken over, although we have driven the guitar body, not the guitar strings.

This behavior is so well-known to all guitarists that they might wonder why this should be a problem. Still, when we find the guitar to consist of a driver and a resonator, the strings and the body, this picture would need to change when we drive the body by knocking on it. But it doesn't! Although, at first, the strings are driven by the body, otherwise they would not move, then the strings start vibrating in their frequencies and then transmit their frequencies to the body. So no matter which part we drive, the strings or the body, in both cases, the strings win the game and force the body to vibrate with their frequencies. Or speaking in the terminology of self-organization and synergetics, the string is forcing the body to vibrate with the frequencies of the string.

Yet this forcing is not happening with the body giving up right away. When again driving a string, so with normal guitar playing, within the initial transient time of the note, so within about 50 ms, the top plate and guitar body start sounding in its frequencies at first. Indeed, within the first 50 ms the knocking sound we hear when indeed knocking on the top plate is clear and loudly present in the guitar sound. Still, most of us do not perceive it at first, as we are so used to a regular guitar sound that we fuse the knocking sound to the harmonic tone following into one single classical guitar tone. Still, without this knocking, the sole sound of the string sounds very dull and boring and does not at all sound like a guitar, but like a plain string without any interesting components. The guitar body sound is therefore making the tone, and we can therefore say that the crucial part of a guitar tone is the first 50 ms, and is the struggling between the string and the body until the body wins the game.

One can train the ear to hear this. Simply knock on the bridge of a string, and right after pluck that open string and try to hear the knocking sound within the very first moment of the plucked string tone. When doing this three or four times, at best with the high e-string, one starts distinguishing the initial part of the plucked tone, the knocking sound, from the rest of the tone, the harmonically sounding tone. Once trained, it is easy to judge guitar tones of other guitars by knocking on their bridges, as the knocking sound will be present in all played tones and, therefore, strongly shapes the character of the guitar.

This holds for all musical instruments, the initial transient phase makes the tone, strongly helps to identify the instrument, tell a violin from a saxophone, and help strongly in terms of articulation.

So the puzzling problem of why the string wins the game and not the body of a guitar is not only a theoretical problem. The precise structure of this interaction is present at each played note and shapes the character of the instrument. Still, there is much more to say about this process with respect to the initial transients building up, as well as with understanding the basic principle of musical instruments at all as self-organized systems, fitting perception, transferring information and form music.

Eigenmodes Versus Forced Oscillations

So there seem to be two kinds of oscillations, the one a string or a body is going into when vibrating on its own, and another which appears when it is forced, like in the case when the string forces the body to go with its frequencies. The first one is called eigenfrequencies or eigenmodes, a term derived from the German word 'eigen', meaning 'its own'. So the eigenfrequencies a body moves in are those frequencies of its own, and the patterns appearing on them, the Chladni patterns, are the eigenmodes.

Such eigenmodes appear in all kinds of vibrating systems and have appropriate names. A loudspeaker is radiating into the room in all directions. If a loudspeaker would not be a box but a very small point, then it would vibrate as a monopole, which looks like a perfect ball. So the pressure wave is traveling from the center in all directions with the same strength. But such a single point may have more complex radiation eigenmodes. The next more differentiated is a dipole. Here the radiation happens as two balls, one to the left and one to the right. In between the two balls, the radiation is smaller, and if one pictures a plane between both balls, the sound pressure on that plane is zero, no radiation at all. This continues, and higher eigenmodes would be four balls, one left, right, front, and back, which is called a quadrupole. The next would be an octopole, and this continues in principle endlessly. A musical instrument is radiating in a very complex way, which is not a pure monopole, dipole, or any higher-order eigenmode. Still, one could add up some of the poles with different strengths, a little bit of a monopole, maybe much dipole and quadrupole, a bit of an octopole, and so an. This ends up in a very complex radiation pattern. The usefulness of the eigenmode idea is that any possible radiation from a musical instrument, indeed any possible radiation, can be built by such a set of eigenmodes with different strength.

The Chladni patterns are also eigenmodes, as mentioned above, but of a very simple geometry, a rectangular plate. Now, musical instruments are much more complex. Guitars, violins, pianos, or dulcimers might have very difficult plates mostly round with cut-offs, sound- or f-holes, they have ribs attached, have changing thickness all over the plate and are normally made of wood, which is a natural material with changing properties also depending on air moist and temperature. In the end, they are covered with varnish and attached to other plates or parts. So their eigenmodes are not simple at all, and the shape of these eigenmodes can only be found either experimentally or with very complex computational methods and algorithms. These methods take the 3-dimensional geometry and compute the eigenmodes on these virtual guitars, violins, pianos, etc. Even with contemporary computers, where multiple kernels run in parallel, these calculations are often very time-consuming and might take hours or even months to run, depending on the complexity of the system and the precision the researcher is trying to achieve.

But these eigenmodes are only present when the body is freely vibrating for itself. In real music, they appear in percussion instruments. When a drummer is hitting a crash cymbal, a snare, or bass drum, after the strike the instrument is free to move and therefore does this in its eigenmodes. Still, most musical instruments are different, as we have seen above with guitars or with violins, pianos, dulcimers, or other stringed instruments. There the string is forcing the body of the instrument to move with the frequencies of the string. This is needed as we only hear the sound radiated from the body, and stringed instruments are constructed in such a way that such a forcing is indeed happening. We take this for granted, as all stringed instruments work that way. Still, this behavior is not straightforward, it could be vice versa, and the resonating body would tell the string in which frequencies to vibrate. So another crucial aspect when constructing a musical instrument is building it in such a manner that one part of it is winning the game. In most instruments, this part is the one which is vibrating in such a way that a harmonic overtone spectrum of the relations 1:2:3... results.

Still, many other cases exist, and over the course of the book, we will find all kinds, where the relation between the forcing part and the forced part might turn around. Also, in cases where the relation is so complex that very fancy sounds are produced with many tones playing at the same time, fluctuating sounds, noise, or the like appear. This is often the case because while playing normally, a simple tone with harmonic overtones is achieved within the instrument geometry, but it might be achieved only just, and only slight deviations from normal playing conditions might drive the instrument in other sounds or regimes, to use a scientific term. Such cases were often used in Jazz and Rock, Contemporary Classical Music, or Free Improvisation there called extended techniques. This behavior is often very much speech-like and reminds of human or animal articulation. It is crucial for understanding the origin of music, as well as the influence music has on us, and why we need it and use it every day.

Explaining musical instruments as self-organized systems, as we will do below to much extend, is a very recent finding. Traditionally, musical instruments are mainly explained by their eigenmodes and eigenfrequencies of the resonance bodies. Still, as we will see below, we are not able to explain crucial parts of musical instruments in such a way. Additionally, it always has been hard to relate musical instrument geometries with their characteristics, the way they are heard and perceived. This relation is indeed complex, and finding musical instruments self-organized systems is capable of relating physics to psychology and culture.

One of the first to ask how such two systems in a musical instrument interact and why one wins the game was Aschoff [7]. In 1936 he published a paper where he discussed the saxophone, an instrument quite new at this time and associated nearly only with Jazz music. These days, Jazz was considered as quite primitive and Aschoff was reminded by the sound of a saxophone on the bleat of a sheep. He tried to understand what caused this bleating and found a strong amplitude change within its initial transient phase. He also found a certain frequency range much stronger pronounced with the saxophone than with other wind instruments contributing to his perception.

Furthermore, he found the saxophone to basically consist of a tube and a reed. This reed is attached to the mouthpiece of the saxophone and is most often made of bamboo. Today we consider the instrument to enlarge to the mouth cavity and neck of the player, manipulating the timbre of the instrument. Still, the basic parts are the tube and the reed, both able to vibrate, both with their respective frequencies.

Still, it is obvious that the tube wins the game. This is why we can play different notes or pitches on the instrument by using different fingerings, so by opening and closing the valves attached to the tube. If the reed would win and force the tube to go with the frequencies of this reed, opening and closing valves would not change the notes, which would make the instrument pretty useless. So indeed, the saxophone is built in such a way that the tube is winning, which we take for granted as we are used to the saxophone working this way. Still, one time in the history of man, one needed to be the first to invent this basic idea of attaching a reed to a tube. Although we do not know who this might have been, we might imagine that this construction was not straightforward at first. The reed needs to meet several constraints, and the mouthpiece needs to be constructed in a specific way to make the whole process working. Go ahead and take a tube and attach a bamboo piece to one end and blow in it, you might experience a sound far away from harmonic or nice, and maybe you would have problems getting a sound out of the instrument at all.

What Makes a Musical Instrument Playable?

When looking through the history of musical instruments, it is interesting to see the hundreds, or maybe thousands of experimental musical instruments have being invented and tried to establish in music performances. Nearly all of them failed. The few we actually use today meet a certain set of constraints that make them usable at all. The most important ones are:

1. loudness,
2. fast attack of tones, and
3. a wide range of possible articulations.

Loudness is needed to be heard at all. As we have already seen with the guitar, a single string without the wooden box attached to it is nearly not audible. The box makes things better, but even today, classical guitars are considered low in volume, which is a constant problem. Electrical amplification has made things better, still radiating the sound via a pair of loudspeakers is missing the complex radiation behavior of the instrument, which is again an important part of the sound.

The fast attack is important to be able to play fast. By fast, we mean that the tone establishes within the 50 ms we found already important with the guitar. The 50 ms is a threshold below which we are no longer able to distinguish between different parts of a sound. So we do not hear if an instrument is sounding maybe noisy for the first 20 ms and then smoother for the next 30 ms. What happens within the first 50 ms is taken as a whole by the brain, and we hear all that happens within this short time frame as one event. Interestingly a similar threshold exists for vision. When we take a time interval $T = 50$ ms we can find the frequency by the simple formula $f = 1/T$, so how often fits T into 1 s. In this case, the answer is $f = 20$ Hz, and therefore 50 ms fits 20 times in one second. Now, this frame rate is about the number of pictures a movie needs to have to be perceived as continuously moving. The old Super 8 format of home cinema, filmed on celluloid, is using 18 frames per second, or 18 fps, which is a bit 20 Hz. Still, the number of frames presented to a viewer per second must not be less, otherwise, the movie is not continuous, and the film is jolted. So, also with vision, there is an integration time within which all events happening are integrated, added together, fused to one event or sensation. In music, if an impulse is repeating 20 Hz, we hear a very low tone. The lowest tone on a piano is 30 Hz. Still, if the repetition rate, the frequency, is decreased even more, so 15 Hz 10 Hz, one does no longer hear a tone but a fast rhythm of single events. This is because the ear and brain are now able to hear time segments with an impulse, and those without them, which leads to the impression of a fast but discrete event flow.

Now it is clear that it has advantages for musical instruments to have the time of their initial transients, their attack time, below these 50 ms. Only then we hear a characteristic beginning which right away becomes a quasi-stable sound (quasi as musical tones are nearly always fluctuating in time in one or the other way, and therefore music consists of only transients, so to speak). If the attack of a musical instrument is larger than 50 ms we find it to come in slow and heavy and with much gyration. It is very hard to play fast with such instruments, as the next tone appears without the previous being fully established. The player has the impression he needs to put much energy and power into the instrument, it is hard to play. Although this effect might sometimes be wanted, if a musical instrument is not able also to play with a fast attack it is of no use in most musical genres.

Yet the third feature a musical instrument needs to have to be useful in most musical situations is a great amount of possible articulations. So the musician should need to play loud and low, hard and soft, scratchy and sweet, and maybe also with extended techniques, so in a way the instrument was not originally intended to be used. Again with most instruments we can buy in a music shop, this feature is present and we are used to that. Still, often, cheaper instruments show a considerable lack of articulatory possibilities, where one might get the impression the instrument has only 'one tone'. So with good instruments played loud and low, the loud tones are not only louder but sound different, and there also might be some thresholds from which on the sound is suddenly changing. With cheap instruments, one might also play loud and low, but it will sound more or less the same all over the place.

Damping Is a Feature, Determining Who Wins the Game

Now to come back to the saxophone, it appears that the reason why the tube wins, and not the reed, in determining the pitch, is that the reed is much more damped than the tube, as Aschoff has already assumed. Of course, things are much more complex, and we will discuss the process in detail in a later chapter. Nevertheless, indeed the reed decays its sound very fast, while a tube or a wave in air is decaying much slower. Damping is a crucial parameter within the fundamental process musical instruments produce harmonic spectra. There will be different kinds of damping discussed below.

With a reed, it is the bamboo which has much internal friction, and therefore when plucking the reed alone, the sound is very short and immediately gone. The precise process of damping in wood and other material is still under debate, and it will be interesting to see any research progress here in the future. Still, the damping itself is easily found and measured by simply bringing the reed into vibrations and see how long it moves until it stops. To do so with the tube is more complicated. Still, we can hit the end of a tube with our plain hand, which makes the tube vibrate. The sound is short too, still longer than that of the reed.

The damping in air is known to be very small, so that this kind of energy dissipation, the energy which is disappearing, is mostly neglected. This also plays a role in room acoustics. There the only way for a sound in a room to lose energy is when this sound is reflected at the walls, ceiling, or floor. Here a part of the sound energy goes into the reflecting wall, and after the reflection, this energy is gone, it is dissipated, it is damped out. The air in a saxophone tube is constantly reacting with a wall, which is simply the tube itself where the wave is traveling along. There much damping appears, and the sound starts to decay. Still, this is less compared to the damping of the reed, and therefore the reed is much more damped than the air in the tube.

So at last for wind instruments, we seem to have found one of the principles which determine who in a vibrating system with several vibrators wins the game and tells the other part in which frequency to go. The winning part is less damped than the forced one. This is clear, as the damped one is vibrating much less, and so after this part has ended its vibration, the other part is the only left vibrating, and is then able to vibrate with its eigenmodes and eigenfrequencies. Then these eigenmodes drive again the other vibrating part which is damped stronger. The saxophone reed is driven by the tube, and therefore the tube will determine the system's frequencies. Again this is not the whole story, and the whole process is much more complex and fascinating, as discussed below. For now, it is good to have identified a first parameter that determines who wins. There are more to come.

Synchronization as Output

We can also say that the two systems synchronize to one common frequency. Synchronization is crucial for musical instruments, and it is constantly present in the brain within neural networks. Synchronization appears in many kinds, like when two frequencies align or when two vibrations are in the same state, in the same phase. It may be that two brain regions vibrate in the same way or that two musical instruments are aligning. During our investigation, we will find different kinds of synchronization and also different causes of them. We will also find when synchronization breaks down and in which way this is necessary in music, in the sound and timbre, but also in the brain, in terms of expectation and even consciousness. We will find that a constant on and off of synchronized and chaotic motions are necessary to make music production and perception the way it is and that this interplay is crucial. In other words, without a constant change from synchronization to noise and back again, music cannot exist, and our perception of music cannot happen.

So in terms of the music production side, we have found some aspects and important findings of many researchers in this field. It is now time to get an overview of music perception and what we need to do to understand and enjoy music. Again we will begin with some major findings but leave the details and the conclusions to later chapters. This is necessary, as while proceeding, we cannot argue in terms of Musical Acoustics or Music Psychology alone but will need to combine both to arrive at a reasonable understanding of music. This will later on also include philosophical concepts that are closely related to psychology. Also, physiological findings are necessary to understand music, especially the ear and the auditory pathway.

Some Fundamentals of Music Psychology

Justinus Kerner was trying to make the invisible visible. He believed in the spirits of dead ancestors still to be around and speaking to people able to recognize and understand them. Although he did not see them himself, he was sure they existed. People not persuaded were asked to spend a night in a small medieval tower in the backyard of his house in Weinsberg, southern Germany. This tower was known as the former prison of the city where sentenced persons were forced into its closed basement on a rope through a small entry on the top of this basement. Normally these people never returned, and therefore their spirits, according to Kerner, were still around, speaking to people spending a night there. Rumor has it that none of them who tried could stand a whole night there and was convinced about the existence of spirits from then on [108].

To make these spirits visible, he needed mediums. In the case of Kerner, all mediums he examined were women to report him about any spirits. His most famous case was a woman from Prevorst, a small village next to Weinsberg, who suffered from the sudden death of his father and got ill. She then reported to see the dead speaking to her and did prophecy, proposing healing medicine and procedures for healing minor illnesses to friends and relatives. She became very nervous and sensitive to light and sound. Today such a state is explained by epilepsy of the temporal lobe.[22] Kerner believed her and met her on a regular basis. In the history of psychotherapy, this is considered as the first case of such a treatment, with documented sessions between a patient and a psychiatrist. Therefore Kerner started what we know today as psychotherapy in the modern sense. Kerner was writing down his findings in the book 'Die Seherin von Prevorst' [154], which became a best-seller.

This regular and serious documentation of Kerner was only an extension of his job as a medical doctor. He was trained in the clinic of Authenrieth in Tübingen, one of the most advanced and respected clinics at that time. His Ph.D. of 1805 was about hearing and displayed a number of empirical investigations not seen before in this area. By extensive measurements, he was the first to discover that the ears of man and women are statistically of different size.[23] He also investigated the frequency

© The Author(s), under exclusive license to Springer Nature Switzerland AG 2021   35
R. Bader, *How Music Works*, https://doi.org/10.1007/978-3-030-67155-6_2

range animals can hear, like that of cats, dogs, or moles, in experimentally creative ways. To estimate if a mole hears lower or higher frequencies, he bent a rod at the mole's tail and let it dig into a box full of earth. He then pointed several musical instruments towards the box, playing low frequencies with the tuba and higher ones with the violin. He estimated that the mole would move if he hears a sound, and this movement could be detected by Kerner through the rod attached to its tail.

Kerner was also solid in many other ways. He was the first investigating Botox (Botulinum toxin), one of the strongest nerve poisons known today. At his times, meat was valuable, and rotten meat was not thrown away but cut to a sausage salad with much vinegar as dressing for not smelling and tasting the rotting too much. Many meat dishes with much vinegar were common these days, and in Swabia where Kerner lived, it was the Swabian sausage salad. Still, many people died when eating it in strange ways, like by suffocation or kinds of paralysis not directly associated with eating. Kerner got a grant from the Court of Württemberg to investigate this strange death and found a poison which was paralyzing muscles and senses. Next to doing animal experiments, he also heroically dropped some poison on his tongue and found it insensitive afterward. This is where the name Botox comes from, the Latin word for sausage is *botulus*. Although Botox is one of the strongest nerve poisons known, it was never used as a chemical weapon in any war. Still, we hear that Hollywood stars inject rotten Swabian sausage salad into their faces to look good.

In these days, a systematic field like music psychology did not exist, which only came into place with Helmholtz notion of perceptual roughness in 1863 [128], Gestalt psychologists like Wolfgang Köhler (1887–1967), Kurt Kofka (1886–1941), or Max Wertheimer (1880–1943), who developed their theories on melodies, or by the work *Tonpsychologie [Tone Psychology]* by Carl Stumpf in 1883–1890, discussing tonal fusion [231]. We return to it later. Still, Kerner provides us with an early application of music therapy. He played the jew's harp, which in his days was a virtuoso instrument in Europe, with professional players making a living from performing the instrument and touring. Kerner reports having used his harp to mediate his patients, also using it in a psychiatric clinic. Such clinics came into place only after the French Revolution, following the ideas of the Age of Enlightenment, which proposed to treat the mentally ill as patients rather than to send them to prison or kill them. Still, in Württemberg, a more liberal spirit was already around in 1753, wherein the local prison in Ludwigsburg, the city where the court of Württemberg lied, a section of 'the mad' was established called *Irrenhaus*, a madhouse. Occasionally Kerner, in his youth, made an apprenticeship as a clothier in a house right next to this madhouse, and he reports that he was visiting the inhabitants, easing their moods by playing his jew's harp. Later on in his life, working in the Authenrieth clinic, he was treating the poet Friedlich Hölderlin (1770–1843), who became mentally disturbed in the middle of his life. If he also used his jew's harp when treating Hölderlin is not known.

Musical Instruments Treating the Psyche

Modern music therapy became more sophisticated over the decades. Calming people by playing music today is known as mood regulation rather than as music therapy. Still, the stage was set, and the psyche began to replace the Christian notion of a soul. In line with this, other musical instruments were used to influence this psyche. The glass harmonica or armonica, invented by the American inventor, author, and diplomat Benjamin Franklin (1706–1790) was one of them. Glass bowls of different sizes are attached to a rod, which is rotated using a mechanism of a sawing machine. When touching the bowls with wet fingers, the friction between the fingers and the glass produces a sweet sound. As all glasses are right next to each other on the rod, the player is able to play melodies much easier as with wine glasses, separated quite a bit from each other. One can also play chords by using several fingers at the same time, again an improvement, as with wine glasses, a maximum of two glasses can be played simultaneously.

The very soft sound of the glasses was considered fascinating and very much influencing the psyche, or better to speak, the nerves. Nerves became increasingly known publicly since the finding of Luigi Galvani (1737–1798).[24] The narrative is that he was preparing frog legs for dinner. Suddenly a lightning truck next to his house and made the frog legs move. He, therefore, discovered electricity, while at the same time found that body parts driven by this electricity were moving. This mechanism was found to works through nerves, which seemed to play a role in movement. Still, nerves were also made responsible for mental and spiritual phenomena.

If Benjamin Franklin was intending spirituality is not very likely, as he was a reasonable man (and part of the story with the lightning, as he was also inventing the lightning rod). Still, the glass harmonica was then used by Anton Messmer(1734–1815), a medical doctor and a contemporary of Franklin, influencing the psyche of patients. He was using the armonia again for mood regulation. Still, he also very much improvised on the instrument, in a time which today is known as the classical period of music, centered in Vienna. Musicians like Ludwig van Beethoven (1770–1827) or Wolfgang Amadeus Mozart (1756–1781) were living, composing, and, not very well known today, were very much improvising on their instruments. Messmer was introducing the 17-year old Mozart to the instrument at Messmer's house in Vienna in 1773, and Mozart was composing for it later in his life. He also mentioned Messmer in his opera Cosi van tutte (1790), where he presented a magnet and attributed it to Mesmer [239].

Messmer was known for a unique technique of moving the hands over the body of a patient in a small distance, not touching the patient, very slowly. Through these movements, the animal magnetism (animalischer Magnetismus) was supposed to be at work influencing not the physical body of patients but their anima, their souls, or psyche. This technique became very popular these days, and because of the great number of people, mostly women, demanding such treatment, magnetic banquets were invented. For such banquets, a bucket full of sand was used, in which the ends of maybe five or seven robes were dug. Each patient held the other end of one of the robes, sitting in silence, waiting for the magnetism to work. In such a way, many persons could be treated at once.

Today Mesmer's technique is considered a light form of hypnosis. He never claimed to have seen any dead or that this would be possible at all. Still, also Kerner was using the 'magnetic lines' (magnetische Striche) with his patients, and late in his life corresponded with the descendants of Mesmer, who had died in 1815, to publish a book about Mesmer's life.

A correspondence between magnetism, music, and health had already been proposed before, most prominently in Athanasius Kircher's (1602–1680) work about magnetism from 1641.[25] Kircher published a melody suitable to heal people being bite by a tarantula, therefore showing strange behavior like wildly dancing the tarantella. He explains that it is a magnetism that attracts the spider's poison out of the body due to that melody. Kircher was a professor of mathematics at the Jesuit Collegio Romano in Rome when asked by Marin Mersenne (1588–1648) to consult his Jesuit colleagues around the world to measure the geographical degrees of longitudes using magnets, which caused Kircher's interest in the subject. The book, with over nine hundred pages, contains many more examples of purely imagined relations between natural phenomena and magnetism to the great amusement of his contemporaries.

To continue with musical instruments: the association between a glass harmonica, as a friction instrument, and the psyche was established, and in the following many other friction instruments were invented, build, and played. We have already found the Chladni instruments, the clavicylinder, the euphonium, or the terpodion to be of such kind. These instruments were found to have a unique sound, which was very different from other instruments. The Chladni instruments were advanced compared to the glass harmonica, as the clavicylinder already had a keyboard, like a harpsichord or a fortepiano, and was therefore playable for musicians trained on these instruments right away. And we already heard about the terpodion.

It is not too surprising though that in some towns, performances of the terpodion were forbidden, claiming that it ruins the nerves [122]. Still, this does not refer to the machine sounds as we might expect, but to the soft, singing, and voice-like sounds. Nerves in these days were publicly associated with a wide range of phenomena. This ranged from hysteria mostly found with women, up to movements of frog legs excited by electricity from a Leyden bottle, the battery of these days. The knowledge of acoustics, physiology, and neuromusicology which we have today was not developed then, and the mentally handicapped were often still kept in prison or being killed instead of getting proper treatment. So there was a fear that such an instrument could ruin the nerves, and therefore the life of people in an unreasonable way.

We have no recordings of the terpodion from these days, of course, as the first recording machine, the Edison phonograph recording on wax cylinders was only invented in 1877. Still, historical records mention the sweet and heavenly sound of the instrument, sometimes reminding on wind instruments like the bassoon or clarinet. When playing one of the few reminding instruments today, such sounds can indeed be produced, although we will never know how the players in those days did the instrument mean to sound.

Still, it is also very easy to make the terpodion sounds differently. Pressing one key might result in two pitches audible at the same time, the basic pitch of the key and a pitch an octave below. Such sounds are known as subharmonics, as they produce a harmonic sound, in this case, the octave, below the regular key. Still, many other sounds can be produced with the instrument, which reminds of wind instruments. They also remind of the human voice, and many kinds of articulation are possible. Pitch glides can be produced where the frequency is shifting in a glissando manner. Hammering sounds can be played reminding on machines with a regular hammer mechanism. In the low range, the sounds are quite impressive, with a heavy, strong bass and additional energy in the higher frequency range. Finally, very noisy and chaotic sounds can be played, varying also in time.[26]

All these different sounds lead to an instrument with tremendous timbre possibilities. As the articulation of these sounds can be varied by the player by either changing the key pressure or the speed of the wooden rotating cylinder, remember the sawing machine, the musician can produce quite expressive sounds, similar to those of a human or an animal. It might be this feature that is responsible for the association of the instrument with spirits, intelligent beings, which are expressing themselves, which are talking, still not using words or building a sentence, but in an articulatory sense.

There is a long list of instruments around the world that mimic animals or ancestor spirits. The 'chicken' is a rod attached to a membrane. When rubbing the rod discontinuously, the sound of a chicken is produced. Again this is a friction process. Another one is the 'frog', a wooden frog with a regular set of wooden peaks at his back, quite saw-like. When rubbing with a wooden stick over these peaks, the sound of a frog appears. Again a friction instrument.

A very rare and fascinating instrument is the *lounuet*, a friction instrument of New Ireland near the Bismarck archipel [22].[27] It is basically a wooden block in which three large chambers are burned. The wood on top of each chamber is cut at one side, making it a plate which is free at three sides. So the block then consists of three plates, underneath each a hollow chamber is placed, open to both sides. The instrument is played by rubbing the hands over the plates. As the plates are of different sizes, three different tones are produced, from low to high. There are different sizes of the instrument mimicking different animals. The largest can produce a pitched sound with a musical note and sounds very much like the hornbill bird, a very famous and symbolic bird all over Southeast Asia. The instrument can also mimic monkeys. The smaller instrument cannot produce a harmonic, pitched sound but sounds very much like a frog. Still, during initiation rituals of young men, which are performed in sheds hidden from the women, the instruments are played and are about to sound like the voices of the ancestor spirits (at least this is what the women are told).

All these articulatory possibilities are intrinsic to the friction process of the instrument, which we will discuss in the following chapters in more detail. The association with the voice is interesting, as the voice is working in a different way, as we will see. Still, both, the voice-like sounds and the voice-like articulatory possibilities, suggest that there might be a similar mechanism at work with both, the friction process of the terpodion and the human vocal fold sound production mechanism. Also, there are other instruments, like wind instruments, which can strongly be articulated, and we will have a look at their mechanism too. On the other hand, there are friction instruments like the violin, which do have some articulation possibilities, but way less than the terpodion. So to understand the way music works, we need to go into more detail with all these instruments, which we do in some chapters to follow.

Music and Trance

The connection between music and altered states of mind is a very close one, and in many cultures, music and trance are strongly aligned. Still, research in this field is not conclusive, mainly because of two reasons. First, in trance rituals, music is only one element, and many other ingredients, like dancing or drugs, are involved too. Also, music which evokes trance with some people fail to do with others. This is good news, as it means that we are free to decide if they want to go into a trance or not.

Still, on the other side, the music inducing trance is not arbitrary, and several features are often found in different cultures. Trance music is most often highly repetitive with a straight beat and rhythmic or melodic figures repeating again and again, often over hours. Another feature is that trance music is often loud, so nobody involved in the ritual can escape it. *Capoeira* music of Bahia, Brazil, has very complex rhythms played by several drums [213]. The *orishas*, the gods of the *capoira* pantheon, all have their distinct rhythm, and when playing a special rhythm, the gods are invited to come, entering the body of a medium and so taking part in the ceremony. Who this medium is becomes clear when someone is behaving very differently and acts typically like the god is expected to do. Then this medium is followed by the other participants of the ritual, people want to know what it is saying and suggesting for the life of the community or about political or health problems.

But of course, there are examples of trance where the music is slow or low in volume, and still, people get into a trance. The shamans of Nepal use the drum *denguru*, next to many other instruments, like conch shells, bone flutes, cymbals, bells, or singing bows to get into trance [216].[28] This music is much less complex and often low in volume, but it is very monotone. Mediums entering a state of trance in Nepal normally start to shake their bodies automatically and therefore are found to be in a trance. Drums beaten isochronous are found all over the world, for example, with the Teton Sioux of Northern America [70].

Trance is a state of solipsism. Someone under trance is nearly only within his or her world, and not reacting to events happening around. Still, our brains do alarm us when things are changing, when we see new shapes or lights, hear people talking or hear noise, feel touch, or smell something which needs to interest us. There are neurons in the auditory and visual cortex which are trained to react only to such changes. When a sensual input is staying the same, these neurons do not react anymore, which is called habituation.

So getting into a trance, getting inside of us, makes it necessary to keep all these disturbances away from us. Now we can close our eyes, but we cannot close our ears. As it is likely that new sounds will appear around us all the time, one way of making the brain not to be interrupted by external events is to produce a sound that is not changing, a regular drum pattern, or any other repeating sound. This sound is masking other disturbing sounds, like talking or any kind of noise.[29] Then our brains decide that it is not necessary to listen to the external world, and is able to deal with itself.

So although there are other features which make trance possible, like extensive dancing to exhaust the body such that also here no feelings from outside matters, in terms of sound, a repeating pattern is obviously of much help when aiming for a trance state of mind. Yet another extreme is silence, and researchers, thinkers, or meditating monks do prefer silence, not to be disturbed by external events. Putting it the other way around, disturbing us by sound or vision, needs a change in the sensations, things going on. Also, speech-like and informative sounds make us listen. Advertisement, which is made to make us look and listen, therefore needs changing events. Also, environmental noise is often a problem when trying to maintain one's own thoughts. The enhanced environmental noise we are all exposed to is, therefore, also a political issue. People are less and less able to follow their own thoughts, building their own opinions, or making their own decisions, as they are constantly interrupted by external events [8].

Still, next to the neurons only reacting to changes, there are those still reacting to events themselves, even if the events stay the same. The information processing in the brain is using both the information itself, as well as the deviations from a mean [218]. This is good news, as it means we are always able to listen to continuous events when we want to.

Trance is also very well known in Western music history. Christian Friedrich Daniel Schubart (1739–1791) was known to have moments where he started performing music, reciting poems, and other forms of art by improvisation and free invention [259]. This might have started while walking through the city market, or in evenings in pubs and restaurants, or in private gatherings. His position was that of an organist in the town of Ludwigsburg, the city where Kerner was performing the very early music therapy sessions. He was also known for his poems, which did not always amuse Carl Eugen (1728–1793), the Duke of Würtemberg in his residency in Ludwigsburg. Schubart criticized the selling of young people from Würtemberg for English colonial wars, and in poems, he did not find the mistress of Carl Eugen very attractive. Therefore he was arrested and imprisoned for a very long time, which was judged as much too harsh by many contemporaries. Still, it gave him the chance to write his music theory on the Hohen Asperg, a prison which was used for political prisoners since the Middle Ages, and especially during the German revolution of 1848, then called the democrats hump (Demokratenbuckel). After Schubart had enthusiastically performed such improvisations, which could last for hours, he was completely exhausted and needed to stay in bed for days. He was said to have been in a trance-like state and then could invent and perform such music and poetry, he would not be able to perform in a regular mental state.

Such performances were in line with the idea of a genius. In the work of Immanuel Kant (1724–1804), or the view of the arts of Friedrich Hegel (1770–1831) among others, a genius has the ability to translate states, ideas, or visions greater than man into music or poetry, to show them the 'heaven of ideas' through a piece of art. Beethoven would later be the prototype of such a genius in Western art music, and many others could be mentioned. Such a genius was often described as acting unconsciously, acting as a medium to transform the ideas of spirits or God into art. Franz Brentano and others were strongly opposing this idea, considering a genius as someone highly educated in an art form, its rules and methods, habits, and customs, along with a strong sensibility of what fits the psychological necessities of perception [53]. A piece of art of a genius is, therefore, more than a combination of former pieces, or a mere perturbation according to rules. It is a sensitivity of what costumes a style has, carefully selecting ideas according to this style, with a clear idea of what fits human senses. Composition for Brentano was a very conscious act, not at all unconscious. He exemplifies this careful selection by a letter from Mozart, where he describes his composition technique in right this way. Later, Julius Bahle (1903–1986) found what Brentano assumed philosophically now on an empirical basis [31]. He asked major composers of his time to write in a diary all steps of a composition and found that, starting from a selection of ideas, a piece of music gradually emerges by selection, according to aesthetic laws.

So in this respect, Kant was wrong. Still, he was not a psychologist, although his work in the 19th century was often interpreted as such. Kant was a philosopher, and to get a better understanding of the development of psychology and perception, we need to make our way through Aristoteles.

Aristoteles and the Beginning of Psychology

Aristoteles (384-322 BC) may be called the founder of psychology, as he was the first to hold lectures about how an external event is transformed by man into a sensation, and from there to an understanding. Today this work is called *De anima*, which is Latin and not Greek. In the Greek original, the word he used was not *anima*, soul or a sentient being, but *psyche*. It was also not a published book, but more a script for lectures, called his esoteric texts, for they were only meant to be used in the inner circle. Aristoteles most likely has also written books, as Plato did, but none of them have survived. Many seem to have been burned by the people of Athens, who did not like Aristoteles and his school, the Lyceum, too much, and only tolerated him, as he was the teacher of Alexander the Great (356-323 BC). After the sudden death of Alexander, Aristoteles had not protector anymore and needed to flee. Only about 100 BC, after the scripts had traveled around and possibly did rot in a basement for decades, these scripts were published by Andronikos of Rhodos for the first time. His exoteric scripts, his books, got lost over the centuries completely. This makes it sometimes hard to understand Aristoteles, as his surviving scripts are written very densely. Also, he invented many fields from scratch, like logic, metaphysics, or psychology. Therefore he needed to use everyday vocabulary to explain his ideas.

Aristoteles' main idea was in contrast to the basic thinking of his teacher Plato (428/27-348/47 BC). Plato was considering all things and beings to participate with respective celestial beings, and the things on earth were, therefore, only 'examples' of these basic things. So all men were only examples of 'the man' in the Platonic heavens of ideas, a word derived from *eidos*, meaning a picture. So thinks, according to Plato, only come into being because they are manifestations of the celestial beings or idea, living in the heaven of ideas. Plato was initiated in the mysteries of ancient Egypt, and it is likely that he brought these ideas with him, as they were not in the mainstream Greece philosophy before Plato. Before him, in the so-called natural philosophy of the ionic sea, today the western coast of Turkey, such a clear description of an idea of a heaven of ideas is not found.

This thinking was challenged by Aristoteles with the objection of the 'third man'. When a man on earth was only a manifestation of a second, celestial man, then this celestial man would again need to have a predecessor, a third man. This would continue endlessly, and all of a sudden, instead of only one man on earth, then there would be an endless amount of men somewhere. This sounded not too reasonable for Aristoteles, and therefore he replaced the idea of a man being caused by a celestial man by his idea of a man created or caused by another man. So the man right there is caused by his father and his mother, who had the possibility to have a son or daughter, a new human being. When these parents choose to use this possibility, they will found a new reality. This new reality again has certain possibilities within it, which might be realized, and therefore continue the process, life on earth. Therefore Artistoteles established a new thinking, using these two motions of possibility (dynamis) and reality (energeia) as causing all things.

This is an understanding fundamentally different from the pre-Aristotelean philosophy, making modern science possible at all. Instead of dealing with an undefinable heaven of ideas, the cause can now be found in the existing things and people, and by understanding the system of causes and relations. Then one is able to make predictions and propose theories. Indeed, in ancient Greek times, the word *theoria* did mean looking into the future, but mostly by interpreting the flight of a bird or the inside of a fish or other animal, fortune-telling. Only Aristoteles paced the way for an understanding of things by examining their properties and laws, their possibilities.

This is how his script about the *psyche* came into place. He proposes two stages of possibilities to become realities in the process of what today we call perception. In a first stage, a human has the possibility to open his eyes. Still, he might also choose to close them. If opened, he has realized his possibilities, and information enters him, making an impression like a picture or figure. Then in a second stage, he might think about the pictures, as he has the possibility of doing so. Still, he might also choose not to do so. If he is reflecting on the perceived objects, he arrives at an understanding, which is the highest form a human can achieve.

Although this is only a very rough description of the process, such a kind of thinking did not exist before Aristoteles, therefore making him the inventor of psychology. Indeed, until today this is still our basic understanding of perception and cognition, and most research done in this field is only an elaboration of the idea of Aristoteles.

Transcendental Aesthetic

The next step in making this rough process more elaborate was done by Immanuel Kant (1724–1804) in his *Critics of Pure Reason*. There he argues that understanding as again a two-step process, where the outside world, the thing itself *[Ding-an-Sich]*, is transcended through a process of sensual perception, what Kant called Transcendental Aesthetic. Within this process, the sensory data are sorted into categories by intelligence *[Verstand]*. Categories were first introduced by Aristoteles, who introduced ten of them, which Kant reduced to four: quality, quantity, relation, and modality. The huge amount of sensory information from the eyes, ears, smell, taste, and touch are transformed using the categories. So categories in the Kant sense are no boxes where something is sorted in. Categories are machines ordering the input. Each such machine, that for quality, quantity, etc. works differently, leading to a judgment about the input. The thing itself, according to Kant, can never be known itself, as all man can know is transferred through the senses and is, therefore, 'second hand'.

So the first step is sorting, which is done more or less automatically. When we hear music we know we automatically understand where the beat is, we perceive a melody which has a beginning and an end, we understand larger musical forms like a cadence or an uprising of musical events up to a four-to-the-floor beat in dance or house music, and we may understand the musical genre. All these features seem to be so natural and self-evident that we take it for granted. Still, if we try to model these features with modern signal processing algorithms found in modern music production software, we often find many errors the software delivers when trying to perform only simple tasks like finding the onset of notes. Indeed, the best onset-detectors available today only find 80–90% of the onsets we hear [1]. These simple tasks are not at all simple, and our brains do a really great job of delivering these features to us automatically.

Indeed, if we would fail too often, we would not be able to succeed in life. Very often, we need to identify a situation very quickly, like in everyday traffic. Indeed, our system of instantaneous judgment from sensory input is failing from time to time. This was already discussed in antiquity. Then a typical example of such a misperception was that of a stick held in water halfway which looks as if it would have a kink or bend, although it is straight. Another example was found by Johann Wolfgang von Goethe (1782–1832), finding snow to once look white and once blue. Explanations of these misperceptions are found in the refraction of light at a water surface in the first example. The changing perception of the color of snow was only found in the 20th century as an automatic white balance of color, known from photography, which is also hard-wired implemented in neural networks in our brains.

So there is much to say about this first 'automatic' perception process Kant was talking about. Still, the phenomenon is clear, obviously, we sort and order perceptual information through our senses very quickly through types of machinery, resulting in a perception like a color, a musical tone, a word, a friend approaching us, or the like.

Now in the Anglo-American language, this is often called intelligence, which means ordering and sorting things. The CIA, the Central Intelligence Agency, does sort and order information it collects. Still, in the German tradition, Kant was raised *Intelligenz* means more than this. Therefore Kant called the first stage *Verstand*. The second stage or process, which Aristoteles describes as the transition from having a sensation up to understanding it, is roughly equivalent to what Kant called *Vernunft*, reasoning. It is a second stage above perception, which later was also called aperception. This second stage is not instantaneous but takes some time. It is the stage where the results of the first stage are negotiated, thought about, balanced to lead to a final judgment on the present situation or phenomena. This second stage leads to a broader and wider understanding.

Now, this second stage is much more individual, and as it is not automatic like the first stage, it can only be performed when one is taking some time to think about the situation, maybe collect more information, ask around, and balance. This can only be done when people are free to do so, and therefore the *Vernunft* for Kant is equivalent to the freedom of humans to take their life into their own hands, rather than just follow a dogmatic view, a religion, or a state doctrine. Kant, as a leading figure in 18*th* century Age of Enlightenment, was therefore calling this enlightenment (*Aufklärung*) as leading man out of his self-inflicted immaturity.

This, of course, had a strong political implication as a revolutionary process towards a state of individuals overthrowing monarchy and religion. Indeed, this was the basis of the modern middle-European states or the USA as states, governed by the people themselves. Still, to be able to govern oneself, the people would need to be able to make reasonable decisions, leading to common welfare. This was before meant to be possible only by people empowered by God, may them be monarchs or priests, who are the only able to make sound decisions as they were thought to be inspired by higher spiritual realms. If common people wanted to rule, they had to prove that they were indeed able to do so, and Kant gave them the philosophical background by claiming that everybody had the faculty (*Möglichkeit*), the possibility of using one's brain of reasoning. Still, this faculty might not be engaged by individuals and would only come into place by education. The common term of 'making a child a human being' was around then, as a child would first need to be educated to be able to build up his faculty of reason. Only then it distinguishes itself from an animal, which has only the first stage of instantaneous perception but cannot think twice about things.

Transfer Effects of Music and Freedom of Will

Kant was a philosopher and not a psychologist. Still, during the 19th century, his work was very often taken as a psychology in a sense discussed above. When science became more interested in brain research, this two-stage process was taken as physiologically separated. The first stage of instantaneous perception was taken to happen in the brain stem, neural regions all animals have, and the second stage of reasoning was found in the neocortex, which only humans have to a considerable extend.

The part of the neocortex which is not the target of the senses streaming their information to the neocortex is called the associating cortex,[30] following the above reasoning that human freedom might found here. Humans deal with the sensory information by recombining them, separating arguments and details, and putting the things together anew. Therefore humans need to be able to associate things not together in the first place. This is fundamentally different from the first stage of perception with more or less automatic perception and instantaneous appearances of entities like musical melodies, visual objects, and so on. This also means that in the associated cortex, everybody can make free associations.

This is a great amount of freedom, still, it is a problem for science, as no rules holding for all people in all situations can be found anymore. In music, questions like if people are able to learn better when music is played in the background, questions about synaesthetics, the associations of tone and color, or the question if learning a musical instrument makes one more intelligent belong to this field. Although much debate has been on these topics, no scientific investigations could give any clear proof that such connections do exist. Indeed, these things seem to be highly individual, and this is good news for the freedom of the individual.

Still, another problem arises from this freedom, the argument that the brain is a physical system, consisting of electric currents and electromagnetic fields representing our consciousness. The physical laws governing these processes are very well known and purely deterministic. This means that when we know a present state, all future states are already determined, and there is no way an individual could change anything here. If this is true, and indeed it appears so, where is our freedom gone?

Philosophers did already ask this question, and a nice example is that of a man sitting on a chair. He is free to stand up at any time and decides to do so. Then, although he was free to have stood up whenever he wanted, after he has stood up, he would not have been able to have acted otherwise. He had no other choice because of the deterministic world. Still, there is no doubt that the man, when still sitting, had the freedom to stand up whenever he wanted.

This suggests that the freedom man experiences is only an illusion. The world is deterministic, and therefore everything which will happen in the future is determined. The reason why man has the illusion of a free choice is that he is not aware of all processes in his brain and in the world. Therefore his view needs to be restricted, and therefore he has the illusion of free will.

Still, this is not a fundamental problem, as having only restricted information about the world and having a neocortex with an associating cortex is a fundamental feature of each individual. Indeed, this is part of what makes us humans, and therefore the experience of free will is present, whether we like it or not. Therefore, the freedom of mankind, although we know it is an illusion in a global sense, is something we need to deal with, and we do not have the option not to struggle in life about it.

Music and Politics

It might be surprising to discuss free will in a musicological book, but indeed the connection between free will music is a topic of politics, still often not on an explicit basis, but on an epistemological one, in terms of musical perception and understanding. Epistemology has started in the Platonic dialog *Sophistes*. It is both, the starting point of ontology and that of political philosophy. So ontology or epistemology is political right from the start. The question of what we trust, our senses, our judgments, and reasoning, or any dogmas and doctrines, is an epistemological or psychological one. And at the same time, it is building states and nations of respective types. So music production and perception are political right from the start in this sense, and we need to keep this in mind when we discuss basic physical and mathematical findings in musical instruments and the physiology of the ear. We will see that this relation is not only one of an analogy but governed by physical laws, making a physical culture theory possible.

The political nature of epistemology was experienced by many philosophers, psychologists, and musicologists during modern times. Franz Brentano (1838–1917) was one of them. He suggested a psychology based on empirical findings in his book 'Psychology from an Empirical Standpoint' (*Psychology vom empirischen Standpunkt*) [54], suggesting to observe oneself performing an inner measurement (*inneres Messen*). From his ideas, thousands of studies of musical perception arose, where listeners were presented some music and asked to fill in questionnaires, judging the music in terms of many parameters, like loudness, roughness, emotional reactions, etc. That this is possible at all was proposed by Brentano, where he claimed that one is able to give judgments about his instantaneous perception, the first stage of perception. Here people are expected to still react very much automatically to the music, and therefore a scientific understanding of the relation between an acoustic input and a psychological perception is expected and indeed often found. But Brentano was also aware that judgment is different from immediate perception. The judgment is something secondary, where the immediate perception is only remembered by the listener. Therefore the listener, when judging about his perception, is in the state of aperception already.

Still, that Brentano was able to formulate such a suggestion was only possible by overcoming Christian dogmas. The main point to overcome for him and for many generations before, especially during the Age of Enlightenment, was the idea that Jesus was the son of God, and not a human. This might come as a surprise to us today. Still, Brentano, who wanted to become a priest at first but also wanted to marry, had a hard life after following the idea of Jesus being a human and therefore following those who wanted to free man from religious dogmas.

Such suggestions about Jesus were not new in the late 19th century. Prominently Gustav Schwab, a professor of philosophy in Tübingen in the 18th century, suggested that there are two interpretations of Jesus, one as the son of God and one as a human, and lost his job immediately. Also, today, this is a struggle within the Catholic church. When Joseph Ratzinger became Pope, he published a book on right this subject, rejecting Jesus as a human. His reasoning is that if Jesus was a human, then the Ten Commandments were an ethic and the speeches of Jesus those of a philosophy or a psychologist. But in these areas, others are much more elaborate than the Bible is. Therefore the core of Christian belief is the dogma that humans might think but have no chance whatsoever to find the truth. God sent his son and gave his Commandments to the human race, and one only needs to live an appropriate life and will go to heaven after death. If humans were able to understand the world themselves, religion would not have any grounding anymore.

It is interesting to see that in the legal case of the Inquisition against Galileo Galilei (1564–1641) the accuse was not about suggesting that the sun is in the center of the planet system. Galileo was accused of doing experiments. Therefore he had tried to find out the truth himself, which was straight against the dogma of the church, that humans are not able to find the truth, and only needed to believe.[31]

These things do not bother a Western society too much today as these days are over. Still, in many regions of the world, the same strong dogmatic exist, like in with the so-called Islamic State, where the core of the dogma is the rejection of experiments and the fining of the truth by humans. This leads right away to a rejection of music, e.g., by the Taliban, where only restricted religious songs are allowed [32]. All other music is a symbol of individualism and a re-combination of new elements into a new musical song, styles, or genres and, therefore, a forbidden permutation of a religious dogma. The reasoning of such religious groups is not so much that music is touching emotions or the senses, as the religious music these sects allow shall do right this, lead to an emotional relation to a dogma or spiritual leader. What bothers them is the use of reason, the association and recombination, experiments, and the search for truth outside the dogmas.

So again, music and politics are related right away again through epistemology or the choice of a system of perception and understanding. Brentano was in a similar situation. To solve the problem, he again discussed Aristoteles, reading him anew, and therefore triggering the reinterpretation of Aristoteles in our time, freeing it from the scholastic interpretation of the Middle Ages. The basic problem was to justify a creative mind, the mind able to decide for itself. This creative mind needed to still be reasonable and not become arbitrary. The creative mind was the freedom to make one's own judgments. Still, as discussed above, this cannot become arbitrary because of several reasons. One is the ethic problem that we understand certain actings as leading to general welfare and others to chaos and inhumanity. Another problem is, again, that the instantaneous perception of the senses is about the same for nearly everybody, it is inter-individual.

The Creation of Creativity

To solve these problems, he took a notion of Aristoteles from the writing called Metaphysics. There, next to other topics, Aristoteles discusses all combinations of the pairs static-moving and active-passive. Obviously, four combinations are possible, some thing is not moving itself but may be moved or may no. Or something is actively moving others while again either being moved by others or not. Such discussions are typical for ancient Greek philosophy, where pairs of opposite notions are used in a reasoning. Now one of these possibilities, the unmoved mover, was taken in the scholastic philosophy of the Middle Ages as God, as having unchanged rules and moving the world. Then humans are moved by God and at the same time are moving other things in the world, and finally, 'dead' things like stones or water are not moving but are being moved. Now the problem applying this system to humans is that if now a human is able to understand the world, he would replace God in this system, making him the unmoved mover. Still, when people have sensual input, hearing music, and seeing the world, they are moved by this world too. They are the moved movers. But then there are no general rules at all, and everything would be arbitrary, all moving all others. So how can one rule the world while being moved?

This question still bothers us today after postmodern times, where many dogmas and rules have been overcome in the Western world. We appreciate the present freedom, still finding that applying this freedom too much takes too many resources or the world, leading to a global ecological collapse. So again, it is not arbitrary what we do, and taking a world as a great shopping mall is not the answer. The same problem holds for music, where today everything is possible, and thousands, maybe millions of artists follow their own creativity. Still, it appears that many of the very experimental art projects are not at all popular, which did not change over the decades, where listeners had the chance to catch up with the new ideas of tonal, rhythmic, or timbre combinations. So there still seem to be some hard constraints to music, and we will discuss several of them in the following chapters.

How did Brentano deal with the situation? Aristoteles adds another term in his psychology, discussing how human understanding works, which is called *nois poeticos*, the creative understanding. This notion helped Brentano out, as it allowed for general rules in understanding the world while at the same time using these rules by creatively dealing with the world. It also explains the new understanding of rules in the world by dealing and experimenting in the world.

Although again a very rough notion, it is the starting point of what today is called creativity and creative arts. Again we will come to this here and there in what follows, but it is necessary to point out that the idea is not to give up the possibility of general rules when we use sensory input and that using sensory input is not leading to a state of mind only depending on the present input so that we can judge and deal with this input. The relation is much more complicated, and still today, different brain theories exist on how the brain works in general, and all these again have political implications.

In modern neural network research, where basic network properties need to be assumed by researchers, as still there are too many neurons to just model the network as a whole, this basic understanding is part of the fundamental choice of the network architecture. Mostly so-called bottom-up/top-down models are proposed, where the sensory data is processed further upwards the brain, and the upper brain is again reacting on this as in a top-down process. Still, going into more detail, it often appears that the situation is much more complex and that the process needs to be viewed holistically as one self-organizing or synchronizing process. Also, adaptations in the brain to physical parameters is so hard-wired that a simple bottom-up/top-down view is not explaining the perception.

Brain Activities

So the brain is actively dealing with the sensory input. To which extend this happens depends on the level in the brain. Low-level neural networks like the nucleus cochlearis or the trapezoid body [204], the next neural networks right behind the ear, are dealing with the input in a much more automatic way than higher brain regions do, like the frontal cortex, where pattern recognition happens, or the Broca region, where musical syntax is processed [157]. Emotions evoked by music [73] trigger typical emotional fields, like the amygdala, associated with negative emotions, or the limbic system, next to others [49]. Still, the relation between music and emotion is highly individual, as expected. Most studies in the field of emotions do follow the localization idea of the brain, that certain tasks are performed in a certain area.

Still, again the situation is not so simple. Low-level networks are also depending on active processes. There is an active mechanism from the trapezoid body back right into the cochlea of the ear, where efferent nerve fibers, nerves which activate action, are changing the basilar membrane, where the mechanical acoustic wave is traveling along. The nerves act on the stereocilia, which are also the cells receiving the sound and transferring it into nerve activity. Cilia are found everywhere in the body, and muscles are made of them. They can contract, making themselves shorter, which is right what is happening if we use our muscles. So the stereocilia do the same thing in the cochlea does, contracting the basilar membrane at very small regions when activated by the trapezoid body neural network. The reason for this is that the regions on the membrane contracted are much better able to distinguish a frequency, to hear it more precisely in terms of what it is. So there is a feedback mechanism. A sound enters the ear and activates the basilar membrane at a certain frequency. This is activating the nerves at the membrane, which transfer the information that there is a sound to the trapezoid body through nerve spikes. The trapezoid body then activates new nerves going down again to the basilar membrane, right at the position where the sound is present, making the membrane shorter there. This improves the precision with which the frequency is encoded into the nerve spike. So the neural network tries to listen more closely and adjusts to an input, they actively listen to the sound.

So the relation between acoustics and psychology is not straightforward, although it is far from arbitrary. Of course, this relation has always been debated. An early example of such a struggle is the critics of the Newton theory of color by Goethe. Newton has discovered that white light is composed of different colors and can be decomposed using a prism. Also, white light can therefore be composed, combining red, green, and blue light, a fact we use in coding color on computers using RGB colors. Today we also know that there are three receptor types in our eyes, for red, green, and blue light separately. Still, for Goethe, the explanation of the world through science was too cold, and he, as an artist, was opposing. He reports hiking through the Harz, a mountain region in mid-Germany in wintertime, and experienced that the show looked white but sometimes clearly blue. This could not be explained by the Newton theory, and therefore he developed his own color theory, a one based on artistic ideas. Although this color theory may have inspired artists over the centuries, we know today that it is wrong. Still, it is pointing to a property of visual perception we know today as color permanence and use it often in our digital cameras as white balance. When taking a picture of a person with a red pullover in bright daylight, the pullover will look red on film. But if this person goes into a room with artificial light, which might be bluer, and takes a picture there, the pullover on the photo might look green. This is because the blue light does not contain the color red very much, and therefore the resulting color is one between red and blue, maybe some green. Still, if we follow the person from daylight into that room, for us, the pullover looks red in daylight and red in the blue room. So our visual perception system understands that the overall light conditions in the room have changed and takes this into consideration when perceiving the color of the pullover, therefore keeping the red color. It is interesting to note that we do not only know that the color is supposed to have stayed red, we indeed have the sensual, visual impression of red.

This permanence of color, of course, helps us through the day where we have many different light conditions. Still, this happens automatically and is indeed hard-wired in our brains as a three-dimensional nervous system of a color cube, where with a simple but robust mechanism, the neuron representing the color red will maintain to be activated under both conditions, taking the environment color into consideration. This was only found in the 50th of the 20th century neurophysiologically and could not be known to Goethe. What he experienced with the ice being white and sometimes blue was an extreme situation, where the internal system of color permanency failed and produced a blue sensation where it was supposed to maintain the white color. So Goethe was pointing to a problem but was not able to understand it, and preferred to refer the phenomena to creativity.

The Split Between Hard and Soft Sciences

Goethe was one of the first to prepare the ground for the split between the hard and the soft sciences, which happened in the second half of the 19th century. In a famous speech of Helmholtz in 1877 [129] he was discussing the growing gap between both, referring to Goethe explicitly. When Helmholtz was younger, he very much opposed Goethe and demanded a scientific explanation of the world. When the two sciences drifted apart, he was trying to keep them together, addressing Goethe, who was also interested in science. Still, he made his point clear that no hard scientist had ever discovered anything on the basis on the philosophy of Hegel, which was very prominent in his days, and added that soft scientists often were more lovers of the 'warmth of feeling', as he expressed it, than in finding the truth. It is interesting to note that again a political discussion, which is still around today, was motivated by addressing epistemology, philosophical standpoints of perception, reason, and the relation between subjective experience and physical reality.

Fusion of Senses

Carl Stumpf was one of the founders of modern Systematic Musicology, then called Comparative Musicology, aiming for finding laws in music by comparing the musics of ethnic groups from all around the world. Stumpf wrote a book on the psychology of tone [241], there discussing tonal fusion. As discussed above, Helmholtz had found that musical tones consist of harmonic overtones of certain frequencies. Still, when listening to a single note only, we do not hear many frequencies, but only one tone, one pitch, one musical note. So the harmonic overtones have been fused to a single sensation.

This does not happen when we listen to inharmonic sounds, like percussion instruments. These sounds also consist of many frequencies, still, these frequencies are not in a harmonic relation. Then we can hear several frequencies separately. But when the harmonic overtones are in perfect mathematical frequency relations of 1:2:3:..., we do only hear one single tone, one pitch. Therefore the single frequencies need to have been fused to one single sensation.

Stumpf found that such a fusion cannot be understood psychologically, which in his days still very much meant philosophically, by basic principles, but need to have a physiological basis. Indeed, even today, the process is not perfectly clear, and many theories of pitch perception, as it is termed nowadays, have been proposed. This means that no neural network has been identified yet, which produces this sensation.

Still, pitch perception is strong, meaning it is present to all of us in the same way, working completely automatic. If this was not the case, we would not be able to sing melodies or agree on who to play the single dots on a musical score as notes. Therefore, tonal fusion is so fundamental to music that it is astonishing to see that still, we do not agree on how it works.

Now, as pitch perception is so universal and not subjective, we would expect the fusion mechanism to be present in lower neural networks, the auditory pathway from the ear to the auditory cortex. The mechanisms which have been proposed here are all computational in a way that, in the end, a single neuron would be active with a certain pitch [59]. The computations most often suggested neural delay lines, which have the periodicity of a certain pitch. If the periodicity of the incoming sound matches the delay time of the delay line, the pitch is perceived.

Such mechanisms are not very robust. Still, our pitch perception is. So there might be a very different mechanism at work producing tonal fusion, a field-notion of pitch. To understand the mechanism, we first need to discuss concepts of consciousness, which is part of the Physical Culture Theory, we are about to discuss in the following.

The musicological discussion of tonal fusion had the search for visual fusion as a forerunner. During the 19th century, many fusion problems were discussed.

One was the fusion of vision, where the two pictures of the two eyes fuse into a three-dimensional picture, our 3D vision. When we close one eye, we can still judge distances, but we do no longer have a three-dimensional picture. When going to the cinema and watching a 3D movie, this becomes obvious. The cinema screen is two-dimensional, and when we see the movie with both eyes, we have the same picture on both of them. Therefore we do not really see the movie as a three-dimensional scene as we do in everyday life. Still, if there are indeed two different pictures on the screen, and one is only entering one eye while the other is only entering the other, we might really experience a 3D movie.

Previously this has been achieved using colored glasses. Some might still remember the cinema glasses, where one side is red, and the other side is green. Today there are two techniques to realize 3D cinema. Either the glasses have polarized filters, which only allow light with one direction in. Then one picture does only consist of light in one polarization direction. The other, shown simultaneously, has a polarization in the rectangular direction compared to the first one. The other system shuts down one of the other glass, making it completely dark for a very short time span. Then the movie projector is showing one picture at the time when one side of the glass is open and vice versa. When this shutting is very fast, the eye does not see the change, and each eye gets another picture. Then the glasses need to be synchronized with the movie projector, which happens through a wireless signal in the cinema acting on all glasses.

At the beginning of the 19th century, such a 3D visual system was already invented in a much simpler way, of course. People in my generation still remember a toy, which is a looking glass, where one can insert cards with a picture on it in a slot. The cards have two nearly identical pictures on them, one for each eye, and the looking glass ensures that each eye is only seeing one of them. If the pictures are drawn in such a way similar to what we see when we normally look around, one can see a 3D picture when using such a looking glass.

This toy was very popular way back then, but was also a subject of intensive scientific investigations. The problem was how the brain fuses two pictures into one, today also known as the binding problem. Of course, when the two pictures do not meet perfectly, the brain starts interpreting them. It either rejects one side completely or partially, and therefore tries to build a single picture, making consistent sense out of two visual input. Also, Helmholtz, when dealing with optics at the beginning of his career, was discussing many experiments and using it for a theory of sensual fusion and perception in general. Today we know that the fusion of the two pictures of the two eyes in the brain of the cat happens as a neural synchronization, a self-organizing process [94]. We will come back to such self-organizing networks in later chapters.

This discussion was a pre-runner for the Gestalt psychologists, Köhler, Wertheimer, or Koffka at the end of the 19th century, which tried to understand perception as *Gestalt*, a German word meaning schemata or bodies [231]. Their first examples were melodies, where they found that we do not perceive melodies note by note, but as a whole, as a Gestalt, as a fused sensual experience, fused not only in space but in terms of music, also in time. Therefore they came up with basic Gestalt principles, laws which constitute a Gestalt like proximity, similarity, or good continuation of events. Then we fuse or join sensual impressions into one, which leads to a better understanding of the input, but also seems to be an interpretation based on a creative process.

Psychoacoustics

The only ancient Greek woman writing about music we know is Ptolemaïs of Cyrene, who lived somewhere between the 3rd century BC and 1st AC. Writing about the Pythagoreans, who were known as mathematicians, a spiritual cult, and as philosophers [136], she wrote

> Pythagoras and his followers wished to treat sense-perception [aisthēsis] as a guide for reason [logos] at the beginning, to provide it as it were with an initial spark, and after setting off from these starting-points to work with reason by itself, discovered from perception. [35]

Aristoxenos of Tarrent (360-300BC), on the other hand, was known for his pure empirical attitude, to understand music purely from experience and listening. So three attitudes were known these days, two from the Pythagorean side of listening first as a basis of understanding, and not listening, but deriving theories from first principle (what later Platon did), and one side of Aristoxenos insisting on using perception as the only base of understanding. The greatest music theorist in ancient times, Archytas of Tarrent (435-410 BC), also a Pythagorean, was with those doing both, listening and reasoning. His discussion on tonal systems is the first to mention microtonal intervals, those smaller than a minor second. He, therefore, divided the tetrachord, a musical fourth (4:3) in three ways, enharmonic (the real harmonic one), chromatic (the colored one), and diatonic (the one with tones left out) [255, 256]. The real harmonic one is that of microtones obviously used in ancient music, which is prolonged by the maqam music tradition of the Middle East, Africa, and Asia until today. The monochord (one-stringed musical instrument) used for these examinations is still around in Subsaharic Africa, and from there was introduced to the Blues of the Southern Appalachian mountains [167]. In the West, the diatonic system is used, a more simple scale with tones left out (the microtones). So Archytas has very well derived his music theory on both, examining the tunings of musical instruments in his time, as well as finding mathematical relations between them. The Pythagoreans used music as an example to understand the world and the cosmos using an aesthetic example, music. Therefore Pythagoreans were the first we know to do music psychology, acoustics, and theory, they were the first Systematic Musicologists.

The relation between the physical reality and the perceived world was formulated systematically again at the end of the 19th century by Ernst Heinrich Weber (1795–1878) or Gustav Theodor Fechner (1801–1887), who termed it psychophysics and, in terms of music, psychoacoustics [231]. The general method of this discipline in these days was to give a judgment about a change in the physical setup. How much weight must one add to a weight in someone's hand that the person will perceive that the weight in his hand has doubled? How much louder must a musical piece be that a listener has the impression the loudness of the song has doubled? So it was not asked about absolute values but about relational perception, the doubling of a sensation.

This is reasonable, as it turned out that all our senses do work as compressors, they compress a very large range of physical input into a small range in perception. For example, one needs to raise the physical energy of a musical piece by ten to obtain a perceived loudness, which is doubled. One needs to double the frequency of the sound to have a perception that the pitch has been raised by an octave. So if a frequency 110 Hz is double 220 Hz, one perceives an octave. Therefore the frequency was raised 110 Hz. But also, when the frequency 440 Hz is doubled 880 Hz, one perceives a rise in pitch by an octave. Still, in the second case, the frequency is increased 440 Hz, which is four times more than 110 Hz raise of frequency in the first case.

So our senses compress the input. The benefit is that this creates a reasonable overview of a large range of a physical parameter. This can be expressed in a logarithmic way, and, therefore, each sensation has its own logarithmic basis. The pitch octave has a basis of 2: one needs to double a frequency to reach an octave. Still, the sensibility of pitch is the question of which differences of pitch we are able to just perceive. These so-called just-noticeable differences (JND), under best conditions, are so small that we can divide each semitone into 100 smaller intervals! This makes 1200 such intervals in the octave. The logarithmic basis then is $\sqrt[1200]{2} = 1.00058$. The sensation of loudness has a basis of 10: one needs to raise the physical energy of a song by 10 to get the impression of doubling the loudness. If we multiply the logarithm by 20 we reach the JNS for loudness to be 1 dB. In terms of pitch, the mathematical formula then is

$$\text{Perceived interval between Frequency}_1 \text{ and Frequency}_0$$
$$= \log_{\sqrt[1200]{2}} \frac{\text{Frequency}_1}{\text{Frequency}_0}.$$

So the logarithmic basis is $\sqrt[1200]{2}$ here. For loudness, this relation reads

$$\text{Perceived loudness between sound pressure}_1 \text{ and sound pressure}_0$$
$$= 20 \times \log_{10} \frac{\text{sound pressure}_1}{\text{sound pressure}_0}.$$

So here, the logarithm basis is 10.

Indeed, pitch interval perception has the smallest logarithmic basis of all our senses, including vision, smell, taste, or pain, and therefore it is our most sensitive sense.

Although this is a very good approximation at first, of course, things are often much more complicated. Loudness perception very much depends on gender, where women perceive the same physical song energy as louder than men. This is caused by the general difference of the outer ear between women and men. The ears of women transport the sound stronger to the inner ear that the ears of men, what Kerner found.

Another example is that sounds with a harmonic overtone structure are perceived very much louder than noise, which has a broad spectrum, like water waves on a beach or wind in the trees. This is a tremendous problem when measuring environmental noise, making us sick, as both sounds, the harmonic and the noise sound, might have the same physical energy. Still, ocean waves might calm us down while the deep of a back-driving van gets on our nerves—as it is supposed to do.

In the 20th century, especially after World War II until now a tremendous amount of research has been conducted in this field of psychoacoustics. Still, many problems are still very much under debate, and it appears reasonable to direct research more into the direction of neural networks to understand perception.

Still, again politically speaking of environmental noise, psychoacoustics is and will be the basis for laws. Unfortunately, the laws in Europe and the US are still based on scientific findings of pre-World War II, where the world sounded considerably different compared to today.[32] Still, restrictions based on today's knowledge would lead to a performance reduction of cars, trains, and aircrafts. Such performance reductions are accepted by consumers when they feel an immediate benefit from it, i.e., when they are opposed to their own noise, but are not too welcome when the benefit is more for the environment.[33] Still, as all consumers are also the environment of the others, one might hope for some changes in the future.

Some Fundamentals of Comparative Musicology

If we find so many things in common in music, in terms of acoustics and psychology, general rules would need to be found in the tremendous diversity of musical cultures around the world. What has Techno music in common with Mozart, the Burmese *hsain wain* ensemble, African xylophone music, the Uyghur *sashmaqam*, or Balinese *gamelan*? What do the choirs of South Africa, Andalusian Flamenco singing, Beijing opera voices, or Mongolian throat singing have in common? Is electronic and experimental music produced in nearly all countries of the world the same? What about Global Hip-Hop, Mainstream Pop, and Electronic Dance Music?

Music is a universal language. This is what we often hear. But when we listen to musical genres of ethnic groups, we do not belong to, and which we do not know, we often face the problem that, although the sounds may be interesting, we do not really understand what is going on. Although we might hear some melodies, we are often not really sure if we understand the language of that music, its intention or emotion. Still, that does not mean that we understand nothing. We might be able to clap or tap to a beat, understand beginnings and endings of phrases, or that the music is getting more or less dramatic. Still, often we are lost, and although unknown music might be interesting to listen to for some minutes, we might get bored quite soon and find the music 'repeating itself'.

Music, DNA, and Sexulality

So maybe the differences in musical styles around the world are caused by special features of single ethnic groups. Relating musical genres to human DNA and genomes of specific ethnic groups have indeed been performed. A correlation between music, language, and DNA in Taiwan showed correlations between music and genes, as well

R. Bader, *How Music Works*, https://doi.org/10.1007/978-3-030-67155-6_3

as language and genes, but not between music and language [55]. Indeed, there are correlations between certain musical features and the genes of the performing ethnic groups. Also, genes correlating with musical talent have been found [246].

To make such investigations, one needs to define musical features to then compare them between ethnic groups, not an easy task. This is most often done by defining music as a set of musical features and examine if these features are present in a music of an ethnic group, a genre, or the like. The tonal system used, the presence or absence of a solo singing style, the presence or absence of polyrhythms, can they be related to a certain gene? Does a certain gene lead to the preference of solo singing or to a tonal system consisting of five or seven tones? So how to measure music along with a set of musical features? We feel uncomfortable with this idea as music is a system and not just adding some features together.

There are many comparative studies looking for such features. A large one is the Cantometrics project, comparing singing styles all around the world according to feature lists like emotional content or solo versus choir singing, with a long list of features describing singing [183]. As an example, when relating the very high pitched emotional solo singing style of Flamenco singing in Southern Spain with the more relaxed and less emotional choir singing of the Spanish North, the study concludes that they are related to the cultural background of these regions. Rigid Catholic moral and sexual restrictions are taken as the cause of an emotionally more dense singing of Southern Spain, while the more relaxed, liberal view on sexuality of the North is concluded to lead to a more relaxed singing style. Although the relation might seem reasonable in some way, is this a strict proof? Would we not need to consider the whole system of relations in the cultures to really decide on this point?

And if there is a connection between a cultural and a musical feature, is this connection necessarily present? Do we expect a certain musical feature to always be present in a culture each time we find the respective cultural feature? Each time a culture shows sexual restriction, can we be sure that there is highly emotional singing performed? There is still free will in place, and of course, we are free to decide if we chose one expression or the other and therefore are not necessarily slaves of our genes or our sexual habits. But maybe we need to argue more freely and say a culture with sexual restriction is more likely to have high emotional singing. Music is often performed for improving the quality of life, is motivated by emotional problems that can be released with music, but also with social interaction. But these might also be motivated by intellectual questions, lifestyle decisions, or the pleasure of building new sonic worlds in music. Then there might be a relation found between music and outer musical topics with a certain likelihood, still always based on free decisions of performing such music or not.

But this would mean that at least the relation between a musical feature and a cultural one is that of the effect music has. In a culture with sexual restrictions, emotional singing will always lead to relief. Still, because of some reason, people in this culture might decide not to use music for this purpose. Then we would find a relation between a musical and a cultural feature to work in all cases present. This can be tested scientifically and is addressed in studies of musical side effects. Does background music turned on when learning vocabulary improves learning? Does listening to music while jogging increase the motivation to continue? Do we become more intelligent when listening to Mozart? The answer is always the same. Some do, some do not. Scientifically there is no relation between cause and effect holding for all people [3]. To have some positive relations does not mean that there is a relation when there are also some negative ones, and the percentage of improves/no improves is a 50/50 chance. So music might be a doctor, but one different for each of us.

But again, music is not present in outer space, we use it every day, and there are many reasons for doing so. It affects us, and as we have seen in the acoustics and psychology chapters before, there are so many things we all have in common in terms of music. So this does not mean that there are no connections, relations, or laws in music over ethnic groups of genres or music, performance styles, or likes and dislikes. It is only that the whole story is much more difficult. Music is a very complex system with some input parameters and many connections in between, leading to an effect. Such a system might be so complex that it is hard to be determined.

Does this mean it is not worth trying? Of course not. This becomes clear when we focus on another system which is extremely complex, the human body. Medicine is still not a science with such strong and systematic laws as is found, e.g. in mathematics or in physics. Billions of cells, thousands of proteins, neuronal connections, and endless molecular interactions in the body make it impossible until today to determine what will happen to the body when we treat it in a certain way. We know this from our doctor consultations. Medical doctors ask us about our symptoms and investigate our body, then associate these findings with an illness and again associate this illness with treatment, some pills, exercises, or maybe even surgery. Still, the relations between symptoms, illness, and treatment are associative. The human body is far too complex to be able to know the precise development of all cells and molecules, which have to lead to the illness. And this situation will continue, as in the near future there is no hope that this will change.

Still, do we stop from going to a doctor, although we all will have experienced some times that a doctor is not always perfectly sure about what is really going on with us? Of course, we do go there because we hope that the associations the doctor does are going in the right direction, and the pills or treatment we get will make things better. So experience, and scientific experiments, have told the doctor what to do, so he is able to give us his best guess.

It appears that the system of music built from the thousands of musical cultures around the world is just this, far too complex to be understood as a system yet, in its relations, its functions to the people, to individuals, and their development. Only the fact that we do not understand the relation between music in the cultures does not mean that there is no such system. The only question we can raise is how do we approach this system. Will we one day be able to input all music into a computer to calculate how music will develop over the next centuries?

There are several approaches around to find such a system, and it is interesting to see how they deal with the problem. Some of these approaches are based in the 19th century, some are recent, and some are only building up now and may give us insight into the problem in the years to come. Throughout the book, we will favor one method which is able to do both, consider music as a highly complex system, while not needing all small details to determine its performance and output. It seems that this method is not arbitrary, and nature has already used it for building life. We used it for building music—and maybe culture in general—for it is both, lively and robust. Still, first, we want to look at the development of Comparative Musicology to study music of ethnic groups around the world in the light of its methods and ideas of what a system is. To warm up, we consider some musical traditions of the world and their development from ancient times to the end of the 19th century.

Diversity and Change

Music was always a fusion of styles, some local, some global. Historically, European music often spread into colonies, like ballroom music, the counterdance, or waltzes, which fused with local styles and developed further. Still, often it was not really mixing up with the music of other countries, not to speak that the music outside Europe was taken too seriously by Europeans way back then. Eduard Hanslick, a music critique, in 1854 was still arguing that the "... wild howling of the South Seas islanders ..." ["... das wilde Geheul der Südseeinsulaner..."] was just this, a howling, but not music. A view which made it possible to postulate a development of music, starting from Middle Age single-line melodic music, like Gregorian chanting, developing into polyphonic music in Renaissance and Baroque times, and reaching its climax in homophonic music of Beethoven, then containing chords and melodies. After that, music was often found to have declined into neo-Romanticism and Contemporary Music as composed by Wagner or Schönberg.[34]

One interesting example of a remote survivor of colonial music was still found in Brazil in the 1940th. Western colonialists in Southern and Middle America would play their contradanzas and waltzes. In the remote region of *Minas Gerais* in Brazil, this music had survived until the 1940th and was recorded there by Corrêa de Azevedo,[35] a rare insight into a lost world. It could survive there because this region is known for diamond production and, therefore, quite detached from the rest of the country. It is fascinating to hear these contradanzas today.

In Nothern America, in the New England states, European classical music was in much favor in the 18th and 19th century, people played guitar, violin or the like. Instrument builders like Christian Friedrich Martin (1796–1873), a European who had learned his craftsmanship in Vienna, settled there, and his company is still one of the major producers of Western guitars. Benjamin Franklin (1706–1790) was bringing forth instrument building by developing the *glass harmonica*, an instrument to ease the playing of singing glasses. All this happened in the tradition of Western art and country music and did not involve the music of the indigenous people. The same holds for the colonies in Asia or Southeast Asia, China, Africa, or other places.

Still, many non-Western cultures prolonged their music tradition, of course. China has a music tradition that had a strong musicological follow-up for thousands of years. During the Han dynasty, 206 BC—220 AC, Emperor Wu (156-87BC) enlarged an already existing institute of musicology *yuefu* (lit. Music Bureau), employing more than 800 people, also including musicians and dancers. Music was state-building as China relating 'everything under heaven' [266]. Until now, China has a very strong and rich ethnomusicology with extensive descriptions about the music traditions of the 56 ethnic groups in China (which are much more, taking all subgroups into consideration).

Music was also used politically. Large music ensembles were playing at Chinese courts. Musicians were slaves in ancient China and sometimes buried with a deceased emperor. But they could be used as presents to another court so that they could play the music of the giving at the court of the presentee. How much political strategy was behind these presents is not always known. Still having the music of another emperor right at one's own house is a political statement.[36]

Today's China summed many ethnic, cultural traditions, theater, acrobatics, stories, and music into a cultural heritage list and managed to make the Uyghur *muqam* a world heritage [118]. The Uygur people live in the autonomous region of Xinjiang in today's China, a large desert land north of Tibet and part of the ancient and today's silk road. These people are part of the Turk family, and Turkish and Uyghur people can understand much of their language right away. Still, the Uyghur are part of China today, although there are many political problems in recent years calling for independence. Still, China, when deciding who to recommend for the World Heritage recognition, favored the *muqam* instead of the Shaoling kungfu, much more known around the world.

So in China, for thousands of years, music was highly developed, very much diversified, and followed-up by musicology.

The same holds for many other parts of the world. In Southeast Asia, music traditions like the Indonesian *gamelan* music with a large ensemble of bronze instruments and often highly complex textures and timbers are very old too. Still, like all music traditions, it was subject to constant change and modernization. The last big change in the music of Bali was after the late conquer of the island by Dutch soldiers in 1904 and 1908, where the noblemen of Bali were walking towards the soldier only to commit suicide before them. The shock following this act was the recovery of old Hindunese epics Ramayana and Mahabaratha in reading contests. The winner was the one reading fastest, and as these contests were followed by *gamelan* ensembles, their music became faster, too, ending in a modern style called *gong kebyar* [247].

Strong music traditions also exist in Thailand, Cambodia, or Burma, only to mention examples. The Burmese *hsain wain* orchestra seems to be a mixture of the gong ensembles known from Indonesia and the drum-and-shawm bands, following the tradition of Turkish military orchestras of the Janizary. These military bands needed to be loud, which is why they use the *zurna*, an oboe-like instrument, and drums, both of huge volume. These bands were introduced mainly through merchants rather than by military interventions, but they are spread all over Southeast Asia and India. The Nepalese *panje baja* wedding ensemble, the temple music of Buddhist Singhalese in Sri Lanka, the band playing live at Thai boxing events in Bangkok and at such events all around the world, as well as many other such bands with highly sophisticated melody and melismas, complex rhythms and drumming are often very old and also very diverse. They are also a good example of the constant interactions of musical styles all around the world.

Indian classical music was also part of constant change. Most Western people know the *sitar*, the stringed instrument Ravi Shankar played at Woodstock. Still, the instrument was invented only mid 19th century, had its peak around 1970/1980 with seven *gharanas*, teaching spots in Nothern India, and has much declined. [115].

The classical Indian dance was reinvented by the Tangore brothers around 1900 in Kerala, South India. They used an ancient writing about music and the arts, the *nadyashastra* of Bharata as a starting point. In the tradition of the *shastra*, a text genre around 200 BC—200 AC, the movements of the body parts, hands, feet, head, eyes, etc. were described in a long list. The Tangore brothers took this list and formed a dance they called classical Indian dance, nowadays still part of many Bollywood productions [115]. So a book preserved much of a cultural performance lost otherwise. Again, music and dance that are supposed to be 'very old' and 'traditional' are subject to constant change.

Many such examples exist all around the world. One last to mention here might be the music of Egypt, which was considered as a reference in the Middle East in the first half of the 20th century. It had a boost in the 1880th when it was modernized by the Nahda, the Islamic renaissance of the 19th century. The modernization was the fusion of what was considered old Arabic music, the *maqam*, with elements of the West, like operettas or musicals. At the famous Cairo congress in 1932,[37] where Egyptian musicologists met Europeans, from the Egyptian side the hope was that the West would recognize the modernization of Egyptian music, while the Western traditionalists hoped to hear 'authentic' or original Egypt tunes, melodies and tonalities. The rise of the 'long song' in the 1940th of artists like Umm Kulthum or Wahab, building up entertainment music from elements of the *Nahda* is now considered as the Egyptian music in the country. Even the event of dance music starting in the 90th did not overtake the Western music completely, but sounds still very much 'Arabian' or Egyptian, and uses many elements found in traditional styles.

The Beginning of Comparative Musicology

All these examples show the constant change of musical cultures and their interactions. Still, this was not the view at the beginning of Comparative Musicology. Around 1900 it was assumed that the traditional musical styles were ancient and old, and nobody in the West knew about their histories. This is not astonishing taking these times into consideration, where the task was to get to know these musical styles at all.

The 19th century saw the emergence of many comparative disciplines. Comparative Archaeology was claimed by Justinus Braun in 1850 in his work *Der Ursprung der Kunst* [The origin of art] [50]. There he compared the art of ancient Egypt, Greece, and that of the Middle East, then called Assyria, and found the spiritually deep but playful culture of Greece to be a product of the very serious spiritual tradition of Egypt and the more playful one of Assyria, a view not continued today. Still, starting from the conquer of Egypt by Napoleon 1798–1801, the number of archaeological excavations had grown to such an extent that it made sense to start comparing the results of different regions. Other such examples are Comparative Paelozoology or Comparative Linguistics, often defining ethnic groups,[38] among others.

The same comparison happened with music only after the Edison phonograph had been invented in 1877, with which one could record and playback music from all around the world on wax cylinders. Before this invention, Europe was only collecting musical instruments from around the world, mainly from colonies, and we have many reports from colonialists and missionaries about the music from Africa, America, or Asia, which are often very rudimentary. If a writer was a trained musician, he would most often try to find a scale similar to a European one or describe musical performances and rites with colorful words, not really understanding what was going on.

As the knowledge about music in these days was restricted to the instruments colonialists, traders, or missionaries brought with them, it was a really new thing to record this music and store it in an archive. The first of such archives was the Berlin Phonogram Archive, with the first recording of a Thai *phi pha* band performing in the Berlin Tiergarten in 1900 [237]. So there were opportunities for some people to listen to 'exotic' music here and there, especially when it came to World Fairs or *Völkerschauen*, views of ethnic groups at fairs and markets. Often the displayed persons were costumed in a way a Western would expect them to be, and so Northern American Indians were dressed with a feather on their heads, a custom they might never seen in their lives before. Still, these occasions were rare, and nearly no recordings were made of them before these days.

The recordings at the Tiergarten in 1900 were done by Erich von Hornbostel and Carl Stumpf, founding fathers of Comparative and, therefore, Systematic Musicology [231]. Together with Curt Sachs and others, they were asking people traveling around the world to record music, sending them empty way cylinders, and collecting the returning recordings in the archive. These recordings made it possible to compare the musics of the world and therefore look for universals, laws, and rules holding for all music around the world. So the idea was a very modern one, not considering ethnic diversity, but exploring what all people have in common in terms of music.

One such attempt was the development of the musical instrument classification system and organology, developed by Hornbostel and Sachs, which still holds today [135]. The idea was to sort all instruments according to their driving mechanisms, whether they are blown or bowed, plucked, or struck. In many subclasses with hundreds of examples of musical instruments from all parts of the world, this classification is able to sort and order indeed most instruments ever built. It works so well because it is based on physical parameters, the driving mechanisms, and not on cultural aspects or specializations. The basic idea of Comparative Musicology was to ask for such universals, music as universal language, rather than to point to special aspects of ethnic groups or countries.

Another example is the book 'Tonpsychologie' [Tone psychology] of Carl Stumpf [241], where he was asking what a musical tone indeed is. We have addressed this question above, where it was not self-explaining why we fuse so many frequencies into one tone sensation. This question is a starting point of modern music psychology and, again, pointing to a universal, the tonal fusion as present in all people around the world.

In later years there were some attempts to misuse the idea of Comparative Musicology in colonial or even in Nazi doctrine. Blume, a music historian, was suggesting to use the recordings of ethnic groups to show the primitive nature of their music and, therefore, the superiority of Western art music. Heinitz was proposing a biomusicology and suggesting a Jewish music style. Although today's DNA-music relations are showing correlations between musical styles, as discussed above, clear evidence for such a correlation is missing. One problem is to determine the distances between two musics in terms of a number. To say South African music is five units away from North American Indian music and seven units away from Japanese *gagaku* music sound weird. Still, it does not mean that one is not able to compare music at all. Of course, one can compare features, like roughness, brightness or complexity, tempo, timbre, and the like. Still, summing all this up into one unit is not straightforward. We will discuss this below with automatic music analysis and sorting using artificial intelligence.

Comparative Musicology, as started by Hornbostel, Stumpf, or Sachs, was indeed not about telling the differences between people, but stressing what we all have in common, by looking for universals and general rules in music holding for all musical styles around the world. This tradition was then taken over by Systematic Musicology as we know it today, consisting of Musical Acoustics, Music Psychology, and Music Ethnology.

Ethnomusicology and Music Archiving

But not all musicologists recording and analyzing music from around the world are looking for universals. Jaap Kunst is known to be the first ethnomusicologist, someone interested in recording and collecting music of ethnic groups without the background of finding universals, and so simply describing what he found [127]. Employed in Dutch colonial service, he was stationed in Bandung on Java, nowadays modern Indonesia, way back in the 20th which until the end of WW II was a Dutch colony. During his weekends, he was taking a boat traveling to smaller islands in the Indonesian archipelago. He arrived there usually at noon, where the people were working in the fields. So he took his violin with him and started playing in the villages. Curious and ambitious, the natives wanted to show the foreigner their instruments and music, and brought them along to play their tunes. Kunst was recording some of their music on wax cylinders he sent to von Hornbostel and the Berlin phonogram archive. He also compiled many writings about the music, the instruments, and the people.

His hobby became popular, and due to the public interest in his work, as well as the benefit it meant to colonial rulers to know about the natives they had power over, he was appointed for the first full-time position as an ethnomusicologist and therefore started this job. So right from the beginning, ethnomusicology was strongly linked to political interests, which still holds today. For rulers of a foreign country, it is important to know about the habits and customs of the natives, where music is a main part of. It is also good to know what to play on the local radios to catch people in their own cultural setting.

Yet another aspect of ethnomusicology needs to be mentioned, which emerged right from the beginning of recordings and had a peak at the World Music times around the 80th, the commercial one. Taking recordings of musicians form countries where no copyright act is present or where it cannot technically be enforced, and adding some Techno beat or bassline to it maybe adding some keyboard pads, this music is often produced and sold in the West with great commercial success. Of course, the musicians and composers should be adequately paid, which is too often not the case. So World Music, contrary to its image of promoting One World and People Unite, is very often a new kind of colonialism with great commercial advantages of Western musicians and recording companies. To which percentage the number of university positions in ethnomusicology in the West nowadays depend on military support or such from the entertainment industry can only be estimated.

Still, there are many ethnomusicologists around which are independent, and we owe them many recordings and insights into musics, which serve us in many ways. One way is to preserve the music of the cultures for later generations to be aware of and reproduce. One such case is the music of Cambodia after the ruling of the Red Khmer between 1975–78, who killed most of the art musicians of the country, as they were viewed as part of the 'decadent bourgeois' as was the standpoint of the Khmer Rouge. Still, they allowed *mahori*, a wedding and party music played by peasants. Although many Buddhist monasteries were destroyed, many monks were killed or forced into a wedding, no systematic destruction of Buddhism was observed [119]. Cambodian Buddhism has a Buddhist chanting called *smot*, which is beautifully melismatic, to please people and attract to Buddhism. Still, texts are mainly depressive in nature, for example, suggesting to ill people to accept their problems as part of their karma [23]. The Red Khmer were still singing these songs but replacing *sanga*, the Buddhist monk fraternity by *ankar*, the political party. Francois Bizot, a cultural historian, working in Angkor at that time and being captured by the Red Khmer, was reporting that they overtook many aspects of Buddhism into their political party and life [42].

So although not all music was destroyed in Cambodia at this time, most of the court musicians and dancers were killed. Thanks to the archives of ethnomusicologists who had recorded the music before, the music and dances could later be reconstructed, where a new generation of musicians learned their own tradition from scratch, from tapes and video recordings.

As in many cultures, the young people are moving into Hip-Hop, Techno, Rock, Pop, or Gospel and are no longer interested in the music of their parents, it is only a matter of time until they want to know about the music of their forefathers and will lean by the recordings made by ethnomusicologists.

There are other cases where phonogram archives help to launch commercial success without exploitation. Harry Belafonte was raised in New York, but his family came from the Caribbeans. Still, he did not know the music of his forefathers. Therefore, as a young musician, he went into the Library of Congress with its huge phonogram archive and listened to folk songs from the home of his parents. Rearranging and recording them, they became world hits in the 50th and 60th [40].

Computational Music and Sound Archiving

Young musicians all around the world have grown up with the internet, and so with music from all around the world. They know about modern production systems like all-in-a-box, where one creates music on a laptop without the need for a recording studio. Still, these musicians are based in their local and traditional music scene, which is clearly audible in many cases. So although the music did change through Westernization, it also kept its roots and fused to very interesting styles. The World Music scene from the 1980th, where Western producers were recording non-Western musicians in a way to fit into a Western market, is more or less over. The musicians all around the world produce their own music, which is a mixture of all they know.[39]

Of course, Comparative Musicology, Ethnomusicology, and Phonogram Archives are modernizing too [43]. The net is buzzing with millions of songs and musicians, musical styles, and genres, expensive productions as well as DIY projects, live recordings made by cell phones, recording of soundscapes, and many other formats. Nowadays, it seems there exists nothing which is not existing, at least somewhere. All this is accessible right away with a click or two. What was big freedom at the beginning of the internet to post an own story, opinion, or news right away, nowadays becomes a thread in many ways. One is that it becomes harder and harder to be visible on the internet, as it is too crowded. Indeed, there are over 600 Phonogram Archives in the world, some large and many smaller ones. Next to that, there are channels on all platforms presenting music from around the world in all formats, with some storytelling, as plain recordings, with presentations coming along, whatsoever.

To address these buzzing, new search engines are needed to suit the needs of people really interested in music recordings from around the world, which are substantial, sound, well documented, and with an interesting background. In terms of such Big Data problems, this task can only be performed by algorithms. A Computational Music and Sound Archive (COMSAR) idea is, therefore, the contemporary technique to cope with all this [12]. Modern algorithms of signal processing take the recordings as input and analyze them in terms of rhythms, melodies, tonal systems, timbre, and other musical features. Then self-organizing maps are used to compare the pieces according to these parameters. The result is a two-dimensional map where the musical pieces are arranged in. Nearby pieces have similar rhythms, timbres, or instruments used, while those further apart are more different. Such a map then shows areas of similar musical styles or other features.

Such algorithms are able to tell quite reasonably from which country musical pieces come from. It might also show pieces with similar rhythms or timbres. Then it might appear that similar pieces need not come from the same country, ethnic group, or musical style, but that these rhythms are used at several places around the world. Therefore the computational algorithms are able to show the similarity of many musical styles in terms of different features, something Comparative Musicology is looking for, universals, and general laws holding for music around the world.

An advantage of such maps is that there is no person deciding about the similarity of features. The computer is doing so as a neutral medium, free from cultural or political bias. Although the algorithms are built by humans, they are so general that such a bias can only be there if a user decides to restrict the search to only a limited amount of parameters or musical pieces. Still, this is easy to determine, and therefore, a much more obvious description and comparison of music from around the world is possible.

Additionally, such a COMSAR platform can serve as a search engine for listeners, musicians, music producers, or recording companies to cope with the buzz on the internet. They do not need to search endless platforms for music similar to what they know and like. They can insert their musical pieces and let the computer look for similar music. Or vice versa, look for music on the opposite side of the map to learn about completely new ways of making music. We know such suggestions from streaming platforms, wherein the easy listening or in the jogging mode, the streaming platform suggests a similar song to come next. Still, these algorithms are quite simple, mainly looking for tempo and brightness and do not consider any musically interesting details. Indeed, most often, they only suggest what other listeners have favored who also listened to the same piece. This is producing an echo-chamber which restricts the view of listeners to always the same pieces.

Part II

Impulses

Musical instruments work with impulses. So does the human brain, it works with discrete neural spikes. The spikes in the human brain are very well studied, as we see later.

The impulses working in musical instruments are not so well-known. Instruments are most often explained using frequencies. Still, the most important part of a musical instrument sounds is its initial transient, the very start of a tone. If we cut this transient off, we might not be able to tell a trumpet from a violin.

Furthermore, music is constantly changing and is therefore transient by nature in all its features like rhythm, articulation, or musical form. While frequencies do more concentrate on a steady-state, impulses are much more appropriate to represent music in its transient, changing ways.

So first, we need a clear picture of what a frequency is and what one can do with this representation while then to find out more about the impulse picture and understand how instruments can be explained in such a way.

Another important aspect is the damping of vibrations in musical instruments, as it determines how the interaction between the instrument parts work. Only then we can discuss the nature of harmonicity and compare it to its opposite, noise. This will lead us to the information and expression we can transport with music, its meaning, its semantics.

Only when we have understood all these aspects, we arrive at the very core of the picture, nonlinearities, synchronization, and self-organization. As we will see, musical instruments are self-organizing systems, where only because of this self-organization, a harmonic overtone spectrum appears. Without this feature, we would not be able to use musical instruments at all, playing melodies and chords.

Also, many aspects of articulation, timbre, or rhythms are only possible due to this self-organizing property of musical instruments. Although one can also try to reduce certain features of a sound into simple, linear systems using the frequency picture, these are only valid for a small range of playing and sound production,

R. Bader, *How Music Works*, https://doi.org/10.1007/978-3-030-67155-6_4

while the whole range of variations is only possible due to the nonlinear nature of musical instruments. Simplified synthesis systems, like most synthesizers use, and simplified signal processing tools, like frequency filter techniques, do indeed produce a sound. Still, most often, this sound is considered unnatural. Although these sounds may be very interesting and musicians produce cool stuff with them, they are only simplifications of a much more difficult and fascinating system of music production.

So the difference between the frequency and the impulse picture is decisive. Although we will find that we can analyze all sounds using frequencies, the nature of the sound production process, and therefore the basis of music, is truly an impulse pattern formulation. The same holds for the impulses in the human brain, where nearly nothing works in a simple, linear way, and most of the time, self-organization and synchronization are the basis of perception, thinking, and action. Synchronization is indeed considered as the basis of consciousness, which will be discussed later.

What Is a Frequency?

But let us first start with the frequency picture, which is indeed very helpful in many ways and often used. Readers familiar with the Fourier transform, filters, or phases might skip this section, still for those who might know what a sine-wave is but are not perfectly sure about the features of a Fourier transform, it might be interesting to follow.

Sound in air is a wave of pressure fluctuations traveling through the air from a source to a receiver, for example, from a guitar reaching our ears. These pressure fluctuations can also be recorded by a microphone so that the pressure is transferred to a voltage. This voltage is digitalized by an analog-to-digital converter (ADC) and then available as a file on a computer or smartphone. When we want to listen to the music by pressing the play button, this digital sound wave data are sent to a digital-to-analog converter (DAC), which transforms the digital data into a voltage again. With this voltage, a headphone or loudspeaker is driven, which again transfers the voltage into the original air pressure fluctuations which enter our ears. An alternative to storing the music digitally in a file or on CD or DVD or any other digital format available today, one can, of course, store it analog on a vinyl record or a tape. Still, the process of reproduction is very similar. These formats need to be converted to sound pressure waves in the air again, which reach our ears.

So no matter in which format the music is present, as air pressure waves, as changing voltage, as a digital wave file, or as analog format, there is a wave that changes in time. Such a wave is called a signal as the most general term. When one is doing changes in this signal, we speak of processing the signal, and the technology of doing so is called signal processing. This again very general term holds for analog or digital signals. It might be a filter or adding reverb to a sound, it might be the transformation of a sound through a loudspeaker, or it is the natural reverberation of an instrument in a room. In all cases, a signal is changed, and therefore processed.[40]

Now for processing signals, one first needs to look for a very general representation for the signal and then apply the processing. With representation, we mean some kind of simplification, a general rule, which allows us to deal with the complexity of sound in a more easy way. The most common representation is that of a signal composed of many frequencies with different amplitudes, so different strength and different phases, so different relative positions in the sound. This is not at all the only possible representation. Theoretically, there are endless others, like an impulse formulation. Still, most people think and talk about frequencies and about a spectrum of frequencies, which is the strength or amplitudes of all frequencies present in the sound.

Now this representation has some basic properties which make it very robust and handy, and which is the reason it is very often used. But before we come to this important point, we first need to understand this concept closer and will find much simplicity and beauty in it.

The term frequency means something is repeating. Signals are time-series, where a parameter is changing in time. Therefore it is straightforward to count the number of events within one second of time. So if such an event happens, say one hundred times each second, then we have a frequency f = 100 Hz. The Hz is short for Hertz, which refers to Heinrich Rudolph Hertz (1857–1894), which was the most prominent student of Helmholtz in Berlin and the first to produce a radio wave. To honor him, we call the physical unit of frequency Hz, which is 1/s, one over second, so the number of repetitions per second.

The parameter repeating is doing so absolutely regularly. So we do not simply count the number of occurrences of an event within one second. No matter when they happened, they need to happen regularly. It could be that one hundred events happen within the first 100 ms of a second, so the first 1/10th of that second, and the rest is silence. This is not the concept of frequency used here, although it is often used in neuroscience, where the firing rate of a neural spike is simply counting the amount of spikes per second, no matter if they come regularly or not. This needs to be kept in mind when reading through respective literature, where we hear of tonotopic neurons and neural fibers that are tuned to single frequencies.

But now, what is it repeating? Of course, there are endless options. Most know that a single frequency is a sine-wave, a sinusoidal. A sinusoidal can be explained geometrically, mathematically, or mechanically. Geometrically it is derived from an arrow rotating in a circle counterclockwise. Imagine an arrow pointing to the right. If this arrow starts rotating with a constant speed counterclockwise, it will produce a sine wave. It becomes clear when following the height of the arrows tip moving vertically. When pointing to the right, we say the height is zero. Then it starts moving counterclockwise, and therefore the end of the arrow is going up until it is pointing right to the top. Then it will move down again until it is pointing to the left with a zero height again. Then it moves down, having a negative height until it is pointing straight down, while only then moving up again to its initial position, pointing to the right. It then has moved a full circle.

Still, sound is a signal moving in time. So we need to project or draw the vertical movement of the arrowhead along a horizontal straight line. By doing so, we need to move along the line at the same speed as does the arrow rotate. Then the up and down of the arrowhead produces a smooth curve. It starts at the very left point of the horizontal, straight timeline moving right. When the arrowhead has reached its peak, this line goes up, moves down to zero when the arrowhead is pointing straight left, becomes negative when the arrow is down and gets zero with the arrow pointing right. It then has moved a whole circle. The resulting cure is a time series of a sine-wave going through one period. The periodicity is, therefore, the time the whole process needed, and we measure it with the variable T for one time one period needed. It might have lasted for 10 ms. If so, and if the arrow continuous to move on that way, we can continue to draw our time series in the horizontal plane, say of one second. Then we will see one hundred of these sine waves filling up that second. So our frequency is $f = 1/10$ ms $= 1/0.01$ s $= 100$ 1/s $= 100$ Hz.

But we can also derive a sine-wave mechanically. Therefore we take a spring, hang it to the ceiling and attach a weight at its lower end. Then the weight will make the spring stretch so that the whole system is resting. Moving the weight down a bit and setting it free, the weight will move up and down with the spring contracting and expanding. When recording this up and down the same way as before, taking the weight as the arrowhead and moving again on a horizontal timeline right, we can record the same curve as before, a sine wave, also called a sinusoidal. So first, we made a geometrical, then a mechanical description of a sinusoidal movement and therefore show that something moving in the real world can be described geometrically.

Still, one formulation is still missing, the mathematical one. Indeed, during the 17th and 18th centuries, this was the most demanding problem where people like Leibnitz, Newton, D'Alambert, and many others were suggesting approximations to this problem, as discussed in the introduction. Indeed, it appears that calculating the sine wave curve mathematically is like doing the quadrature of a circle, which is not possible. Still, it is possible to give approximate solutions, and this is what has been suggested. Mathematically, a sine wave is an endless series of terms, where adding each additional term makes the solution a bit better:

$$y(t) = \sin t = \frac{t^1}{1!} - \frac{t^3}{3!} + \frac{t^5}{5!} - \frac{t^7}{7!} + \dots = \sum_{n=0}^{\infty} (-1)^n \frac{t^{2n+1}}{(2n+1)!} \,. \tag{1}$$

For those not too familiar with mathematical equations, this may look complex. Still, it is interesting to look at the details, so stay tuned! We have one parameter called t, which stands for time, as the sine wave moves through time. In other publications, you might find an x instead of t, which is a variable often used as that of a horizontal axis. Still, here we are in the time domain and therefore use the t. The whole thing starts with y(t), where y means the amplitude, which is the height of our moving arrow or the height of the sine curve. So y(t) only means that we can choose a time-poiint t and get an amplitude y at right this t. We can do so for all possible t. But how to calculate this?

Here was the real mathematical problem and the answer is that $\frac{t^1}{1!} - \frac{t^3}{3!} + \frac{t^5}{5!} - \frac{t^7}{7!} +$ .... The '!' means the faculty of the respective number. We see that the amplitude y(t) for a certain t is an endless amount of terms denoted by the ... at the end, meaning these terms will prolong in this way forever. Still, the terms have higher and higher powers (starting with 1, 3, 5, then 7, etc.) but are divided by greater and greater numbers (the faculties 3!, 5!, 7!, ect.). Each new term will make the solution a bit better, but we will never reach the perfect value. The last term in the row with the sum symbol ($\sum$) is only a short version of the terms before. Note that the sum runs from zero (n = 0) till infinity (∞). While the rest is a formula resulting in the respective terms when inserting n = 0, 1, 2, 3, ... (give it a try and insert one or two yourself!). This whole process is much to write, and, therefore, it is reasonable to give it a short symbol, which is the 'sin' at the beginning. So the sin is only a convention representing the endless series.

This sine-wave repeats as we have seen. So it is interesting to see for what t the wave starts all over again. Again the solution can be found in the rotating arrow. As the arrowhead moves along a circle, the length of the line it moved through is right the length of the horizontal line we projected the circle to the time series. Now, as we use an arrow length to be 1 for convenience, the distance of the arrowhead traveled is the circumference of a circle, which is $2\pi r$. Its radius is r, and as a unit circle has r = 1, the circumference is simply 2π.

So the sine wave starts at zero (y(t) = 0)) at t = 0 and it ends again at y(t) = 0 when t = 2π. This also means that it is again crossing zero at t = π and has a positive maximum at t = $\pi/2$ where y($\pi/2$) = 1, and has a negative minimum at $3/4\pi$ where y($3/4\pi$) = -1. Then the sine wave repeats. So we can restrict our description of the sine wave within this range of $0 \le t \le 2\pi$. During this time, the sine wave goes through different phases, going up, reaching a maximum, going down, reaching a minimum, and ending at zero again. So we say, for each time-poiint t the sine wave is at a certain phase, which can only be between zero and 2π. The phase is another very important property of a sine wave.

Yet a third important parameter is the maximum amplitude of the wave. At the moment, the maximum and minimum we had is ± 1. But of course, the wave may have a different maximum amplitude. Adding this to the wave is very simple. The whole wave y(t) only need to be multiplied by a maximum amplitude A. Then we have for a sine wave

$$y(t) = A \, \sin 2\pi \, f \, t \, . \tag{2}$$

Here the maximum amplitude, the overall strength of the wave is denoted by A, which is multiplied with the whole sine wave. The argument of the sine wave, which was simply t, is now $2\pi \, f \, t$, taking into account that one turn of a wave lasts for 2π. If we take t to count in seconds, we can add the frequency f to the argument and get the final version shown above. After one second (t = 1), the sine wave has gone through the whole wave 2π for f times.

So we see that a sinusoidal can be formulated mathematically, geometrically, and mechanically or physically as a vibrating body on a spring. This vibration has a frequency. It repeats a certain amount of times per second.

How to Deal with Many Frequencies: The Fourier Theorem

Or course, most time series, like those in music, is much more complex than a simple sinusoidal. This is because it consists of not only one but of very many frequencies, many sinusoidals. So taking several such sine waves with different frequencies and adding them all together gives very complex vibrations and, therefore, very complex time series. Musicians doing analog synthesis know this very well. A common method of producing a sound in electronic music is to simply add an amount of sine waves.

The Fourier theorem is the one to use in such cases. But it is more than just adding sinusoidals, it states that all existing time series, no matter how complex they might look, are composed out of a number of sine waves with their amplitudes and phases. This is a much stronger assumption. Suppose, as a musician, you want to create a sound and think that additive synthesis is your synthesis method of choice. You have the sound in mind and are adding sinusoidals together, hoping to arrive at about your imagined sound. This process of creating a sound does only sense if you can be sure that with such a method as additive synthesis, you are able to arrive at this sound at all. Maybe the sound you want to have is so complex that it cannot be achieved by additive synthesis.

Still, the Fourier theorem guarantees that any possible sound can be composed out of an amount of sinusoidals, each having its own amplitude and phase. Mathematically this reads

$$y(t) = \sum_{f=0}^{\infty} A_f sin(2\pi f t + \phi_f) \, . \tag{3}$$

Here A_f is the amplitude, the strength of frequency f and ϕ_f is a number $0 \leq \phi \leq 2\pi$, so between 0 and 2π. As we have seen above, this phase shifts the whole sinusoidal wave along the time axis back or forth. So each frequency f has its own amplitude A_f and phase ϕ_f.

Theoretically, f goes from zero to infinity (∞), which means that theoretically, an infinite amount of sinusoidals are needed. Still, if we have a musical sound, which might last from some milliseconds to some seconds, or a piece of music consisting of several minutes, we can restrict the amount of frequencies needed. Also, frequencies above our hearing range of about 16–20 kHz, so 16.000–20.000 Hz can be neglected.

This is the crucial point why we use the Fourier theorem and the representation in sinusoidals at all. It guarantees that all possible musical sounds can be decomposed into a number of sinusoidals with their frequencies, amplitudes, and phases. If this would not be the case, adding sinusoidals together may make fun, and we still might be able to do fancy things with this method, still, we would always know that there is an endless amount of sounds which could not be constructed with this method.

If we do not want to synthesize new sounds but want to understand a musical sound we find interesting, we can analyze it with the Fourier theorem and decompose it into its sine waves by calculating the amplitudes and phases of the time series with this theorem. Again we are sure that we represent the sound completely by this method, leaving nothing behind.

Yet a third advantage is that we can go both ways without loss. We can decompose a sound into its frequencies, amplitudes, and phases, and then again reconstruct this sound by using these parameters exactly. This is very helpful when using signal processing algorithms like filters, reverbs, delays, or codecs (coding-decoding algorithms for data transfer via the internet or other data formats). Although there are many methods of performing filtering or the like, a very robust and straightforward way is to decompose a sound into its sinusoidals, then strengthen certain amplitudes, and reconstruct the sound again. The resulting sound will be much like the original, still with some frequency range stronger than before. So if a vocal track or a guitar has a frequency region that is not strong enough, with such a method, one can make the sound energy in these regions stronger.

But also, the transient nature of sounds is within a Fourier spectrum. So if a sound decays or starts with a strong slope, like a drum or a guitar tone, this overall amplitude envelope, first silence, then suddenly a tone onset, then decay, is coded into the Fourier spectrum, in the amplitudes and phases of the frequencies. This fact indeed enlarges our understanding of frequencies as pure sinusoidals, which are giving a frequency a strength in the sound. As the Fourier theorem is a hard constraint, it represents all possible time series, no matter how complex they are. So they also code very complex overall amplitude envelopes. Still, then we cannot simply call all frequency amplitudes as sine waves which we hear, but need to consider all frequencies as taking part in both, the sounding of frequencies and the development of sound envelope developments.

Consider the spectra of two sine waves, as shown in Fig. 1. Two sounds are shown here, one sinusoidal with constant amplitude over time on the left and one which starts suddenly and decays afterward on the right. In both cases, we hear a single frequency. Still, the first is constantly there, while the other suddenly appears and decays afterward. Now below the two time-series the Fourier spectra are plotted. The first spectrum of the sinusoidal in its pure steady-state has only one frequency with a certain amplitude as expected. The other spectrum shows much more frequencies with respective amplitudes. These additional frequencies code the overall amplitude envelope of the sound. Still, all we hear is only one sinusoidal, not a more complex spectrum of frequencies.

So we see that in temporally changing time series, what music is all about, the spectra are not perfectly representing frequencies we hear. They also code the transient development of the sound. From a signal processing standpoint, this is not a problem. Indeed the hard constraint of the Fourier theorem, namely that it represents a time series precisely, places us on the safe side. Still, when we want to understand a sound, we need to keep in mind that a spectrum does not only correspond to the sound we hear but also to the sounds envelope, its rise and decay.

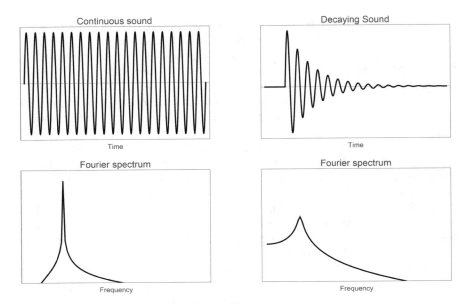

Fig. 1 The decay of a frequency is also decoded in a Fourier spectrum

Beyond Sinusoidals

So all possible time series, or sounds, can be represented by a spectrum of frequencies with respective amplitudes and phases, and the Fourier theorem guarantees us that we have represented the sound completely. We also know that we can rebuild all possible sounds from a given spectrum. Still, these features do not only hold for sinusoidals but also for other waveforms. Everybody familiar with electronic music production knows sawtooth, rectangle, or triangle waves. In Fig. 2, such waves are plotted in the top row (left to right, respectively). On the very left is our sinusodial wave again, next to a wave looking like the tooth of a saw. Such a wave exists when bowing an instrument, so with violins, cellos, or other bowed instruments like the Chinese *erhu* or the Uyghur *satar*. We will discuss the reason for this waveform further below, as it is a beautiful example of the self-organizing nature of musical instruments.

The waveform second from left in the bottom row is a so-called wavelet, a 'small wave' often found in signal processing today. Actually, this wave was first used in geology. Here geologists wanted to investigate how the earth is layered by different kinds of stones. A simple way to find out is to 'knock' on the earth with a heavy load. This produces a wave very similar to such a wavelet. These waves travel down into the earth, and each time the layer of stones changes, there will be a small reflection at the boundary between the layers. This reflected wave travels back to the surface again. Each time such a boundary exist, a new wave travels back. So when recording the back-coming waves at the surface, one can estimate how the earth beneath is constructed.

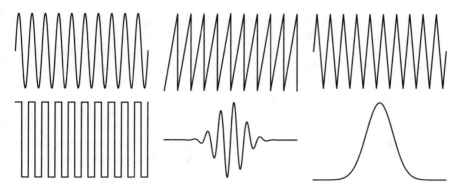

Fig. 2 Different waveforms which run through musical instruments. Top (left to right) sine, sawtooth, triangle, bottom: square, wavelet, gauss impulse

The first wave from the left in the bottom row in the figure is a rectangle wave. It is found with wind instruments, where an opening and closing of a reed happens, like with saxophones or clarinets. Such a wave also appears when musicians are using their lips, with instruments like trumpet or trombone. Again we will discuss this further down when we are considering turbulence of an airflow, which is causing such waves. Rectangle waves are also a standard waveform for electronic music sound production.

The waveform at the very right of the figure is similar to a rectangle, triangle, or sawtooth wave with corners more smooth. Mathematically it is a so-called Gauss, pulse-shaped, or Gauss impulse wave, honoring the mathematician Carl Friedrich Gauss (1777–1855). Mathematical is reads

$$y(t) = A \ e^{-(t-t_0)^2/\sigma^2} \ . \tag{4}$$

Again we have a time series y(t), which has a maximum amplitude A. The bell-shaped curve comes from the exponential function, where e = 2.71828... is the Euler number, a number very useful for exponential functions, and called to honor the mathematician Leonhard Euler (1707–1783). The exponent of the function leads to a bell-shaped impulse that has its maximum at $t = t_0$ and has a width σ, where larger σ leads to a broader shape.

This shape often appears in musical instrument sound production too, especially in string instruments like guitars, pianos where we will show some examples below. It is also very close to the impulse of a mallet acting on a xylophone bar or as a clapper acting on a bell. In a way, it is also much like a smoothed rectangle, triangle, or sawtooth motion. Indeed, these 'perfect' shapes never appear in musical instruments but are smoothed out a lot. It also comes close to a neural spike, which is also a very short impulse of a current in the neurons.

All the more complex waveforms can be analyzed using the Fourier theorem. This often makes sense, as we hear frequencies and so we can find out about the spectral content of a sound by transforming a time series into a Fourier spectrum. Still, when it comes to the mechanical production of musical sounds in musical instruments, the temporal development of the waveform might be of importance much more than its spectrum. This is especially true when nonlinearities are present in musical instruments, as we will see below. Also, the human ear and the neural system do not use sinusoidals, but neural spikes, short sharp impulses of electric charges. Such peaks are very sharp, and between two such peaks, more or less silence is present. In such cases, it makes much more sense to switch from the spectral, Fourier picture back to the time domain.

This leads to a representation of a sound as a series of impulses of different shapes following each other, maybe overlapping or repeating, maybe constantly changing their precise waveform shape, but still as a series of time-poiints when the impulses happen. This series of time-poiints may be very regular, it may have a very irregular shape. A quasi-steady-state of a played note, like a guitar tone after its initial transient phase, will show a quite regular appearance of nearly the same waveform.

The situation changes very much if we consider initial transients, where the impulse pattern is not regular at all. Indeed, transients are the most important part of musical instrument sounds where much of the articulation happens like hard or soft attacks. Also we identify instruments mainly by their initial transients. Still, to get to a more complex picture of how impulse patterns in initial transients are built, we first need to get deeper into the sound-producing process of musical instruments in the next chapters.

Damping

So far, we have discussed the possibility of taking musical sounds as a series of impulses, where we have a time-poiint the impulse is happening and a waveshape of the impulse, which may be one of those discussed above, or it may also look even more complex. So we can also view the vibrations of musical instruments in this way. Although we will go into details of guitars, saxophones, or the like only later, for now, we need to discuss some more properties of these impulses before we can formulate a method describing the interactions of them.

Maybe the most fundamental parameter which influences the development of the impulses is damping. Damping means that energy is lost over time, which decreases the strength of a wave or impulse. The energy may be lost due to the radiation of sound of a guitar into the air. Of course, the sound leaving the instrument is decreasing the energy in the instrument, and therefore the vibrations on it are decreased, they are damped. We might call this damping an external damping. The other kind of damping is internal, an energy loss because of internal friction within the instrument. This kind of damping is very strong with tonewood.

Think of two plates, a wooden and a metal plate, both of the same size and quite thin. Our first-hand experience when knocking on a wooden plate and knocking on a metal plate is that the sound with wood decays much faster than with metal. When knocking on a wooden plate, we hear a short knocking sound. With metal, the sound might continue for quite some time. Both have about the same sound level, so external damping due to radiation is about the same for both, wood and metal.

So indeed, overall, the damping in wood is much stronger than that in metal. This is crucial for musical instruments, as guitars or pianos would sound completely different when they were made of metal. Experiments at the end of the 19th century on metal soundboards of pianos showed that metal is an inadequate material for pianos, as the piano sound is very much 'metal-like' then [117]. The impact of metal on the sound is so considerable that we cannot accept metal as a material for musical instruments. Still, basically, a metal plate vibrates the same way as a wooden plate, except that the internal damping of metal is much less than that of wood. So internal damping is indeed a crucial parameter, and it seems better for an instrument to be build out of a material that is heavily damped than out of one, which is damped only slightly. This finding might be counterintuitive at first, as we could expect an instrument to sound better if it is able to vibrate much easier without much damping of the material. Still, this is not the case, and understanding this, we first need to consider the interactions of the impulses traveling in the instruments.

Although for our further discussion it is not of much importance, internal damping of wood but also of other material is only poorly understood. We hear of internal friction, energy loss, damping parameters, but nearly never about the physical reason for such a loss. But consider a sound wave in air traveling in free space. Say you stand on a mountain and make some loud sounds. These sounds travel very far and can be heard at many kilometers distance. This means that the internal damping of sound in air is very low. Indeed, one could wonder why there is damping of sound in air at all. Sound in air is the movement of particles of air in space. If a molecule in air, say an oxygen molecule has a neighbor, which is moving into its direction, both molecules interact through their electrical fields. So one is pushing the other in its direction. The second particle will start to move and again push another molecule next to him in the same direction too. This pushing takes some time as each particle needs to be accelerated, and so we end up in a speed of sound of about 343 m/s at room temperature.

Still, where is the damping? Each particle gives some energy to the next one, and there is no loss at all. Damping of sound traveling through air is only caused by an effect based on temperature. If a sound wave moves, there are regions where the molecules are closer together and others where they are farther apart, as they would be if no sound was there. This is caused by the pushing of molecules and their movement towards or away from neighbors. But particles pushed suddenly together increase their temperature, although only slightly. Contrary, if particles are suddenly less dense and further apart, the temperature in this region is getting less. But in such a case where at one region in space the temperature is higher than in another one, a balancing mechanism starts, which tries to achieve the same temperature in both regions. It is the same mechanism why one needs to heat up a house in wintertime. Inside the house, the temperature is higher than outside, which causes the temperature inside the house to reduce and increase the outside temperature slightly. Now, temperature is also the movement of molecules, next to these movements being sound, still very much faster. Nevertheless, a little bit of this movement, which is sound, is used to level out the temperature difference between two regions in air. So some energy of the sound wave, some movement of the molecules, is lost into the temperature leveling, and therefore the strength of the sound wave is decreased, it is damped internally.

Indeed, this is the only mechanism of a sound wave being damped in air. We need to go to the level of molecules to understand it. Damping of this kind is also called dissipation of sound energy into thermal energy, where the word dissipation is pointing to a transfer of kinds of energy, the sound, and the thermal energy. Of course, the first law of thermodynamics, the field in physics dealing with energy and heat, claims that energy is always preserved, it only might change its kind. So the sound energy is not lost, it is only lost for acoustics. It is interesting to note that again Hermann von Helmholtz was one of the first to discover this first law of thermodynamics, but as a young physiologist dealing with nerves, mainly those of frogs, and not as a physicist. His motivation for doing so was to argue against the idea of vitalism, very much around during the mid 19th century. Vitalism claimed that physics cannot explain the appearance of life out of 'dead' matter, and therefore an unknown divine power, maybe an aether was around to give life to dead matter. Helmholtz argued that if this was the case, the energy which moved a frog leg was not physical, and therefore moving it, again and again, would not reduce the internal energy in the leg. Still, if internal processes, like the stimulation with voltage, caused the movement, the internal structure of the leg would stay intact, no matter how often the leg is moved. But when measuring the legs mass before and after many movements, it appears that the leg was lighter in the end. So there is an energy loss depending on the movement. Therefore the finding of the first law of thermodynamics was an important step to show that the appearance of what we call life is not caused by a divine, external force. Still, this does not say what it may then be that causes life then. This might only be found when considering self-organization of dead matter, a mechanism also present in musical instruments and in the brain.

Returning to internal damping of wood, the thermal leveling is an internal mechanism that needs to be present in wood too. Still, we might find it to be of minor importance, considering again the difference between a wooden and a metal plate. If it would be the main cause of internal damping, this damping would stronger with increased thermal conductance. Thermal conductance means the ability of a material to transport heat. We use woolen cloths to prevent us from freezing because wool has a very low thermal conductance. The body heat is only very slowly moving away through the cloths into the surrounding air. Other materials have a very strong conductance, so if we sit on a stone, our back might become cold very soon. Stone is transferring energy away from us very fast. Now wood has a small conductance, able to preserve the energy well, while metal feels cold, caused by its ability to transport the body heat away very quickly. So a wooden plate has a much smaller thermal conductivity than a metal plate. Of course, the strength of thermal conductance will determine if a material is transferring acoustic energy into heat fast or slow. If thermal conductance were the major part in internal damping, wood would vibrate much longer than metal. Still, the contrary is the case. Therefore the thermal leveling mechanism does not seem to be very prominent with internal damping in wood.

So there needs to be at least one more reason for internal damping. One very well known from polymers, such as plastic, is called anelasticity. There is some kind of plastic material which one might stretch in such a way that it is getting considerably longer. When setting the plastic free again, it might not move into its original form right away. Some material take some time until they reach their original size again. In such a case, there is a time delay between the application of a force and the stretched or unstretched state. Such a time delay is called hysteresis. If the way from unstretched to stretched is different than the way from stretched to unstretched, we have a hysteresis loop. Such a loop does not appear under perfect conditions, and if present, makes things much more complicated. One can imagine how much more demanding it would be to apply a vibration on a material that has strong inertia. Some of the energy applied to it goes into internal damping. Now metal is known to have nearly none of such behavior, while wood is a very complex material consisting of many polymers, like cellulose or lignin, which are long and complex molecules composed of the one of only a handful of basic molecules chained together. So with this phenomenon, the main reason for internal damping might be expected, although research has not understood this fully until now.

Still, it is interesting to see that these processes of hysteresis in large polymer molecules are an effect that can only be understood based no longer on classical mechanics, but where we need the theory and framework of quantum mechanics. Although, when understanding the basics of the mechanism one might be able to formulate a more simple mathematical model without the use of quantum mechanics, which might also be more handy compared to quantum mechanical modeling, only the latter holds the reason for this kind of damping. Basically, what is happening is that a molecule under pressure might change its shape fundamentally, but without breaking apart. Such a shape change of a molecule has its reasons in the quantum mechanical equations, which try to minimize the internal energy of the molecule under the present conditions. Such a minimalization of energy may make a different shape of the molecule favorable, and therefore it will undergo this change. The new shape has a different overall energy level compared to the first one. Now the difference between both shapes will be lost by the molecule as it is sending out a photon, a very small light particle, which has the energy of the molecule energy difference. This emitted photon is then the portion of energy lost in the acoustic wave, it is the internal damping.

We are fortunate that we do not need to go into more detail here for our basic reasoning. Still, it is fascinating to see that understanding the maybe most important parameter of musical instruments, their internal damping, one needs to go down to quantum mechanics.

Now we can also understand why different frequencies are damped differently. A higher frequency has more cycles per second than a lower one. So a high-frequency vibration will undergo all these mechanisms more often. Therefore it is expected to have more energy dissipation per second than a slower wave, and therefore we expect it to decay much faster. Indeed, this is happening, although, especially with wood, there are so many other mechanisms going on that the damping of wood is very complex, making the liveliness of musical instrument sounds composed of wood compared to instruments built of plastic, which has internal damping much less complex.

Dispersion

Another phenomenon well known in acoustics is dispersion, which we need to discuss briefly. If one of the impulses we discussed above is traveling over the surface of a plate, like a guitar top plate or a piano soundboard, it changes its shape not only because of damping. If no damping were present, the shape of a Gauss impulse would undergo a sharpening at its front side and a flattening at its tail. This can be explained using Fourier's theorem, where we think of this impulse as consisting of many frequencies with their amplitudes, which are right in such a way to form a Gauss impulse. Now on plates, different frequencies have different traveling speeds. High frequencies are faster, while lower frequencies are slower. Putting all the frequencies together again, forming an impulse, while traveling over the plate, this impulse will start changing its shape gradually, it will raise steeper at its front and decay slower at its tail. If waves travel at different speeds on a medium, this is called dispersion.

Another nice example of this phenomenon exists with brass instruments, like with trumpets or trombones. While listening to a swing orchestra like that of Duke Ellington or a Funk or Fusion band with a brass section, in the loud passages, this section often plays small phrases and melodies in a very rhythmic way, which have a sound called 'brassy'. This is cool and groovy. Still, these sounds are very different from those sounds played by the instruments when they play with low volume. Played softly, the instruments sound smooth, only when played loud, this additional brassiness appears. The reason for this is the high sound pressure level of the impulses traveling through the instrument's tube when played loud. Remember that loud music may have sound pressure levels of around 100 dB, which may already lead to damages in the ear. This is definitely the case when listening to music with 120 dB over some time. Now the sound pressure level (SPL) within brass instruments may be up to 140–160 dB! Such levels will not be radiated because only a small part of it is actually leaving the instrument. So within wind instruments in general, sound is much louder than around of them.

Now the sound in air does usually not show dispersion. If an impulse is leaving an instrument, the impulses in brass instruments are mostly rectangle-like, as discussed above, and this impulse is traveling through the air to a listener, it will not change its shape. Air has nearly no damping and no dispersion, and therefore the impulse will only be distorted when it is reflected at a wall in a club or concert hall. Still, this is no longer true for SPLs of 140–160 dB. There the speed of sound in the air is faster. This is because the molecules of the air are closer one to another and are, therefore, better able to push each other. In such cases, we have found before that the temperature is raised. Indeed, the higher the temperature of air, the faster a sound is traveling through it. For SPLs of about 80–120 dB all frequencies travel by about the same speed, which for normal air temperature is about 343 m/s. So within about 3 s, our impulse would have traveled about one kilometer. But if the SPL is higher, this speed increases considerably.

Now picture our impulse. At its very beginning, it has a low amplitude and, therefore, a low SPL. Still, it is raising up to its peak, which may have an SPL of 140 dB. Then it is decaying again to zero amplitude and, therefore, 0 dB. So within one impulse, the SPL changes dramatically. Therefore again, within the impulse, the speed of sound changes. As higher SPLs move faster than lower ones, obviously, the peak of the impulse, which was in its middle at first, will move towards the front. Again this leads to a steepening of the impulse towards its front and a larger decay towards its tail. Such waves are called shock waves. They also appear with supersonic airplanes. Now when again doing a Fourier analysis of this steepened impulse, we find more energy and, therefore, larger amplitudes in the higher frequencies. This rise of the energy in higher frequencies is what we hear as brassiness. It is a kind of distortion, just like with distorted electric guitars. But here, the reason is dispersion of air. This is only happening within the tube of trumpets or trombones, but once raised in amplitude, these louder frequencies will radiate from the instruments. This behavior of air being faster with higher SPLs is a typical example of a nonlinearity, which we turn to now.

Linearity or Nonlinearity

To decide if a system is linear or nonlinear, we need to draw a plot comparing two physical quantities. We will come back to the brassiness as a nonlinearity in a second and first want to discuss another example, a guitar string. The reason will become clear within the next pages.

Clearly, to make a string vibrate, one needs to pluck or strike it, the harder one strikes, the larger the amplitude of the vibrating string, the louder the sound. Now guitarists know that there seems to be an upper limit of loudness. This has several reasons, but one can be experimentally determined when taking a string between two fingers and stretching it. Clearly, the more the string is stretched, the more force the fingers and the arm needs to apply. Still, there is some point of displacement of the string where it becomes considerably more difficult to displace the string further. One needs to enhance the force very much and will have a fear of breaking the string. But how can we describe this mathematically?

We can do so by using a graphical plot of two physical quantities, as shown in Fig. 3. The force we apply to the string versus the displacement the string undergoes with this force is shown in the left plot. Here on the horizontal axis, we draw the plucking force on the horizontal axis versus the string displacement on the vertical axis. Now we do measurements. We apply forces F to the string, and for each force,

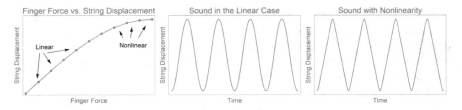

Fig. 3 When a string is plucked with only a little force, the string displacement follows the plucking force linearly. This is the part of the curve in the left figure on the left lower side. Still, when the plucking force is getting stronger, the displacement is not following up linearly anymore but in a nonlinear way, shown in the left figure on the upper right. When the string would only move with its fundamental frequency, in the linear case, the vibration would be a sinusoidal, as shown in the middle plot. Still, when the nonlinearity appears, the string vibration is very different, it is distorted as shown in the figure on the right

we measure the string displacement. The result of every single measurement is a pair of numbers, the force we choose, and the displacement we find. Therefore for each measurement, we can draw a point in our plot, which are the blue dots. When we have done so for many cases, we can draw a line connecting all these points. What we see is that with a small plucking force, the displacement doubles when the string force doubles, the string behaves linearly. Still, when the plucking force is increased strongly, the string displacement is no longer able to follow, as can be seen in the upper right part of the left plot in the figure. There the curve bends and therefore is nonlinear. If we would even increase the plucking force beyond the point on the very right, the string would break.

The two plots in the middle and on the right of the figure then show how the string would move (in its fundamental frequency only) in the two cases, the linear (middle) and the nonlinear (right). The middle case shows a regular sine wave for the linear case. The nonlinear case on the very right still shows a vibration, still, the vibration is very much sharpened. This is as when the vibration is coming to a maximum, the force acting on the string to move back is getting much stronger compared to the linear case. The string, therefore, is forced to return much faster from the maximum, causing this kind of vibration. Then the vibration is no longer a sine wave, it is distorted, and therefore more overtones appear.

Indeed, when strings are displaced strongly, the nonlinear effect causes higher frequencies of the string to be enhanced. This is similar to the brassiness of bass instruments, where also higher frequencies are becoming stronger with nonlinearities present. Indeed, this is a general first finding of nonlinearities, that they enhance higher frequencies are therefore add brightness to the sound.

Generally, the difference between linear and nonlinear appear in such figures as the one above when examining the line in them. This line might show one of two possible shapes, it might be a straight line with no curving, or it might show any kind of curvature. In the first case, where the line is perfectly straight without any deviations from this straightness, the law governing the relation between any two parameters, like force and displacement, is linear. In the case where there is any deviation from the straightness, we would no longer have a linear law governing the relation but a nonlinear one.

Now obviously, there is only one linear case, but there are endless possibilities of curved shapes that are not linear. This is why we do not call the nonlinear case with a name on its own but only say what it is not, it is not linear, it is nonlinear.

Now, telling a linear law from a nonlinear is like changing worlds. This is because of many reasons which we will discuss further down for musical instruments and music perception. Still, there are some general differences.

In the linear case, we have a simple law between two quantities, here force and string displacement. This law can mathematically be written as the fraction of differences in these two quantities. This fraction needs to be constant, no matter where we are measuring. So if we increase the force, starting from a certain force F_0, which can have any value, and increase it by a fixed amount ΔF, in a linear case, we can tell right away the increase of a displacement ΔY above a previous displacement Y_0. The fraction

$$k = \frac{\Delta F}{\Delta Y} = \text{constant} \tag{5}$$

will be constant no matter where the initial force or displacement has been, adding the additional force ΔF. This is wonderful as we can insert this constant in mathematical equations and can deal with it very easily, as it is a constant and, therefore, simple to handle.

This is no longer possible in the nonlinear case. The fraction then depends on the force F_0 from where we have applied the additional force ΔF, and therefore the above equation now reads

$$k(F_0) = \frac{\Delta F}{\Delta Y} \neq \text{constant} . \tag{6}$$

Here the fraction k depends on F_0, and therefore, we write that it is a function $k(F_0)$ thereof. Inserting this into mathematical equations describing the vibration of the string is possible, but solving these new equations is tremendously more difficult and often only possible using complicated algorithms.

This is why we first try to explain the relations in physics in a linear way. This is often reasonable also with nonlinear systems when we are not interested in the precise behavior of the system, but only in what it is doing roundabout, considering its general laws. Then we skip the nonlinearity and replace it with a linear approximation.

Now doing this in musical acoustics is often not a good idea because of two reasons. First, our ears are extremely sensitive. They are able to distinguish physically only slightly differing quantities in terms of timbre, loudness, roughness, and many other parameters. Musicians and composers know this and play around with it. So musically often not the very basic sounds are of interest, that a drummer is producing a rhythm or that a singer is singing a melody, what makes us shiver, what is fascinating about music, what makes us tell musical styles one from another, or what musicians do when developing their personal sound is very often happening in the fine structure of the sound. So when only approximating musical acoustics to linear laws we might miss the very core of musical articulation, style, personality, or interpretation.

One example of missing the interesting part when leaving out nonlinearities is the brassiness example from above. There the constant in the linear case is not a fraction of differences like the ΔF and ΔY of the string case. Here the constant is the speed of sound in air, which is about c = 343 m/s. In a linear case, this speed holds for all sound pressure levels (SPLs). So in our plot SPL vs. speed, we again would have a straight line up to about 120 dB, but this time, the line would be a horizontal line no matter which SPL we are examining, c is always 343 m/s. Only when coming to SPLs of more than 120 dB, the behavior becomes nonlinear, and the curve would go up in c with higher SPLs. Still, the principle is the same, the only difference is that this law is not comparing differences like with the displaced string, but stays at a constant only with higher SPLs to increase.

So brassiness also belongs to the first kind of behavior added to musical instrument sounds when including nonlinearities. Indeed, when we would take trumpet sound traveling within the instrument to only be linear, we would miss all brassiness and, therefore, a very important feature of the instrument!

Still, there is a second reason for including nonlinearities in our analysis, which is even stronger. Nonlinear systems are not only adding some details to the sound musically relevant, they are also about to determine a completely different behavior of whole systems, here of musical instruments and of music perception, which we will discuss further down. This is a second major difference why the linear and the nonlinear cases are indeed different worlds. Only when a nonlinearity appears we have emergent behavior of the whole system, which is not present in a linear case at all. Such emergent behavior is caused by self-organization and synchronization, and we will see that this is the reason why musical instruments work at all, producing harmonic overtone spectra what we call musical tones or notes. So even playing a melody is only possible because musical instruments are self-organized systems.

These two basic differences between a linear and a nonlinear system, the adding of musically relevant details, and the emergent phenomenon of making music possible at all, will guide us through the book in many details.

Impulse Pattern Formulation (IPF)

To get closer to the second and even more important aspect nonlinearities add to music, we first want to discuss the global picture before going into details of musical instruments in the next sections. Therefore we need the notion of impulses as discussed above. Here we are not interested in the precise wave shape of the impulse, but in the time-poiint it appears, and its amplitude, its strength.

Still, not to become too abstract, we want to discuss the very basic process by looking at the bowing mechanism of violins, cellos, or other bowed instruments. Here a bow is moved over a string and makes it move, makes it vibrate, producing a musical tone. Still, all musicians playing a bowed instrument know that bowing is a very difficult task that needs to be practiced a lot until it sounds good. This is very different from a piano, where everybody is immediately able to play a tone by simply pressing a key. Bowing is much more difficult, and with beginners, scratch sounds may easily disturb neighbors and friends. Also, sounds that are too thin or not really having a clear pitch can be produced. Indeed, when pressing the bow to the string with much pressure, the note may change completely and is more determined by the pressure than by the fingering on the fretboard. Also, a problem for beginners is a pitch increase when releasing the string from the bow. When played with normal pressure, a stable pitch is heard, a stable tone. But when ending that tone by simply taking the bow from the string, the string is still vibrating a bit longer. Then the pitch of the played note is still heard but sounds a bit higher than it was when playing the string with the bow, an effect called flattening effect.

What everybody playing such instruments is also recognizing easily is that a violin string is heavily damped. When plucking violin strings with the hand, sometimes demanded in musical scores called pizzicato, the note is very fast decaying. This is very different from guitar tones, which sound for a very long time after plucking. So the damping of violin strings is very strong, which is needed for the bowing process. Indeed, it is very hard to bow a string with low damping. Although it is not impossible, you might try to bow your guitar once.

Every player of a bowed instrument is also very familiar with putting rosin on the bow. A bow without rosin is hard to play and sounds thin, often without a clear pitch. Rosin is a very sticky material that is right what is needed, as the bow needs to stick to the string to displace it. This is how the bowing process takes place, the bow sticks to the string, and by moving the bow with the arm or hand over the string, the bow is taking the string with it a little bit. But this will not work for long, and after some displacement, the bow tears off the string again, and the string is moving back towards its original position. It is moving back as a displacement means that the string is no longer in its equilibrium, its resting position, and want to start moving back to this equilibrium.

So the string is moving back, gliding over the bow. This can happen as there are two kinds of friction, a sticking and a slipping friction. Friction is very strong during sticking, and it is very much decreased when a body is slipping over a surface, here the bow slipping over the string. So the string has no hard time slipping over the bow.

Still, when moving towards its resting position, at a certain point, the restoring force pulling the string back to this equilibrium position becomes small, and therefore the string's velocity is decreasing. Then it is not slipping anymore but sticking to the bow again. Therefore the friction changes once more from slipping to sticking. Sticking friction is very strong, and therefore the bow is able to displace the string again, the process starts anew.

So far, so good. But why is the bow torn off the string, changing from sticking to slipping and sliding back? We may suggest that this is because the further away from the equilibrium or resting position the bow is, the stronger the force that is driving the string back. This is indeed the case. Now could it not be that this force, which is getting stronger with larger displacement, is reaching a threshold from where on the bow is no longer able to make the string stick on it, and therefore the string is torn off and sliding back? Indeed, this might happen, but it is not what is normally going on. It is easy to see why this cannot be the usual mechanism.

Violin players know how to change the loudness, the volume of a tone. When bowing fast, the tone gets louder than when bowing slow. So moving the bowing arm and, therefore, the bow fast over the string results in a loud sound, moving slow gives one of low volume. Still, imagine what is happening if a string is displaced by a bow up to a point where the force driving the string back is too strong for the bow to still hold the string. If so, there is a certain displacement where the string is torn off. Still, when playing with a low bowing speed, it takes much more time to reach that point than when bowing fast. But this would mean that the time it takes between two tear-offs depends on the bowing speed. If a player bows slowly the string takes more time to reach the point of tear-off compared to when playing fast. Now the time between two tear-offs is the repetition time of the sound, it is its fundamental periodicity, which is the inverse of the fundamental frequency of the sound, its pitch. Therefore the played pitch would depend upon the speed a player is bowing. This is obviously not the case, the pitch when playing violin or cello is determined by the fingers of the left hand on the fretboard and not by the bowing speed.

So although the idea of the string force being too strong to still hold the string on the bow is possible, and indeed sometimes happening while playing, it cannot be the main process during normal bowing. The main process needs to have to do with the fact that the basic pitch of a tone is indeed determined by the fingering on the fingerboard. So we need to have a look at the string itself and what is happening after the string is torn off the bow.

If the string leaves the bow, a kind of impulse is applied to the string. It is very similar to when one would strike the string with a small hammer at the bow position. Suddenly the situation changes at the bow-poiint. If this is happening, two waves are traveling away from the bow-point in both directions of the string. These waves are traveling with constant speed, as waves on a string do so. So we can determine where the impulse has traveled along the string after the tear-off in both directions. Although, again, the precise shape of the two impulses may look different due to the constant traveling time, we can clearly say where the peaks of these impulses are at any time-poiint after the tear-off.

Now impulses reaching the end of strings are simply reflected, although always a small amount of motion is traveling into the endpoints. With the violin, the impulse traveling towards the bridge is reflected at the bridge but is also moving the bridge a little bit. This need to be this way, as the bridge is again moving the top plate of the violin, transmitting the string movement to the top plate. This top plate is then radiating the sound into the air, and only then can we hear a violin tone.

But we want to concentrate on the bowing mechanism for now, and so for us, it is important to follow the impulse traveling along the string, being reflected at the bridge point, traveling to the nut, is again reflected there, and traveling back towards the bowing point. The second wave is taking the other way, first traveling to the nut, is reflected there, travels to the bridge, again being reflected, and traveling to the bow position again.

Still, there is a difference between both traveling waves. The bow is normally much closer to the bridge than to the nut. Therefore the first wave traveling to the bridge arrives there very fast and travels to the nut through the bowing point normally at a time when the bow is still slipping. Therefore this impulse is not really affected by the bow. Still, the other impulse first traveling to the nut takes much more time to travel back to the bow-poiint and reaches it later, after the bow has again begun to stick to the string. Therefore it is effected by the bow very much and has two options. Either it is strong enough to tear off the bow from the string. Or it is not, and in this case, the bow acts as a boundary or wall, like do bridge and nut, reflecting this impulse back to the nut.

Now during normal bowing, this second impulse is not strong enough to tear off the bow and does not play any further role in the bowing process. Still, the first impulse is strong enough, and when it is returning to the bowing position, it is able to tear off the bow from the string. So during normal playing, the reason for the tear-off is a returning impulse acting on the bowing position.

This is very reasonable in terms of the playing pitches because the time interval between two tear-offs is then determined by the time the impulse takes to travel one round over the string. This corresponds to the fundamental frequency of the string length, and therefore the pitch is determined by the fingering on the fingerboard, as it is supposed to do (Fig. 4).

It is clear that such a process has many points where it can fail. The impulse first traveling towards the nut may indeed be able to tear off the string from the bow. It does so when the bowing pressure is very low. Then another pitch would appear, but the impulse traveling towards the bridge first is still underway and might tear-off the string again. Then we have two pitches. Such a situation is called a double-slip, in case the bow is able to stick to the string soon after this second tear-off. The sound of such a double-slip is one of a constant pitch, but with a blurred and shimmering timbre, violin players know very well when playing with low bowing pressure, sometimes also called surface tone.

Another behavior of the motion appears with very strong pressure. Then no returning impulse is able to tear off the string from the bow anymore, and the very first mechanism mentioned above holds that the force driving the displaced string back becomes too strong at a certain point to still hold the string. But then the pitch is no longer determined by the finger position on the fingerboard but by the bowing pressure and speed. Such sounds are called subharmonics. There the parameter determining the pitch of the string has completely changed! No longer does the finger on the fingerboard determine the played note, not it is the bowing pressure.

So, all in all, bowing is a very complex mechanism, vulnerable, and therefore difficult to control. Still, it has many possibilities, and therefore bowing has a wide range or articulatory options. The mechanism is a nonlinear one, as there are two distinct phases, sticking and slipping, which are fundamentally different. If we plotted the bowing force against displacement, as we have done above, we would get a strong kink at the point where the bow tears-off from the string. The curve would no longer be a straight line, not linear, but strongly nonlinear.

Fig. 4 The motion of a violin string, bridge, and top plate over one cycle or periodicity of a played note. On the left, the bowed string shows the sticking and slipping and the two impulses or waves traveling from the bowing point after the string tears off the bow. In the middle row is the displacement of the bridge and on the right that of the violin top plate. These are results from a violin simulation as a computational algorithm. From: [21], where more details are discussed

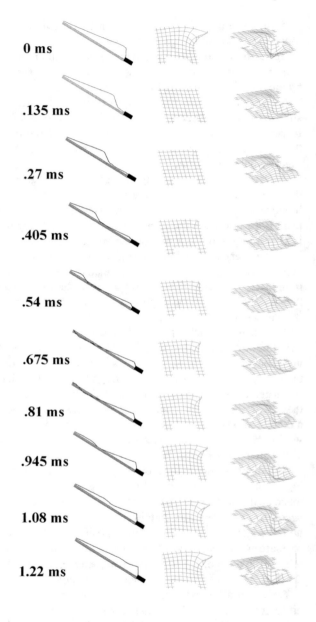

0 ms

.135 ms

.27 ms

.405 ms

.54 ms

.675 ms

.81 ms

.945 ms

1.08 ms

1.22 ms

Bowing is also a very nice example of what is meant by using impulses to describe musical instruments. When only discussing the position and strength of the two impulses traveling along the string, the full range of possibilities of bowing can be described and understood. The situation is simple in this view, an impulse is leaving a place, here the bowing point, is traveling along a string, and is decaying, decreasing its amplitude, is reflected at some point, and travels back to its origin, interacting with this origin, and causing a new impulse to leave the place.

Such a system can mathematically easily be described with the following equation:

$$g_{t+\Delta t} = -\ln\left(\frac{1}{\alpha}g_t\right) \tag{7}$$

Here g_t is our impulse, and $g_{t+\Delta t}$ is the next impulse to come. Looking at the right-hand side of the above equation, we see that g_t is multiplied with α first. This means it is stronger or weaker depending on the number α, which is determined by the physical system. Secondly, a logarithm is applied to it. A logarithm is the other side of an exponential function, where

$$a = e^{-b} \text{ can be rearranged into } b = -\ln a\ . \tag{8}$$

The first equation means an exponential decay. This is the basic damping behavior of a wave traveling along a string, a plate, or a membrane. It means that it is decaying exponentially, very much at first and slower and slower after a while.

So a logarithm is the other side of an exponential decay. As we have a strong exponential decay in the string of a violin, we need to use the logarithm in the first equation of the impulses to describe the system.

So g_t is a series of g's, where each new g is calculated from the previous one with the above equation. Each g is representing a new impulse traveling along the instrument, here the violin string. As impulses have both time to travel and amplitude, the g's tell us about the relative times between two impulses and their amplitudes. The relative time is the played pitch, and the amplitude is its loudness.

So for normal violin playing, the g's do not change after an initial transient phase, and therefore the time series is quite boring. Indeed, the interesting thing is the initial transient phase. Here the g's are struggling and producing a very complex pattern of impulses, an impulse pattern. Indeed, when constructing a musical sound out of the above equation, it sounds very much like the initial scratching a violin bow produces when starting a sound. By varying the bowing pressure, which would be varying the parameter α, arbitrarily complex and varying initial transients can be produced, representing the whole spectrum of violin articulation.[41] The impulse pattern formulation (IPF) is indeed able to produce very complex initial transients of cello bowing, which are those measured.[42]

But the above equation is also able to produce complex behavior during the steady-state of the sound. We would need to expect this from a model of violin bowing, as there are double-slip motions, high-pressure sounds, and much more. The above equation produces all these behaviors, although it seems so simple at first. So when choosing $1/\alpha$ right, there will never be only one value of g, which stays constant after an initial transient phase. This is shown in Fig. 5 on the left side of the upper plot, each $1/\alpha$ has only one value of g (because of the derivation of the equation, we use $1/\alpha$ instead of only α).

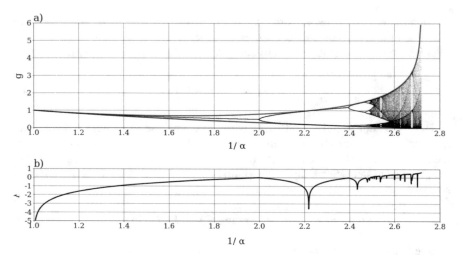

Fig. 5 The very simple Impulse Pattern Formulation (IPF) equation is able to display all possible regimes of violin bowing, from regular sawtooth motion over subharmonics, bifurcations to noise. The figure a) shows how the system parameter g is depending on the bowing pressure $1/\alpha$. Up to about $1/\alpha \sim 2$ (from left to right), g has only one value per $1/\alpha$, and therefore is a regular Helmholtz or sawtooth motion. More to the right, each $1/\alpha$ has two values, which is the most simple bifurcation, musically a so-called 'surface sound'. From about $1/\alpha \sim 2.4$, the situation is getting more and more complex, entering noise. The plot b) below is a Lyapunov exponent. When it becomes larger than zero, the system is chaotic, which is the case at the right part of the plot. From: [178]. This behavior can also be found with wind and stringed instruments, showing that the IPF is a method able to model all musical instruments

One thing that happens is that there are two alternating values, g_1 leads to g_2. When again inserting g_2 we get g_1 again, which produces g_2 again, etc. Such behavior is called a bifurcation, meaning one line is splitting into two, becomes 'bi' like a fork. In Fig. 5 we see from about $1/\alpha > 2.4$ to the right that there are now two values of g for each $1/\alpha$. Such behavior continues, where, by increasing α even further, each new value of g is bifurcating again, resulting in four values alternating when inserting one after the other into the above equation. Then, from a certain value of α, there is no clear pattern anymore, and we enter chaos, each new g is different from all previous ones (on the very right of the figure). Musically chaos is noise, and all violinists or cellists know how a bow without rosin on a string sounds like, it is pure noise.

This all happens with only one such a simple equation! Indeed, the equation is not new, it is very similar to a logistic map, discovered by mathematicians like Benoit Mandelbrot and others in the 1980th, following a long tradition of mathematicians and physicists dealing with nonlinear systems, like Poincare, Cantor, Julia, and others.[43] So the mathematics is very well studied, still that it is holding for musical instruments was not clear until recently.

This has much deeper implications than it might appear at first. Indeed, it means that at least the bowing is a self-organizing system. Self-organizing here means that out of a very complex and highly nonlinear system, a very simple output can arise, here the beautiful tones of violins, cellos, and the like. Still, the system can also be driven into regions where no such simple output appears, leading to a rich variety of articulation, a fundamental aspect of music.

In the following, we will see that this self-organizing nature also holds for all musical instruments and that without it, there would be no music as we know it. We have seen a first example here. The tone of the violin is only possible due to a highly nonlinear interaction between bow and string, and only because of this complexity a regular motion happens, a musical tone or note. We have also seen how fragile this process is, when we change parameters, we easily enter chaos and noise. But without a regular periodicity, no tone would be present, we would not be able to play melodies with a violin or cello. Therefore the self-organizing nature of the system results in musical notes. We will see this happening with wind instruments below, still, the mechanisms there are very much different.

This self-organizing nature is also what seems to drive music perception and neural networks in the brain. Although this is an ongoing debate, evolutionary it may be that this self-organization of musical instruments, and most likely also of hearing, is a crucial aspect of communication, semantics, and emotion,[44] and that we formed culture as a self-organizing system. Until now, we have seen that at least one cultural thing we have invented behaves that way, bowed musical instruments.

Environmental Noise

To view sound not only in terms of frequencies but as impulses answer some questions discussed today in environmental noise as traffic noise produced by cars, trains, planes, large diesel engines or, wind turbines. Many of those noise sources are annoying, and noise became a considerable health risk only during the last decades. Still, the legal regulations, as well as the measurement techniques, are most often not capable of dealing reasonably with the present kind of noises. Each country has its own legal regulations, and often, each noise source is treated with a different legal act. Still, there are some general problems which are common to all regulations around the world.

In 1933 Fletcher and Munson published measurements concerning the subjective loudness of sinusoidal tones [90]. Therefore their study was in the realm of psychoacoustics, relating a physical parameter, a steady sound consisting of only one single frequency, to a perception, the loudness as heard by listeners. They found that a single tone played to a listener is perceived with different loudness, depending on the frequency that sound has. When they choose a fixed physical amplitude with which all sounds were played and only changing the frequency of the sound, people reported that the low-frequency sounds were much less loud than sounds between 1–3 kHz. Again very high sounds above 3 kHz were lower in volume again.

So they asked a simple question: Which physical amplitude, which physical sound power a single frequency sound need to have at different frequencies, to have the same subjective loudness compared to a reference sound of 1 kHz? After playing a sound of 1 kHz with a certain amplitude, they played a second sound of a different frequency and changed the amplitude of the second sound until a listener would say the second sound has the same loudness as the first one. Doing so for many frequencies, they got a curve of equal loudness going through all frequencies. When playing the first sound with low volume, say 40 dB, the curve was lowest around 1–3 kHz and raised towards the low and the high frequencies. In other words, if one wants a bass sound to sound as loud as one of 3 kHz, one would need to tune up the amplitude of the bass sound very much.

Still, this does no longer hold for strong amplitudes, loud sounds. If a sound is played with sound pressure levels around 100 dB, all frequencies sound more or less the same. So the curve of equal loudness for loud sounds is very different from those of lower amplitudes.

To characterize this general finding, the different curves of equal loudness were given names. That of low sound pressure level with a strong dependency of loudness perception depending on frequency is called dB(A). That of practically no dependency is dB(C). There is a middle curve for medium pressure levels called dB(B), but this was never much in use.

Now, noise is mostly loud, and therefore needs to be measured with dB(C). Still, the legal regulations always use the dB(A) scale. Practically this means that if one measures the loudness of an airport, a train, car noise, or wind turbines, using dB(A) gives much lower values as dB(C) would do. So making the legal change today from dB(A) to dB(C) would mean that tomorrow nearly no cars, trains or planes would be allowed to operate anymore. They would simply be too loud to be considered not to threaten the health of people. This situation has been this way right from the start of sound measurement. This is why a change to realistic measurement techniques nowadays would be that dramatic.

It is true that the necessity to use dB(C) with most noise measurements is known since 1933. The reason for this is hotly debated for decades, as the error is too obvious for everybody, and the impact of environmental noise on us is well studied. Generally, there are two kinds of health impact of noise. The first is when being exposed to high levels around 104–120 dB (C-weighted, which means not weighted). This may happen when attending a concert, may it be in the popular music or the classical domain, or when dancing the night away to electronic dance music or dubstep. This might also happen in loud environments in the workplace. In such cases, the hair cells of the cochlea are not 'blown down' because of high pressure. Such sounds would even tear the eardrum down first. The reason for hearing loss in such cases is that the transformation of sound through the cochlea to electric signals into the brain is very demanding to the ear. The ear needs a constant inflow of vitamins, minerals, and several other molecules to keep going. When the ear needs to work hard through a club night, and the body is also dealing with alcohol and maybe several pills, there will not be enough supporting substances to maintain the hair cells, and they will literally lay down. When not exposing to such extensive noise during the next two or three days, they will stand up again. During this time, one hears tremendously less and may have some tinnitus-like sounds. But then the system will recover. If this is not the case, the hair cells will not stand up anymore, and the ear is irreparably damaged. This is an increasing problem, especially for young people nowadays in the Western world.

The other health impact of noise which is not permanently on that level is stress, an increase of blood pressure, heart rate, and innervation of the autonomic nervous system. Increased stress is reducing life expectancy, and therefore it can be calculated how many people die because of noise. A WHO report from 2018 finds only in Europe that 1.6 million healthy years of living are lost through noise each year, which corresponds roughly to about 30.000 noise deaths each year [262]. Of course, noise does not kill immediately, you hear a loud sound which is the last you ever heard. The situation is like with other health impacts like alcohol or sugar. They also rarely kill directly but cause health problems, which leads to a reduction of lifespan. Still, health problems are also a problem during lifetime of course. People exposed to noise and, therefore, in a constant state of stress are easier tiered, less robust, and may have increasing problems at work and in private life. This also leads to costs on a public level, like increased health care or absent times in the job. So the noise is also expensive for the public.

Noise and sound are also used for torture and warfare [104]. During the Irak war, the American troops set up large PA systems sounding towards the enemy, playing music of Metallica or Britney Spears day and night. Torture rooms use noise, flickering lighting, and the shaking of the room itself. But also bagpipes, horns, and drums have been used by military bands to frighten the enemy. But also, on an everyday level, noise has been discussed as a form of suppression. As the world became louder after World War II, many people found it more and more demanding to maintain their own thoughts and ideas, constantly annoyed and sidetracked by environmental noise [8].

Indeed, from the industry side, each larger company producing loud devices doe extensive noise research to make suggestions reducing the noise. Still, rarely these suggestions are followed by the company later on, and only those are implemented needed to meet the legal standards. The new jet engine of the Airbus A380 airplane, for example, had extensive noise investigations during development, which would have led to a considerable reduction of the engine noise. Indeed, only one of them was implemented, the engines cover towards the front is not build of two halves put together, resulting in a small gap anymore, but of one closed circle. Such a gap leads to turbulence, and does both, increase noise and reduces the performance of the engine. Still, the noise reduction is not really much. All other suggestions which would have made the engine considerably less noisy were not implemented. They all would have meant a reduction of performance. Indeed, noise reduction is often possible to an astonishing extend. Standing next to a train or an airplane may blow your ears, still, inside the plane or train, noise volume is very low. Bit for those not paying the tickets or buying the car noise reduction does not pay off. The economic pressure, therefore, makes many companies reject most of the noise reduction.

So it's on the legal side to take action, and the first thing to do would be to change the laws in terms of using dB(C) and no longer dB(A).

Still, then the legal regulations for noise would be in the state of 1933. Since then, noises have tremendously increased, and a modern industrial society produces sounds nobody would have dreamed of in 1933. In Germany, the gap between the legal situation and reality was felt even by officials already in the 1960th, which lead to an administrative instruction, not a law, introducing some more elements of noise (all still using dB(A) instead of dB(C)), called *TA Lärm*, Technical Instruction Noise. This mainly includes three special cases for noise as especially annoying, low sounds, pitched noise, and impulsiveness. In case a noisy sound has such a nature, the measured dB(A) sound pressure level gets a 6 dB plus on top and then needs to meet the legal limits.

Comparing the state-of-the-art of science in the field of sound, noise, health, and the social aspects of the environment, modern lifestyle, working places, etc. the legal situation seems still to better fit the pre-industrial times, taking noises of horses and blacksmiths as a model. So in terms of annoyance, noise needs to take the content of the sound into consideration, not only its loudness. This is recognized in the *TA Lärm* and other instructions, still only too briefly.

Pitched and low-frequency sounds are more disturbing than so-called broadband noise. Broadband noise would be that of a waterfall or a fountain, which has many frequencies, a broad-band of frequencies, and therefore no single frequency appears as a pitch, a tone one could identify and sing along. This is not too astonishing, taking typical alarm signals into account. Alarm clocks, the signals back-driving trucks, and other diesel engines need to make according to legislation or other signals which are to attract people's attention. The inventors of these signals clearly, and most intuitively, favor pitched signals like peeps or rings. A waterfall-like noise signal may not be too effective, indeed some people would find this more relaxing. There are neurophysiological reasons for this, which much still to be investigated. But we will also find a reason for the evolution of human communication, which has to do with intelligent beings, information, and entropy.

Another very annoying kind of noise is impulsive noise. This is considerably different from a noise that does not change in time. As we have discussed impulses before, it is interesting to note that impulses also play a special role in noise. Again considering alarm clocks or the sound of back-driving diesel engines, in both cases, pitch and a repeating sound are combined. The sound is not steady but interrupts, starts again, obviously to even increase the annoyance of people, asking them to take care of a situation. So impulsiveness is also an issue in longer time spans. Before we discussed impulses only a few milliseconds long, here we talk about several hundred milliseconds or even intervals of around a second. Again the changes in sound are very much different from a steady sound.

In terms of music, this is trivial. Music is changing all the time, and there rarely is a steady-state. Rhythm, articulations, effect manipulations, filters, or other effects make music move constantly. The idea to measure music in terms of frequencies, which are assuming steady-states, is therefore not at all straightforward, and an impulse pattern formulation comes much closer to the real thing. We discussed the Fourier theorem above in some detail, and we found that it was, first of all, a mathematical tool, a transformation of a time series in a spectrum consisting of amplitudes and phases of some frequencies. We also found that a transient, a non-steady sound, can be perfectly represented by a Fourier spectrum, still many amplitudes and phases of the frequencies are not real-sounding frequencies but represent the transient development of the sound, its increase or decay, etc. Therefore, interpreting a Fourier spectrum in terms of a perceived sound as a bundle of frequencies for a transient or impulse sound is not correct. So although mathematically everything is correct, the representation is not fitting musical perception very well. Sound engineers know this very well when they try to manipulate music using this theorem. When music is in a steady-state, filtering, adding reverberation, or delay effect using the frequency representation is going quite well. Still, when the sound is changing, the algorithms produce what is called a phasing effect. Everybody immediately hears this as a spacious-like sound, which sounds interesting in terms of musical composition, but it does not sound naturally anymore, and the manipulation failed, it cannot be used. This phasing is the result of interpreting all amplitudes and phases of frequencies in a spectrum as steadily sounding music, where many of them are only representing the transient or impulse development of the sound. These components are manipulated by the methods, too, leading to an unnatural sound.

Indeed, Fourier introduced his theory in 1807, discussing problems of heat transfer in a solid body. His approach won a price in 1807 at the Paris Academy. And indeed, this representation is very powerful and can be used in many fields, Fourier himself developed it to describe heat flow. Still, it seems not to be straightforward in music, although the notion of frequencies is all around in the business. That does not mean that we need to discard the idea completely, still, it stands in our way when trying to understand how music really works, and therefore it is worth considering representations which naturally appear in musical instruments instead, the interaction of traveling impulses.

Musical Noise

There is a musical genre called 'noise' which developed in the 80th, but had forerunners starting at least in the 60th on [124]. It is a music working in a live music club environment like Hip-Hop or Rock does, but includes ideas from art, philosophy, or contemporary classical music, with bands like Throbbing Gistle, Caspar Brötzmann Massaker, or Merzbow, earlier examples in free jazz are Ornette Coleman, Derek Bailey, or Alexander von Schlippenbach. Many of the sounds used are supposed not to have too much of a pitch nature, but contain many partials not harmonic one to another, are rapidly changing with strong impulsiveness, broadband noise, and all kinds of sounds normally called annoying. These sounds are produced by using musical instruments in uncommon manners, like distorting a human voice or bowing a cello body instead of its strings. When playing saxophone with very high blowing pressure or with complex fingerings, wind instruments can play several pitches at once, so-called multiphonics. Musicians like Even Parker or Michel Doneda.[45] also use other techniques, where the saxophone produces several inharmonic tones at the same time. Of course, electric and electronic music use synthesizers, driving them into noise-like, highly impulsive sounds. All kinds of things might be taken as musical instruments. The Einstürzende Neubauten used construction machines, performing in spaces right below motorways. Survival Research Laboratories used explosives and heavy machines they had developed for blockbuster Hollywood movies in concerts in the Mojave desert.[46] In the domain of contemporary classical music, these uncommon playing techniques are called extended techniques. In other contexts, they have no special names. The field of noise music is very broad and has many different notions, some elements are found in free jazz or free improvised music, some were taken over by techno, electronic dance music, or hip-hop, and even pop music, still very much embedded there with harmonic sounds, more typical for these genres.

It is interesting to see that many musicians in noise do not really call their music 'music', but often find it more information-like.[47] Although many artistic standpoints are taken, of course, nihilistic ones at first, there is a highly intellectual background in many of these works, in terms of reproducing modern civilization soundscapes, getting deep into understanding human communication, freeing old conventions to allow new textures and sounds, representing pure power, getting into a trance or altered state of conscious reflection, envisioning new ways of living, or enforcing to face reality in a Zen Buddhist sense, next to many others. They discuss problematic and boundary aspects of human life like transsexualism, death, or other existentialistic states.

This is not new. In many cultures, noise is used for trance rituals, funerals, or as signals transmitting information. People living an alternative life have found a home in the arts, especially in restricting societies and express their experiences there. The arts are also very suitable for expressing highly emotional states during life-crises and personal problems. In a more soft sense, this is called mood regulation and is implemented as such in streaming platforms, suggesting musical pieces fitting to an emotional state.

In the history of the West, many stories are known where the unordered is fighting the order, in the hope order wins, still being conscious that without chaos, the world cannot exist either. The Phrygian people in ancient Greece, in today's south of Turkey, had a myth of a god Hekate, which was very fertile, but regularly was destroying her garden in a rage, also castrating her husband [198]. Most likely, this is to represent nature developing over the year, where in spring, plants and flowers start growing and getting green, and in fall, most plants lose their leaves and decay, just like a god had run through the garden destroying her work and ending fertility. When this cult was institutionalized by priests, they were still castrating themselves, which might have been the origin or eunuchs in the Osman Empire. It has also been proposed that the whirling dervishes of the Mevlevi Sufi order in today's Turkey are derived from this tradition.

In terms of music, in ancient Greece, it was Chiron the centaur who taught people to play the kithara, the ancient lyra [257]. Still, the centaurs were known to be pretty wild, entering a wedding by killing the guests, raping the bride, and drinking until they get their fill. One interpretation of this behavior is the impact of African people from Libya entering the Grece world while riding on their donkeys. These Libyan donkeys are quite wild, represented in the wild nature of the centaur. Still, these people also brought along the African tradition of playing stringed instruments, mainly harps like in Egypt or the West-African kora, while singing epic stories. Homer might have been one of them, indeed we can argue that there were many Homers, singing epics, a tradition found until the 20th century around West-Africa, the Arabian world up to Afghanistan and Pakistan, Russia, Tibet, and way down to Nepal, where the Gaine sang stories to transport news from other areas, mainly from Nothern India. The Ilias and the Odyssee are indeed not the larges epics. The story of Gesar, the epic foundation story of Tibet, also sung in Russia or Turkey, is by far the largest [159] with more than one million verses. Music theatre most likely has developed from these epic singers, therefore, the opera most probably comes from Africa.

Initial Transients

In musical terms, chaotic or noise-like sounds are present all the time. At the beginning of each played note, there is an initial transient phase that is chaotic. This is because when playing a tone, one part of the instrument, often called the generator, is forcing the other part, the radiating part, to go with the movement of the generator. So a string forces the guitar body to go with the frequencies of the string, although the guitar body would not move this way on its own. Therefore at the very beginning of a sound, there is a struggle between the two parts on who will win the game. This struggling is causing very complex behavior of vibration and is therefore chaotic. The amount of this chaoticity is important in terms of articulation and expressiveness, where high chaoticity is perceived as especially characteristic and expressive. This initial transient phase is indeed so important that when cutting it off artificially, e.g., by computer software, a violin might sound like a trumpet or a piano like a guitar.

The first to investigate this was Carl Stumpf in 1926 [242], who had a musician playing in one room and a listener in a neighboring room. Between the rooms was a hole in the wall, which was closed when the musician started a note and was opened right after the start of a played note. So the listener in the other room could not hear the initial transient, but only the steady-state of instruments like trumpets or violins. When then guessing which instrument it was, the listeners very often failed. Still, they were guessing correct when they also heard the initial transient of the instruments, so when the hole in the wall was open all the time.

Of course, there are additional features that make listeners identify musical instruments next to initial transients. Violins are often identified when played with characteristic vibrato. Saxophones and clarinets have formant frequency regions, at which the sound power is always strong, no matter which tone is played. This is due to the reeds of the instruments vibrating on their own a little bit, independent from the played tone. Of course, plucked instruments like guitars, pianos, or percussion instruments decay in sound over time, while wind or bowed instruments do not. Still, next to these particularities, the initial transient is the most crucial part when perceiving a sound.

This transient phase is very fast. Piano tones may start within only about 1 ms, guitars may take 20–50 ms, and low organ pipes maybe even slower, up to 200–300 ms. The organ case is too much and no good, as then the transients are perceived as a separate part of the sound and are no longer fused to a single event anymore. This fusion can only take place within about 50 ms, as this is the time the auditory system uses as the smallest grain size of perception. 50 ms is the inverse 20 Hz, and so if an event repeats in longer time intervals, say 100 ms, corresponding to 10 Hz, we do no longer hear this repetition as a sound but as single events. We then call it a rhythm. So there is no fundamental difference between rhythm and sound, it is only that all rhythms faster than 20 Hz for use are sounds.

This interval of 50 ms, at which all events coming to the ear are taken as one grain or unit, is also present in vision. A video is normally played with 24 frames per second. The old Super 8 film format was even slower, with 18 frames per second. If this is decreased even more to maybe 15 fps or lower, we do no longer see a continuous movie, but single pictures, and have the impression the movie is not working anymore. This is because also the visual system takes the 50 ms time interval to sum up all visual events happening. When something is moving, we expect to see a different picture each 50 ms at last. If at two adjacent time-poiints the same picture appears, there is something wrong, we do not see a continuous film, but something unnatural. The reason for this 50 ms integration is not clear yet. Still, it makes sense to reduce the amount of information coming in and sort it into chunks, grains, or units that we can deal with.

So much of the identification of musical instruments is within this initial transient phase, and therefore chaotic vibrations are crucial to music.

Information in Music

The identification of musical instruments, the articulation of tones, the tone onset points, stressed by the chaoticity of musical sounds, all this is information contained in music. Within initial transients, there is much more information contained compared to the steady-state of the sounds. Also, treating these transients in terms of music analysis is much more complex. When using music streaming via the internet, the system is proposing similar pieces of music compared to the fast pieces downloaded. They are classified by emotions and style, but also in terms of speed, mainly for those using music for jogging and workout, or other reasons.[48] Such systems often use the information contained in the sound of the musical pieces by algorithms that analyze the music and retrieve information out of it. These engineering tasks are known as Music Information Retrieval (MIR) and is a huge ongoing field of research. The algorithms take the digitalized sound file, and from this data are extracting onsets of notes, tempo, pitches, melodies, chords, instruments, and many other features.

So when a streaming platform is suggesting a similar piece of music compared to the one just heard before, this similar piece has similar features in terms of rhythm, tonality, being in major or minor mode, and many other features which are also clustered to emotions. The emotional content is most often simplified into two dimensions, arousal and valence [107]. So a piece might be calm or lively (arousal) and, at the same time, bright or dark (valence). Very roughly, this results in four possible states, lively and bright is happy, lively, and dark is powerful, calm, and bright is chill out, and calm and dark is sad. As both dimensions can be present in a musical piece, more or less subtle differentiations are possible, and neighboring musical pieces in this abstract emotional two-dimensional space are expected to have about the same mood.

Of course, such a very rough emotional description is not able to distinguish between more complex emotions. From a musicological standpoint, much more can be said about emotions caused by music.[49] All these are extra-musical emotions, emotions not present 'in' the music but caused by music. Therefore they are highly subjective and most often only valid for a certain group of people, a certain age, ethnic group, with its preferences, needs, and musical educational background. Above, we have discussed psychoacoustics, where also emotions were discussed as sensations, sensual reactions to music, like brightness, roughness, etc. These are universals, everybody hears them, therefore, we can say they are 'in' the music. Still, they are felt, and therefore emotions. A very loud or rough sound is emotionally very different from a soft and low one. Other musical feelings, like tension, is in between. A chord progression of subdominant, dominant, tonic, e.g., F-major, G-major, C-major, which is a cadence, causes a tension bow for all listeners familiar with a cadence. Still, with untrained people who had never heard a cadence or had any Western musical training, this will not be the case at first. So there are musical universals inborn, part of our musical nature, and there are trained feelings, nurtured. To this nature-nurture field come those emotions which are very subjective, depending on daily moods and events, like the chill-effect. Although such effects have a strong physiological basis, to music they are only loosely connected.

Simplified approaches to music, like the arousal-valence field, are most often engineering ones, which are about to make music accessible to a large number of people, and in the end, earn money. Still, it is interesting to follow the music being composed for such platforms, they increasingly try to fit the algorithms used by the streaming agencies to increase commercial success. So a new musical style is created by possibilities, constraints, and restrictions of a technical invention, the streaming industry. Such developments are found very often in the history of music, and it is interesting to see how musical styles are formed by the techniques and inventions of musical instrument builders, music theorists, electrical engineers, sound designers, informatics, or social or political constraints.

It is interesting to note that extracting even very simple information out of music is often that complex, that up to now they are not working perfectly at all. The simple task to detect if a note has been played at all, so if and when it started, works only up to about 80% of the time, depending on the musical piece. So a percussion piece with very strong amplitude changes at each strike is easy to extract the note onsets from [1]. Still, a piano piece with only very soft changes in amplitude at each note is so complex that no algorithm up to now is able to detect all notes in such pieces correctly. Still, we as listeners do. Taking the amount of research in this field into consideration, this is really astonishing.

Things become even worse with more complex tasks, like extracting the pitches and, therefore, the melodies out of a piece of music, a task involving musical skills of a listener too. Here the percentage of correct answers very much depends upon the piece, may be quite good with simple melodies but can also fail completely with complex polyphonic pieces.

Extracting features of higher compositional dimensions, like larger forms, is also not straightforward. Especially in terms of modern music which does not follow a song structure like Electronic Dance Music, Techno or Hip-Hop, which are the most prominent music styles today, but also of ambient, chill-out, free improvised, experimental, or electronic music often no clear pitches are present, and very complex sounds are presented. So other compositional ideas are used rather than a chord structure of a 4/4 rhythm. Here algorithms taking the musical density and complexity are suitable to explain the compositional or instrumental ideas. Algorithms calculating the fractal dimension of musical sounds as the number of dimensions it exists in are very suitable to display long term forms [21]. Also, algorithms displaying entropy or fluctuations are very close to musical thinking.

These algorithms calculate the chaoticity of the sound as the number of harmonic overtone spectra of a sound, adding all the inharmonic frequencies. So a single tone C has the fractal dimension one, a chord C-E-G has a fractal dimension of about three. If chaoticity in the initial transient phase is present, during this transient phase a ll additional frequencies not harmonic to the fundamental are added. Single guitar tone initial transients have fractal dimensions between 3–5, violins might have up to 8. Percussion instruments mainly consist of inharmonic spectra, still, many are tuned to sound with a pitch, like tom-toms or tablas. Their fractal dimensions strongly depend on the amount of their inharmonic partials. Pure noise has a dimension of infinity.

It appears that when one is calculating the fractal dimension over time for a whole piece, it very often follows the musical form, maybe starting with soft and loose sounds which are getting more complex and dense, reaching climaxes, and returning to a soft sound. Or a piece is constantly shifting between complex and simple event densities. A typical tension built-up of a Techno or EDM piece up to the climax, where the four-to-the-floor bass drum comes in, shows a constant increase of the fractal dimension up to that climax.

Therefore the fractal correlation shows the amount of information in the musical piece. As everybody readily hears complexity and density changes in music, this is an important compositional tool. It is able to produce or reduce tension, differentiates musical parts, or causes more or less attention.

As the increase of the event density means more or less noise, information in terms of the strength of the fractal dimension corresponds to the amount of information present in a musical piece.

Turbulence

Turbulence is still an unsolved problem in classical physics. When David Hilbert (1862–1943), one of the leading mathematicians at the turn to the 20th century, gave his famous lecture in 1900 at the Sorbonne in Paris about the remaining big problems in mathematics, the solution for turbulence was one of them. In 2000 the Clay Mathematics Institute was repeating the Hilbert problems, now with a 1 million dollar prize for the solution of one of seven problems to be solved within the next century (Hilbert had presented ten problems at his lecture, although he as 23 in mind.)

From the original Hilbert problems, only two have been solved yet, one during the 20th century, and one only a few years ago. Kurt Gödel (1906–1978), with his famous incompleteness sentence, showed that number theory could not be proofed within the domain of numbers alone. Additional assumptions from outside the number domain are needed [133].[50] This proof was taken as a general proof of incompleteness of proofs. It challenged the idea that mathematics is the only discipline that can fully be understood by man, as it was made by man. Hilbert took this standpoint vehemently that all mathematical proofs need to be consistent (without contradictions), complete, and decidable (starting from Euklids book *Elements*, where proofs are derived from axioms, simple, undeniable sentences given, and no longer proofed themselves.). Still, Hilbert could not prove the completeness assertion. After Gödel had shown that completeness did not hold even in such a simple case as numbers, that there cannot be a full proof of numbers from within the relations between numbers, the whole idea of completeness had to be given up. This proof also spread out to science as a whole, with its suggestion that no discipline, no physics, philosophy, biology, etc. can be fully understood within its own domain, laws, or rules.

The example of the Gödel theorem shows the enormous impact the solution to the problem posed by Hilbert had on the development of science and our understanding of the world. Now the problem of turbulence is much alike, as although there was tremendous research in this field for about two centuries now, still, there is no solution to the basic equations. So the turbulence problem again appears in the Clay Institute problems posted in 2000.

R. Bader, *How Music Works*, https://doi.org/10.1007/978-3-030-67155-6_5

So it is no wonder that turbulence is a challenge for musical instrument research too. Still, it is decisive to understand wind instruments like saxophones, clarinets, flute, or the organ. There have been many attempts to solve these problems without the help of turbulence, by reducing the problem to simplified air pressure and velocity treatments. Still, although much can be understood with so-called lumped models, many basic properties of wind instruments cannot be explained by them. Especially the very basic fact that we expect wind instruments to sound with a harmonic overtone spectrum, and therefore can be used as musical instruments at all, is missing when only explaining the sound-producing mechanisms with simplified methods. This can only be explained by taking turbulence into consideration. Therefore we need to discuss turbulence in some detail, where we will understand many basic properties of synchronization and self-organization found further down in the book and in our overall reasoning of how music works.

Turbulent and Laminar Flow

Blowing into a saxophone does lead to two basic kinds of flow or wind in the mouth-piece and the tube, a turbulent and a so-called laminar flow. Laminar flow means a smooth traveling of a gas or a liquid, in our case, the air through the tube, caused by the musician blowing into it. It is like a constant stream, where the whole flow travels roughly at the same speed in the same direction. This is the easy part of the flow because it can simply be described by a velocity and the direction the flow is traveling. So when a saxophone player is blowing into his instrument, the laminar part of the flow may travel with a speed of one meter per second (m/s) with very low blowing pressure and up to 20–30 m/s when blowing with very strong pressure. This might seem high, still, this only holds for the small region in the mouthpiece and is much slower and much less intense in the adjacent tube where the finger holes are placed.

Indeed, the main thing about a blown instrument is the transfer of wind into sound. So when blowing into a tube, the wind produced by the player is driving the system, it is supplying energy to it. Still, wind, in the laminar form, is no sound, as sound is a change of velocity, of pressure back and forth. A tone 100 Hz means the air is traveling 100 times per second back and forth. This is sound. As we have seen, laminar flow on the other side is traveling with a constant speed in only one direction. So laminar flow is no sound, and therefore cannot be heard. So the initial laminar wind produced by the player needs to be transferred into a different kind of wind, a one which is changing in time rapidly to become sound, to become music.

For this to happen, the laminar flow needs to be disturbed. Disturbance of flow most often leads to a turbulent flow, which is the case in all wind instruments. We will discuss the process of sound production in wind instruments below in much detail. Still, to really understand what is happening, we first need to get into turbulence itself and see how it works.

A basic experiment with turbulence is the flow through a tube. The tube is straight and round and not too long. At one end of the tube, a liquid or gas is put in with some pressure and some velocity. If both pressure and velocity are not too high, the flow will be laminar, it flows through the tube smoothly and leaves it at the end. Now when increasing the pressure of that flow, it will go faster through the tube, but still in the same laminar way as before. Now, one could expect this to continue on and on for even higher pressures acting at the beginning of the tube, the flow would become faster but still flow smoothly through the tube.

Still, this is not the case. At one certain pressure threshold, the flow is no longer smoothly flowing, but travels in all directions in the tube, creates vortices, smaller and larger eddies, and therefore becomes very complex and very different: it is turbulent. This flow is considerably different from a smooth laminar flow, which is unexpected at first. Also, it cannot be simply described by one velocity and the direction it flows to. At each point in the tube, the flow now has its own velocity and its own direction and is, therefore, much more difficult to describe. Mathematically a horrible situation, as there is no simple formula to describe such a complex flow field.

Additionally, this change from laminar to turbulent flow happens during a very small region of velocity increase. Over a large range of low pressure acting at the beginning of the tube, the flow was laminar, no matter which pressure was applied. Still, at a certain pressure value, there is only a very small increase of pressure necessary to make the transition from laminar to turbulent flow. In physics, the state a system is in is also called a phase. This is different from the phase used in acoustics, where it describes the position within a sinusoidal cycle a vibrating body is in. Phase in the case of flow means a phenotype, an overall behavior, or state. So flow can be in a laminar state, a laminar phase, it can also be in a turbulent state, a turbulent phase. The transition from laminar to turbulence is then called a phase transition, a fundamental change of the phenotype of the flow.

This transition has another property, making it even more complicated. There is a pressure where the flow is still laminar and where a small increase in pressure makes the change to turbulence. Still, when we increase the pressure so that the flow is turbulent then, we might think we could make the flow laminar again by simply decreasing the pressure to the old one, where the flow was laminar. Still, if we do so, the turbulence continues! Only when we reduce the pressure to a value which is below the original one, the flow becomes laminar again. Again, if we then want to return to the turbulent state, the turbulent phase, we can do so with the old transition pressure as we did in the very first place.

This is a basic phenomenon found in many complex, nonlinear systems, a memory effect, where the state or phase a system is in does not only depend on the present state but of previous states in time. Therefore, the system 'remembers' the previous states and behaves accordingly. Such behavior is called hysteresis.

One can also interpret this behavior by saying that a system tries to stay in one phase, in one phenotype as long as it can. It tries to preserve its state and stretches the region of parameters, in our case, the pressure at the tube beginning. This is a basic property of self-organizing systems, they organize themselves by maintaining their state they are in and try to stay that way as long as possible under changing conditions. Wind instruments have this too, as we will see later.

Why is all this happening? Today most researchers will agree that most kinds of flow can be modeled with an equation called after its inventors, the Navier-Stokes equation. Although many more people were involved in the development of this equation, which took about a century to fully write down and[51] although there are many version of this equation according to special flows,[52] research questions or methods of solution, the very core of it is the same. To understand turbulence let us have a closer look at this equation.

Navier-Stokes Equation

The basic idea of the Navier-Stokes equation is that of a balance between several physical parameters, the speed of the flow, its pressure, its density, and its viscosity, which is how liquid or how viscous a fluid or gas is. The equation states that if one of these parameters changes at one point in the flow, one or several of the other parameters need to change too to keep the balance. So if the flow changes, the pressure or the density needs to change too. The viscosity cannot change, as it is defined by the kind of fluid or gas if it is air, water, or syrup. Still, this balance, what does change, and what will take place under which conditions is the basic problem which took so long to find.

The Navier-Stokes equation holds for both laminar as well as turbulent flows. Still, as we have seen above, turbulent flow has a different velocity and pressure at each point in the flow field. Therefore the Navier-Stokes equation needs to assume the pressure, the velocity, and the density to have a separate value at each point in the flow field. Mathematically speaking, pressure, velocity, and density are functions of space and time, at each point in space and time, they have a distinct value.

Still, an equation needs to give the general rule for the behavior of flow at all points in space under all conditions. So it formulates this rule considering a single point or region in the flow but says that this general rule holds for all places. Therefore one can choose any point in the flow, apply the equation, and find it to hold. The other way round is also true. As the whole flow field is governed by the rule, the equation states, no single region can act on its own but is connected to all surrounding regions. As again, these surrounding regions are connected to their neighborhoods, indirectly each region is connected to all others, and so all velocity, pressure, and densities in the flow field change only according to all others.

Such an equation assumes everything being connected to everything else. To account for the relations between the regions, the changes or differences of velocity, pressure, or density between regions are the fundamental parts of the equation, not the values themselves directly. When considering differences between regions, the distance between the regions considered may influence the result. Still, mathematically there is a way to overcome this problem by assuming very small distances between regions, transferring the equation from one considering differences into co-called differentials, making it a differential equation. But the basic idea is the same, the equation assumes that everything is connected to everything else and needs to behave accordingly and that the equation is not considering the values of pressure, velocity, and density directly, but only as differences, or differentials, between adjacent regions, taking this idea into consideration.

Differential equations are fundamental to most problems in physics, and nearly everything in this world can be understood using them. The basic idea of everything-connected-to-everything-else has been applied successfully in electromagnetism, quantum mechanics, classical mechanics, flow fields, and many other applications in sociology, economics, or biology. The calculus of differential equations has been developed by Gottfried Wilhelm Leibniz (1664–1716) and Issac Newton (1642-1626) in the 17th century, independent one from another, still the formulation used today is that of Leibniz. It also assumes a field, a region where all behavior in the field depends on all other regions in that field, a fundamental description of nature in general.

So it is no surprise that also for flow differential equations hold, and we can understand flow as a balance of the differences of these parameters between adjacent regions.

Still, most other differential equations that describe phenomena in music, like the vibrations of plates, of membranes, of drums, or in room acoustics do not have such behavior of suddenly changing their phenotype from one phase into another. The reason is that most other differential equations are linear, in a sense, discussed in the previous chapters already. Still, the Navier-Stokes equation is nonlinear. Therefore it is necessary to go deeper into the equation to find out where this nonlinearity is and where it comes from.

Equations consist of terms, which are all parts of the equation, and are added or subtracted one from another. It makes sense to call them terms because most of the terms can be associated with different physical behavior, and are therefore units on their own.

Below the Navier-Stokes equation is shown. Its first term (Term 1) is simply the change of the velocity v in time. This is a differentiation of the velocity with respect to time. Again there is a different velocity change at each point, still as we have seen above, it is enough to look at the general rule holding for all points. In the equation, this is the only term that changes in time. All other terms change in space only, and therefore we can say that when the flow will be different a very short time point later, this must be caused by the other terms in the equation. If we calculate the values for all other terms and they become zero, then we know that the flow will not change in time. This does not mean that it will stop flowing, it can have any flow velocity. But it would mean that the flow will not change this velocity.

$$\frac{\partial v}{\partial t} \quad + v\,\nabla v = -\frac{1}{\varrho}\nabla p + \frac{\mu}{\varrho}\Delta v$$

Term 1 Term 2   Term 3   Term 4

Another term (Term 3), one of the three terms that can cause a change of flow v in time, is a differentiation of the pressure with respect to space, times the density ϱ at the position we are looking at. The Δ means a differentiation with respect to space in all three directions, in x, y, and z. This is also reasonable. If at one point in the flow, the pressure is larger to the left than to the right, this pressure difference would make the flow change. Only if the pressure at both sides of the point we are looking at is the same, then the differentiation of the pressure with respect to space, the pressure gradient as it is also called, is zero, and therefore the velocity will not change in time because of the pressure.

Maybe it is getting clearer now what was meant above when saying that the equation is not using the physical parameters pressure, velocity, and density itself, but only their changes. The *change* of velocity in time can be caused by a *change* of pressure in space, no matter how strong both values might be. This consideration of changes is what makes an equation a differential equation, talking about differences rather than absolute values.

The next term we want to have a look at is introducing the damping of the flow, Term 4. There are two versions of this term, one for the compressible and one for the incompressible Navier-Stokes equation. They are basically the same, although the compressible version is much harder to calculate. The above equation is the incompressible version of the Navier-Stokes equation. The difference between compressible and incompressible is important for our problem, but we will discuss this in a minute, it need not disturb us now. Still, we already find that in the incompressible version the density ϱ is a constant, while in the compressible version is may change in time and space.

The incompressible version of the damping term, Term 4, is a second-order differentiation of the velocity with respect to space, times the viscosity μ over density ϱ, which are both constants in the incompressible version of the equation. Now a second-order differentiation is applying the idea of a change to a parameter twice. This means we talk about the change of the change of the velocity in space. If this is happening, then the flow is experiencing a kind of friction, which is strong if the fluid is viscous, and is not so strong if it is more fluid, which is the case for water or air. For us, this term is very small and, therefore, not important for acoustics in general and in our discussion.

Coming to the last but most important term (Term 2), we have a change of the velocity v in space, still, this time multiplied by the velocity at this position. This is fundamentally different from the previous terms, as now, contrary to the rules of only looking at changes rather than at absolute values, there is the velocity itself involved. As both parts of the term are multiplied, the change of velocity and the velocity itself, it is clear that this term is zero if only one of them is zero. So if there is a strong velocity but no change of it in space, this term is zero. Also, if there is a large velocity at the right side of the point we are looking at and a velocity in the other direction on the left side, with zero velocity at the point itself, this term is also zero (this second possibility is quite unusual, still possible in principle, the first one is often the case). Nevertheless, often both parts, the change of velocity as well as the velocity itself, are not zero, and therefore the term as a whole is not zero and will cause a change of the velocity in time.

Still, much more than this, this term is composed out of several terms that interchange the flows in different directions. A flow is always in a three-dimensional space and is therefore flowing in all directions. The term now combines the change of flow in one direction with the flows (not their changes) in the other directions. Therefore all directions are interacting and are no longer independent one from another.

The reason for this term to appear can be derived mathematically from other equations, which is even more complicated to explain. Still, it can be understood quite intuitively when remembering that a flow is moving particles of air or of a fluid. These particles do not only move in one direction, they might change their direction partly or wholly, and this possibility needs to be modeled in the equation. Of course, this makes this term very complex and represents the interaction of the flow directions.

Still, the really interesting thing about this term is the involvement of the velocity itself multiplied with the change of velocity. This makes it a nonlinear term, and therefore the Navier-Stokes equation is a nonlinear equation. As we have seen, nonlinearity can produce many kinds of unexpected behavior, like sudden phase transitions, hysteresis, and here turbulence. Indeed, the reason why flow can become turbulent is present right in this term of the Navier-Stokes equation. In other words, if these terms would not be part of the Navier-Stokes equation, no turbulence would appear.

So, in summary, the flow changes when there is internal damping, when there is a pressure difference, or when there is a flow difference with a flow itself present. Still, there is another parameter left, the density. This density changes with time too, and this is what we call an acoustic wave, a sound we can hear. Such a density change will also be a pressure change and a changing velocity. Still, the pressures and velocities we have discussed so far are only very slowly varying in time and are of a flow-type. So although theoretically, they can be sound, in practice, they are far too slow to be sound, and so far, we have only described the blowing, the wind.

A density change, on the other hand, changes fast, influencing its surrounding also fast, and therefore travels fast in space. It is the moving sound wave that will leave the musical instrument and enter our ears where we can perceive it as sound. So we need to have a look where such density differences appear in the flow. These points are the sound sources, the points where the flow is transformed into sound, where flow becomes acoustics, the reason for which these instruments are built.

Theoretically, such density changes can appear everywhere and indeed do so. In the next chapter, we will have a look at wind instruments in more detail to see how this is happening in the most important cases. Still, there is some general behavior of turbulence to discuss first.

What Does Turbulence Lead To?

The first is heavy damping. In a standard experiment already discussed above, a flow of water is traveling through a tube of maybe one meter in length and maybe five centimeters in diameter. If the flow is slow, it will be laminar and flowing smoothly and straight through the tube. If the flow is getting faster, the laminar flow will continue, still faster. Then, at a certain speed turbulence will appear, and the flow will go into many directions, all over the tube, in a very complex way. Still, then at the end of the tube, there is nearly no water flowing out anymore, and at its beginning, the new water can no longer enter the tube. This is clear as the flow inside is constantly changing directions, and therefore need to travel long ways to get from one side to the other. So suddenly, the fast, laminar, and simple flow from one end of the tube to the other stops nearly completely. It comes to a halt. There is something damping out the flow in the tube. Still, this is not damping caused by viscosity of the fluid, as discussed above. This viscosity damping is very small in air or water. The damping causing the flow to stop in the tube is caused by turbulence, and is therefore called turbulent damping. Still, it has the same effect as all damping, it brings the flow down to a halt. And indeed, it is very strong and able to stop the flow.

Now turbulent damping is the crucial point for sound production in wind instruments, as at the points where the flow changes direction, there is a very big density change. Remember that density changes are the sound sources in a flow, and therefore produce the sound of the instruments. This mechanism differs between different wind instruments slightly, still, the basic mechanism is the same as we will see below.

Another important aspect of turbulence is when a flow is leaving a small tunnel, or slit, and flows into a large room. If the flow in the small tunnel is still laminar, the flow leaving to tunnel or slit is laminar too. Still, after leaving the tunnel with its rigid walls making the flow stay in the tunnel, in the free field behind the tunnel, the flow has no such walls or boundaries around it anymore and becomes a flow stream. So it might change directions easily and does so very easily when only very small differences in the surrounding air above or below the flow stream are present. A very small difference is a very small cause. But such a small disturbance increases to a cascade of disturbances because of the nonlinear behavior of the Navier-Stokes equation. The flow will change direction slightly right after the slit and more and more while traveling into the free field. It changes direction exponentially, and after a while, it turns around itself, forming an eddy or vortex, a rotating circle.

This first rotating cycle or vortex is the beginning of a cascade of more and more vortices, then caused by the first one. These new vortices are again causes for new vortices because after about one rotation in the first cycle, the speed of the flow has decreased, and, therefore, it is even more vulnerable to further influences. Each of these smaller fluctuations leads to other vortices. So the first, big vortex causes several smaller ones, where each smaller one again causes a couple of even smaller vortices. This continues until a smallest vortex size, where the flow is so slow that it more or less comes to a stop.

Such a vortex street looks so complex and chaotic that we call it turbulence. It brings the main, fast flow to a stop by so many changes in flow directions as found already above. Still, here we do not have a fast flow through a tube but a small laminar flow leaving a small slit into an open free field.

There are many examples of turbulence, each with slightly different conditions and outcomes. One might know this from the smoke of a cigarette, which leaves the tip moving upwards smoothly, and after some time becomes more complex, ending in turbulence. Here the temperature of the smoke is very important, which is not a basic feature with wind instruments, although the wind flow of a musician is at body temperature and becomes room temperature in the instruments. Still, here the temperature difference is small and not causing a basic mechanism.

There are also other places where turbulence appears, for example, at surfaces, walls, or boundaries. This is of great importance to musical instruments when a wave is traveling through the tube of a flute, saxophone, or trumpet. Within the instruments, there still is some laminar flow left from the blowing of the player. This flow is very small and not of much importance. Still, of course, the sound waves travel inside the tube. As we have seen, slow flow caused by blowing and fast sound waves is both present in the compressible Navier-Stokes equation. The difference is that flow can exist without any density changes, while sound waves are density changes by their very nature. But as we have found above, a sound wave in air is the movement of small particles of air back and forth, which is the density change. But then, the particles also need to indeed move back and forth, they have a velocity. Again, this is different from the flow velocity, which is only going in one direction and changes direction only slowly. In a sound wave, the particles move much faster, but also within much smaller distances.

Still, particles in a sound wave flow too, and therefore may also be influenced by neighboring particles or walls, and end up in turbulence, still on a very microscopical level. This happens extensively at the walls of wind instruments, where the particles are disturbed by the walls themselves. This is especially true for instruments build of wood, like clarinets, which may not have such a smooth surface compared to instruments built of brass, like trumpets. Flutes are often built of bamboo, like the Japanese *shakuhachi*, the Balinese *suling*. Western recorders are also built of wood. This effect of turbulence at the wood surfaces is even stronger when the tube has a small diameter, and therefore a larger fraction of the total wave traveling through the tube is influenced by the tube walls.

When we remember what we have found with turbulence, especially that it is heavy damping, it is clear that the interaction between the sound wave and the walls causes large damping. This damping causes the appearance of small turbulences at the walls, and therefore a transfer of a part of the energy of the sound wave into this turbulence happens causes damping. It also means that the laminar flow, which is still present in the middle of the tube, is affected, drawn back, and therefore the velocity of the tube is reduced. It is clear that this effect is stronger when the walls are rough, and therefore the sound wave will be disturbed even more.

Practically this means that a wind instrument player needs to blow into the instrument much stronger to make it sound when the surface walls within the instrument are rough or when the diameter of the flute is small. Normally the material a wind instrument is build of does not play any role in the sound, as the wall vibrations are so small that they cannot be heard. An exception might be with trumpets or trombones, where the sound wave traveling through the instrument might stretch it a bit and therefore influence its length and so indirectly influence the sound [155]. Still, the wall vibrations themselves cannot be heard, and so the material a flute is built of, may it gold or platinum, does not matter. Still, this is considerably different when the surface of the material is rough. Then the player needs to use more lung pressure, playing in a different way. Also, the damping will affect higher frequencies more than lower ones, and therefore the flute itself will sound different. Yet a third reason may be the haptics. Musicians get feedback from the instrument through their fingers, feeling the instrument's vibrations. If these vibrations are different, it might change the performance of the musician and, therefore, the sound. Still, flutes with rough interior surface are so hard to play that these instruments can often not be reasonably played at all, and, therefore, instrument builders need to take the internal turbulence into account.

Indeed, turbulence is not yet understood in full. To most mathematical equations that describe vibrations, sound, and music, there is a mathematical solution. As discussed above, the solution to the Navier-Stokes equation is still missing. A solution would mean that we can predict precisely how such a complex turbulent flow field would look like after some seconds when we start blowing with a defined pressure and speed into a flute, saxophone, or somewhere else. Such a mathematical description has not been found yet.

Still, there are some general rules. The cascade of vortices of such a turbulent flow can be found to follow a scaling law, as do many things in nature, we have seen this before. The relation here is the rotating speed of vortices compared to their size. If we make a graph, a plot, where one axis is the vortex size, plotted logarithmically, and the other axis is the rotating speed, again plotted logarithmically, we find that a straight line describes their relationship with a slope of $-5/3$.[53] This is a scaling, as this law holds, no matter at which size or rotating speed of vortices we are looking at. It, therefore, is a general rule of turbulence, which sometimes helps to understand or predict the basic behavior of flows.

Another finding is that when a flow is leaving a small slit or tunnel, a case we have discussed above, then the flow starts to spread out. So its diameter becomes larger, the larger its distance from the slit is. This spreading is an exponential one, up to a certain point where the first large vortex starts. Also, this general rule sometimes helps to predict what will happen globally in a flow.

But all this depends on the geometry of the instrument. Blowing into a saxophone mouthpiece is different from blowing onto a labium of a flute or an organ pipe. We will find that both mechanism are different, still have basic properties in common, which are fundamental to musical sound production.

To start with, we now turn to single-reed instruments, saxophones, or clarinets. Then we come to flutes and organs. Here we will find that these instruments are self-organized systems by their very nature and need to be to work as musical instruments at all. Only then do we turn to guitars and violins, where we will find the same thing.

Still, with these instruments, it is not so obvious, with the exception of the bowing mechanism with violins, which is complicated but straightforward in terms of self-organization. But it is important to also consider the interaction of a string and a soundboard as a self-organizing one, to understand things like the initial transient or the robustness of tone production of these instruments. Furthermore, when linking the tone production to music psychology and evolution, it is crucial to first understand musical instruments in such depth.

Saxophone

The saxophone is showing most of the features we have discussed so far. It has many articulatory possibilities like most wind instruments have. Playing soft and low, hard and intense, rough and dirty, sliding tones in and out, changing pitches slightly, its sound possibilities are tremendous.[54]

Still, one can also play several tones at the same time, so-called multiphonics, consisting of two, three, or up to five or six pitches at the same time. There are two ways of achieving this. Either one plays at the threshold of tone production with very low blowing pressure or on the other side with very strong pressure above the usual blowing. Or one can use complex fingerings, placing the fingers not in a straight way but leaving gaps between them. In normal tone production, all finger holes are closed by the valves on the saxophone tube down to a lowest closed finger hole. The length of the tube, roughly from the player's mouth down to the first open hole, then determines the pitch. But if one opens one or even several holes in between, such multiphonics can appear.

So there seems to be a great variability of the played timbre, and therefore in Jazz there is the notion of a speaking saxophone tone, one which comes close to human voice articulation due to this tremendous variability.

Saxophone Pitches Are Not Self-evident

We take it for granted that we can play musical notes, pitches, melodies with a saxophone. Still, this is not at all self-evident.

In normal playing range, when producing single notes and pitches and no kinds of multiphonics, the saxophone tones have a very precise harmonic overtone structure. Then the sound, no matter if soft or hard, rough or sweet, consists of frequencies, of partials, which show the simple ratios of 1:2:3:... When playing a low A with f_0

R. Bader, *How Music Works*, https://doi.org/10.1007/978-3-030-67155-6_6

= 110 Hz, the partials fit nearly perfectly the 110, 220, 330, 440, 550 Hz, etc. No matter which timbre a player chooses, the partials are all harmonic one to another. This is absolutely necessary for using saxophones as a musical instrument at all. If this was not the case, the sound would not have a clear pitch. A clear pitch can only appear when the partials are all harmonic one to another. If there is no clear pitch, the human ear and brain will not detect a musical note. But most music is about pitches, which are the notes in the score. A musical instrument that is not able to play pitches cannot be used playing a regular score. Percussion instruments often do not have harmonic partials and are seldom used for playing scores.

Still, saxophones, as well as all other wind instruments, show such a clear harmonic spectrum, they need to. Most musicians, music lovers, and instrument builders do never really consider this as something special, as it seems self-evident. Still, the fact that these harmonic partials appear with such high robustness, no matter how the played timbre is, is astonishing and will lead us to a fundamental understanding of the tone production of these instruments. Indeed, it is astonishing when measuring the frequencies the tube of the saxophone vibrates in, so when it is not played with a player blowing in a mouthpiece, but by an artificial driving mechanism like a loud-speaker. When driving the saxophone tube with a loudspeaker with many different frequencies, the air in the tube will start vibrating strongly when the driving frequency meets a frequency the tube is able to vibrate in. When the driving frequency of the loudspeaker is off such an eigenfrequency of the saxophone tube, the vibration in the tube due to the driving is very low in volume. This is a simple way to measure the frequencies the tube will vibrate in on its own when not driven by a player blowing in a mouthpiece.

These eigenfrequencies of the saxophone tube are no longer in a harmonic series like 1:2:3:... The lowest frequency is often much too low, often by a musical fourth or fifth, and also many higher partials are off too. When combining such a series to a sound, which can be done, e.g., by a synthesizer, just to check the resulting timbre, there will not appear a pitch but a kind of a rough sound or noise, much like that of a percussion instrument. Still, it is the same instrument which, when played by a player blowing in the horn, produces a simple pitch.

When you are familiar with the vibrations present in a simple straight tube, this might come as a surprise, as when considering the saxophone and all other musical instruments in principle as a straight tube with some length, one is indeed expecting a harmonic series. Still, saxophones are no straight tubes but are conical with some curves and end in a horn where the sound comes out. On the other end is the mouthpiece, which is narrowing the tube towards the slit between the mouthpiece and a bamboo reed, which has a height of about 1 mm when resting. This changes the frequencies the tube vibrates in on its own tremendously compared to a simple tube.

Indeed, this is very well known to instrument builders when it comes to horns that are overblown. It is important to discuss the difference between the harmonic overtone spectrum and overblowing an instrument. Overblowing is achieved by increasing the blowing pressure until the horn is vibrating with a pitch based on the next eigenfrequency the horn is able to vibrate on its own. This pitch has a harmonic overtone series on its own, which is still astonishing. But the fundamental pitch of overblowing is the eigenfrequency of the horn. So we can indirectly measure these eigenfrequencies, the ones we measured so far by driving the tube with a loudspeaker, by overblowing the instrument. And indeed, constructing horns which show a perfect harmonic overtone spectrum while overblowing is very difficult and a continuous struggle for instrument builders. No perfect solution has been found here until now, and therefore players try to adjust the overblowing pitches by changing the lip or blowing pressure. This indeed means that the saxophone horn itself does not have eigenfrequencies, which are perfectly harmonic and deviate considerably from the simple harmonic series. But, as discussed above, when playing every single tone, although the fundamental pitch does not fit into the harmonic series, its partials, all frequencies above this fundamental, are again harmonic to this very fundamental nearly perfectly. So we find that the series in the horn itself is not harmonic, but when playing single tones, their overtone series are harmonic, synchronized to such simple harmonic ratios of 1:2:3:...

So when the instrument is played by a player blowing in a mouthpiece, where the off-partials align to a simple harmonic series, it must be the driving mechanism itself that aligns or synchronizes the partials to such a simple series. Still, this would mean that such a synchronizing mechanism would be the crucial mechanism for the instrument to be used as such at all. Otherwise, no pitches could be played, and therefore the instrument could not be used playing melodies or scores.

Blowing in a Saxophone

So we need to take the blowing mechanism into closer consideration. The player blows into a mouthpiece that is attached to the tube of the saxophone. The mouthpiece is a small tube that narrows towards the player's mouth. At the bottom of the mouthpiece, a reed is attached, most often made of bamboo reed. The reed is quite thin so that it can move. Where it is attached to the mouthpiece with a metal clamp, it is about 2 mm thick and cannot move here because of the clam. Then it becomes thinner and ends at the end of the mouthpiece, where the reed is only about 0,2 mm thick. The way it becomes thinner changes the sound of the saxophone, and many different kinds of reeds are produced, resulting in a Jazz, Classical, or Rock sound. Also, saxophonists rasp their reeds to obtain a special sound.

The reed is as wide as the mouthpiece, about 2 cm, and its tip ends at the tip of the mouthpiece. Still, a small gap above the reed and below the top of the mouthpiece is left, which is only about 1 mm in height. The reed is flexible enough to bent upwards and theoretically could close the gap completely, therefore closing the air channel between the player's mouth and the mouthpiece. Still, when normally blowing, the reed never really moves up so far. Of course, there is also a small gap between the reed and the mouthpiece at the sides of the reed, which also becomes smaller when the reed is moving upwards.

The player takes the mouthpiece into his mouth, closing the gaps at the top and the sides of the reed/mouthpiece to the outside air. He then increases his lip tension, which starts closing the gap a little bit. Lip tension is a crucial part of saxophone playing where the player can change the pitch a little bit. He also needs to increase the lip tension with higher notes to make them start sounding at all. Contrary, lower notes are much easier to play with lower lip tension. Next to the lip tension, the player can blow stronger or weaker into the mouthpiece, and there are the two mechanisms with which he can articulate the sound. Of course, with the hands, he will change the fingerings closing or opening the valves at the tube, changing the pitches of the played notes.

As we have discussed the Navier-Stokes equation in a previous section, we are now able to take a closer look at the process of tone production taking place in the mouthpiece. First of all, the player does blow into the mouthpiece by increasing the pressure in his lungs. This pressure causes a pressure increase in his mouth. Still, then the pressure inside his mouth and inside the mouthpiece is different. There is a spatial pressure gradient between mouth and mouthpiece. The Navier-Stokes equation tells us that this makes the situation unbalanced, and to balance it again, there needs to be a velocity of the air, the blowing starts. As the pressure inside the mouth is larger than that in the mouthpiece, the direction of the blowing is clear. The airflow goes from the mouth into the mouthpiece.

Now, as the slit between the mouthpiece top and the reed is only small, all flow needs to go through this slit, and therefore the speed of the flow is high. Also, the flow going into the mouthpiece is only a thin layer of air, simply because the gap between reed and mouthpiece is small, only about 1 mm. Now the mouthpiece is a tube that narrows strongly towards the player's lips. Therefore the airflow, entering deeper and deeper into the mouthpiece, has more and more space above and below it. Therefore there is a strong velocity gradient between the area where the flow is going and above and below it, where the air is still at rest. Again according to the Navier-Stokes equation, this will produce a pressure gradient between the flow and the air above and below it.

This large gradient will now produce friction on the air stream. The air stream is scratching and rubbing against the still resting air. Now it is unlikely that the friction at the top of the stream and that at its bottom will be the same, will have the same strength. More likely, one of the two is a bit stronger. This means that the pressure gradient on one side is stronger than on the other side. But if on one side the pressure is larger than on the other, the whole airstream will be moved towards the direction where the pressure is lower.

As the air stream is moving into the mouthpiece, this movement will become stronger the deeper the air stream is in the mouthpiece, simply because the air stream, which is deeper, has experienced more of this pressure gradient during its travel through the mouthpiece. This means that the air stream will start changing its direction more and more towards the up (or down) direction. Indeed, this directional change is exponential, so after a very small change, this change becomes increasingly larger

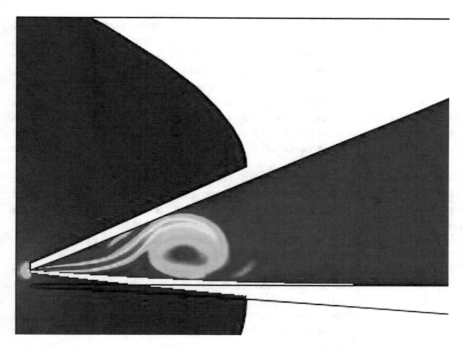

Fig. 1 When blowing into a saxophone mouthpiece, the air stream or air-jet forms a circle, a vortex, or eddy. On the left is the saxophone player's mouth, on the right is the mouthpiece. This vortex has a strong pressure gradient, what we call a musical sound. So in the vortex is the transition from blowing into sound. The flow is simulated using a computer model. From: [21].

and will end up in a complete direction change up (or down) in the mouthpiece. This will continue beyond the direction change up or down, and the air stream will go in a larger cycle, form a vortex or eddy (Fig. 1).

This first large eddy is crucial for the whole tone production, as we will see in a minute. Still, to continue with the process, of course, the airstream slows down during this process of direction change and becomes wider. Still, then again, only small deviations of pressure or flow are necessary to make one part of the stream move faster or slower, and the air stream starts to split-up. It divides into smaller circles, smaller vortices. These vortices experience the same pressure gradient as the large one and will, therefore, again change direction, again building smaller and slower vertices or eddies, and so on. This process continues until the vortices have become that small and slow that there they will practically stop and diffuse. So there will be a cascade of vortices which are smaller and smaller and by that get less important.

So within the mouthpiece, the same process happens as we have discussed with a tube, wherefrom a certain flow velocity on, a laminar flow suddenly becomes a turbulent one. In the turbulence within the mouthpiece, the flow nearly comes to a complete stop, and only a very slow airstream will travel through the rest of the saxophone tube. This stop is turbulent damping, and the energy of the flow is nearly completely sucked within this turbulence in the mouthpiece.

Still, the first large eddy has an important function. As it changes direction, again, according to the Navier-Stokes equation, this direction change needs to be balanced by a pressure gradient. This pressure gradient is at the very front side of the airflow, where it changes direction from straight on towards up or down. A pressure gradient means a new pressure in the mouthpiece. So due to the turbulent process in the mouthpiece, a pressure gradient builds up, the pressure rises over a very small time span.

During this process, the reed is closing. It closes due to the larger pressure inside the mouth compared to the lower pressure inside the mouthpiece. But when the reed closes, the air stream inside the mouthpiece can no longer continue with great strength, and therefore the first vortex will be the only one built. Now, as the vortex becomes weaker, it can no longer maintain the pressure gradient at its front end. Therefore the gradient will vanish again. This means that over time a pressure gradient has build-up, had its peak, and decayed again. This produces a small pressure impulse over time, which is indeed measured when putting a small microphone inside the mouthpiece and recording the sound in there.

So until now, we have the player's lung pressure becoming a flow through the gap at the tip of the mouthpiece, which again becomes a pressure gradient within the mouthpiece. Now a pressure gradient is what we call a sound. And indeed, the production of sound happens right there, at the front end of the vortex in the mouthpiece during turbulent motion of the air stream.

Generally, short sound impulses are traveling through air, and therefore also this pressure gradient is moving through the tube. It is reflected at the next open fingerhole or the bell of the horn and will travel back to the mouthpiece. When it comes back, it will change the pressure situation within the mouthpiece, the pressure will increase. Still, this will make the reed move again, it will open a bit more.

When the reed opens again, a new airstream can enter the mouthpiece. This airstream will feed the old first vortex, refresh it, or even create a new one. Of course, this will again produce a pressure gradient according to the Navier-Stokes equation. This pressure gradient will then take the same way as the first one, travels through the tube, is reflected, comes back, and the process will again be started.

The old pressure impulse coming back from the tube to the mouthpiece is strong enough to open it again. Still, after this action, it is not strong anymore and will not travel through the tube again and again. This impulse is damped very strongly and therefore dies out very fast. This becomes evident when the player stops playing. The tone is gone right away. We take this for granted, still this is not true with other musical instruments. With guitars, a plucked string is vibrating for many seconds. A plucked violin string is much shorter and dies out very quickly as the string is strongly damped. This is needed to make the bowing process possible. With wind instruments, a played tone is damped out even faster, it is gone right after stopping the blowing process. The reason is heavy damping at the inner walls of the tube, the loss due to sound radiation from the tube, as well as the interaction process with the reed. As with the violin string, the tube needs to be damped strongly, as otherwise, the whole process would become chaotic with too many impulses traveling through the tube, and the system would not work the way it does.

So we have a system which reproduces itself, it is recursive, it is self-organizing. The system is organized as the process is repeating again and again, therefore producing a periodicity, a musical tone. Indeed, the input to the system, the constant pressure in the player's lungs, which produces a constant blowing, is not vibrating at all, it has no periodicity, there is no structure or organization in it. The resulting pressure gradient moving through the tube on the other side is highly organized, it is periodic, vibrating, a musical sound.

This is caused by the strong nonlinearity of the Navier-Stokes equation, where a directional change will lead to a pressure gradient that is acting back to the flow again. Without this strong nonlinearity, the whole process would not work, there would not be such a thing as wind instruments, at least not as we know it.

The process can also work vice versa, in some cases, the pressure gradient is an underpressure rather than an overpressure. In these cases, the reed stays open most of the time, producing a strong pressure gradient nearly all the time as the vortex is producing it, and only when the impulse from the tube is coming back as an underpressure impulse, the reed closes, reducing the pressure gradient which is then the impulse produced by the system. Still, the mechanism is right the same.

Only now we can understand why the overtones, the partials of a saxophone sound, are that straight in a harmonic relation one to another. The mechanism is producing a new pressure impulse each time the vortex is moving. This pressure impulse is the same at each instance. The old pressure impulse coming back from the tube is only needed to make the reed move again. As an impulse, it is more or less 'thrown away'. Therefore, if it has changed during the travel through the tube in such a way that higher partials would deviate from a perfect harmonic spectrum, this is of no importance as this back-coming impulse is thrown away anyway. Therefore, a perfect harmonic overtone spectrum is guaranteed.

This throwing-away and reproduction of an impulse make the robustness of the process. No interaction between the energy of different partials is needed or any other mechanism which is balancing the system. The harmonic spectrum is caused by a new production of an impulse each time the previous one comes back, and therefore the new impulse is right the same as the old one. Only with such a robust process players can lean on producing harmonic spectra over a large range of blowing pressure.

Blowing Out of Normal Range

So we find that the tone of a saxophone, the periodicity of the vibration, is nothing simple. Indeed, it is a very complex process with two strong nonlinearities, that of the flow as described by the Navier-Stokes equation and that of the closing of the gap between the reed and the mouthpiece, which reduces flow considerably. The periodic vibration, which is the result of this complex process on the other side, is very simple, it is a simple repetition of the same thing. A sound with such a simple periodicity is what we call a musical tone and is what we expect from a musical instrument. So a very complex, nonlinear process leads to a very simple output, a musical tone.

Indeed, such a musical tone is not easy to produce. When we go into a music store and buy a musical instrument, it is very easy to produce a sound with them. Still, this is because these instruments have been build and tuned to do so, which in some cases took centuries to develop to such a high standard. Try to build your own musical instrument from scratch, and most probably, this will not be simple at all. Maybe one can blow on a bottle and get a sound, still how to change the pitch. Also, the sound of the bottle is low in volume and very noisy. The musical instruments we buy are high-tech instruments which often had a very long development of try-and-error to get the most out of it. And still, they are often not satisfying in terms of playability, loudness, or articulation, and professional musicians spend quite some time with their personal instrument builder to improve them.

And indeed, the stable periodicity or pitch of saxophones or other wind-instruments are only present when playing them within a normal range of blowing pressure and lip tension. When decreasing the blowing tension, the sound will stop, and only noise is heard. It is interesting to see that this is happening at one particular pressure threshold. Above that threshold, the saxophone sounds with a pitch when just passing the threshold, the pitch stops, and only the sound of wind blowing through a tube is heard. This means that the complex process is no longer able to produce a periodicity, it fails, and the system all of a sudden shows completely different behavior.

Such sudden changes of the behavior of a system are called a phase change. This notion comes from physics, where there with phase a phenotype or typical state of meaning. So, e.g., water, when going from a freeze to a liquid state, is said to make a phase transition between the phases 'frozen' and 'liquid'. We adopt the notion because it is quite common in the scientific scene, still in acoustics 'phase' is normally used for the state of a sine wave.

So a sudden phase transition is happening with the saxophone sound from pitch to noise at a certain pressure threshold. The reason is that the process described above depends on the production of a new impulse each time the old comes back from the tube. It does so by moving the reed. If the blowing pressure is decreased, the impulse is weaker, of course, and therefore cannot move the reed with much strength. In a linear system, this would simply mean that the new impulse might be weaker. Indeed, when blowing with less strength, the sound is lower in volume. But this is not because the back impulse is weaker but because the inflow of air into the mouthpiece is not that fast with lower blowing pressure. Therefore the newly produced impulse is weaker, which makes the sound lower in volume. But even this weaker impulse, when it comes back to the reed, is able to move the reed again. But below a certain threshold, the movement of the reed is so small that there is no longer a considerable inflow to maintain and reproduce the first vortex. But if there is no longer a clear vortex, no new big impulse is sent out, but only the smaller impulses created by the turbulence inside the mouthpiece are present. Still, these smaller impulses are irregular, and none of them is able to take over the reed. But if non of them is able to dominate the reed, there will be no periodic interaction between the vortex and the reed and, therefore, no periodicity in the sound, no musical pitch.

This happens suddenly, as there is a struggle between the reed and the back impulse. Of course, the reed does not move on its own very much, it moves when it is driven. The reed itself is heavily damped, which one can find when taking the mouthpiece from the saxophone and clicking the reed on the mouthpiece simply by a finger snip. The resulting sound is very short, which means that the reed is heavily damped, its vibration is nearly immediately sucked by internal energy loss and therefore cannot vibrate longer. So when an external impulse is moving the reed, the reed will follow but will not move very much on its own afterward. So it will not try to add an own vibration to the system. This is important because if this happened, the saxophone would sound in the pitch of the reed!

This is again not expected from saxophones, but this is a crucial point of instrument building. We will get back to this with other instruments, like the guitar or other stringed instruments. For now, it is important to see that the reed need to be damped very much otherwise, the impulse making it move would also make it vibrate in the eigenfrequencies of the reed, and this might become the pitch of the instrument. Indeed, we want the tube length to determine the pitch of the instrument, not the reed. Otherwise, we would not be able to change the pitch of the saxophone using fingerings, and therefore would not be able to play melodies at all.

So a saxophone can also be considered as a system consisting of two systems, which can basically vibrate, the tube in its frequencies, and the reed. Both struggle to take over the other system and force it into its vibration, its eigenfrequencies. Now with the saxophone, the tube wins because it is less damped. The tube is air vibration, which is damped very low because the internal damping in air is low. We hear sounds from miles away. So we found another important aspect here. One of the subsystems, the tube or the reed, is forcing the other into its frequencies, and the one less damped wins.

Still, as we have seen with the lower blowing threshold, this does not work in all cases. It only works if the energy of the impulse coming back and acting on the reed is strong enough to move the reed considerably. Otherwise, this interaction fails to produce a periodicity, and therefore the whole process changes its fundamental behavior from pitch to noise.

It is also interesting to see that if we go the other way and start with a low blowing pressure, which we increase, therefore getting from a noisy sound into a stable pitch, the blowing threshold at which this is happening is higher than the one when changing from pitch into noise. This is called hysteresis and happens typically with such highly nonlinear self-organizing systems. The reason is that if a system is in one mode or phase, like in the pitch phase, everything is up and running, and driving it out of this phase is not too easy. This is another important aspect of self-organization, a system doing such self-organization tries to maintain this organization as long as possible. Only if the situation is getting too worse for this phase, it breaks down, giving way to the other state, here the noise. The same holds the other way around, also the noise state tries to maintain itself as long as possible and only changes into a pitch state when it can no longer hold its present behavior.

So self-organization really means a process of maintaining one state and defending it as long as possible. This is also true for other musical instruments, as we will see later.

Multiphonics

We have seen that the mechanism which is building up a regular vibration is very complex. If all parameters fit together, it is very stable and robust within itself, and a player can rely on it. Still, if the parameters are no longer in the right range, the simple periodicity breaks down, and other states occur. These states are more complex than the simple oscillation, still, they are not perfectly random. A random behavior produces noise. Noise is easy to perform with a saxophone when blowing into it with only small lung pressure and lip tension. Still, then increasing this pressure and tension, right at the transition from noise to a regular motion, strange sounds appear. These sounds are called multiphonics, as there are multiple musical notes present in them at the same time.

Multiphonics are part of what is called 'extended techniques' starting around the 90th and used very much in Jazz and Free Improvised music since the 60th. Extended techniques are those extending the regular sound range of musical instruments experimentally. Many of them were effects, like striking strings of a guitar not before but behind the bridge or bowing the cello body instead of the cello string. Others were surprising. So when sticking a thin nail between the strings of a classical guitar near the bridge, the instrument sounds more like a marimba than a guitar. In the case of multiphonics, two basic techniques are known. Either one plays the instruments with a lung pressure at the tone onset threshold as discussed, or with extremely high pressure. Or one can use complicated fingerings, opening finger holes above the lowest closed valve. Hundreds of such patterns have been invented and documented in text books.[55] In all cases, extended techniques helped the player to extend the sound possibilities of the instrument. In some cases, like the multiphonics or with undertone violin bowing discussed below, these techniques give an interesting insight into the way instruments work.

So when playing a multiphonic of any kind, it is quite complex to discuss what is happening in detail, as it depends on the special case, which there are very many. Basically, we are in the regime of bifurcations, where there is not only one system state, producing one periodicity and, therefore, one musical tone, but several of them. The system switches between these states constantly. So in one case, an impulse produced by the vortex in the mouthpiece is sent out and is reflected partly at one open tone hole down the tube. Still, as a multiphonic fingering may have additional closed sound holes further down the tube, a part of the impulse will travel further down and will be reflected only at the last open hole. So two impulses travel back and interact with the reed. Still, this does not mean that the resulting sound would have exactly two pitches, one of the end finger hole pitch and one of the other open finger holes in between. This might happen in rare cases, but most likely, the first impulse is too small to move the reed in such a way that the reed will open enough to produce a new vortex. But still, it will make it move somehow, and this means that when the main impulse comes back, the reed is in a different state as it should be. Therefore, the later back impulse interacts with the reed differently, it finds it somehow open, and therefore the interaction can happen faster, but the process itself took a longer time, first the small impulse acting a little bit on the reed, then the larger impulse making the rest. Still, when the process is longer, the vortex builds up smoother, and therefore the pressure gradient, the new sound impulse going out is smaller, lower in volume. So the next back impulse from the first open finger hole might be too small to make the reed move at all, and only the second impulse does. But then the opening process of the reed is faster again, the pressure gradient steeper, and therefore louder again. This means that the whole process starts over again.

When playing with a regular fingering but with a lung pressure right at the threshold between noise and a regular sound, a similar situation appears. The returning impulse is not perfectly able to move the reed properly, and the resulting new vortex is shorter, resulting in a sharper pressure gradient, and therefore a stronger new impulse is going out. This new impulse is reflected, able to move the reed stronger, leading to a longer opening of the reed, and therefore to a smoother gradient, a weaker impulse. But this means that the weaker impulse takes a shorter time to produce a new vortex than the stronger one, and therefore two periodicities are present at the same time, leading to an inharmonic sound.

This is only one possible scenario. Still, it shows that an additional open finger hole can disturb the process at one cycle but not on the next, the system is bi-stable, there are two interactions happening one after the other. The resulting sound is often no longer harmonic, and additional inharmonic frequencies appear in the sound. These make it more complex and more 'dirty', some might say more interesting, sometimes speech-like, and give additional expressive possibilities. Such processes can become very difficult, where there are four, six, or eight stages through which the system goes until it repeats, making the sound more and more complex but not perfectly random. Therefore the sound is no longer periodic, but it is also not pure noise.

Modeling Saxophones with Impulse Pattern Formulation

These states wonderfully appear when using the Impulse Pattern Formulation. In the case of normal playing, there is only one impulse coming back. Still, the interaction between the back impulse and the reed, as well as the new vortex formed, are nonlinear. When taking the usual nonlinearity of a logarithm, the saxophone sound becomes perfectly stable when choosing the right parameters for $1/\alpha$, here the blowing pressure and lip tension, and the system parameter g, the periodicity of the sound is stable, a normal pitch is played.

When decreasing the blowing pressure, the sound suddenly gets into the chaotic regime and becomes noise. This is what a player hears when using a low blowing pressure. Still, in between both regimes, that of normal playing and of noise, many possible bifurcations appear, and multiphonics are heard. Still, this intermediate state is short, and one needs to choose the parameters at will to get it. This is analog to regular saxophone playing, where the lung pressure needs to be very precise to get into this state.

When adding additional reflection points to the system, like additional open finger holes, which make impulses travel back at different times and with different strength, all kind of wired behavior can be found, and endless possible multiphonics are produced.[56]

The Beginning of the Tone

A similar situation appears when a tone is started. This is because, in the beginning, the system is not settled and, therefore, not in a regular state. Indeed, when using the right parameters, a saxophone can get into a regular motion right from the start, an ideal situation. Still, in real playing, this very hard to realize.

When a tone is started from scratch, that there was not tone played before, then the player needs to increase the lung pressure first. Even if he has a fixed lung pressure present in his mouth, he needs to hold it, possibly with his tongue. Still, releasing the tongue to start the tone takes a small amount of time, which is again an increase of the pressure. So the pressure goes through a range which need not be one of a periodic pitch, it will be one of noise or bifurcations. Therefore, during an initial transient phase, several regimes may be passed through, and so it takes some cycles of the tone to reach a stable periodicity.

If the new pitch played is part of a melody, there has been a previous pitch. So when opening or closing finger holes to reach the new pitch, both periodicities are present. Strictly speaking, the instrument is in a multiphonic state in between. Therefore irregular periodicities will appear between each pitch transition.

The beautiful thing with saxophones and reed instruments alike is their high range of articulation. So a player can choose to enlarge these stages and therefore get a larger part of inharmonic sounds during tone onsets. This is heard as expression, as it extends the pure periodic regime. He might also choose to make these parts as small as possible. Indeed, in high-speed camera recordings of the reed motion using artificial lips and a compressor as the player's lungs, it is possible to produce sounds that start periodically right away. Still, in practice, this is not realizable, and therefore players wanting smooth sounds often start the tone with very low volume to make the inharmonic part nearly inaudible while only then raise the volume fast when the periodic regime is reached. This is a kind of a trick, still, as it is nearly impossible to start a tone without some kind of noise, this is a practical solution.

Indeed, the initial transient of musical instruments is a very important part of the sound. When cutting the initial transient out and only playing the rest of the sound, the so-called quasi-steady-state, in many cases, a trumpet cannot be told from a violin, a guitar from a piano, a voice not from a synthesizer.[57] The amount of identification still possible strongly depends on the kind of instrument, the pitch range, and other articulations. So in a high pitch range, the instruments are mostly indistinguishable without the transient. Still, if a violin is played with a typical violin vibrato, everybody immediately identifies it as a violin. But in most cases, without the initial transient, it is hard to tell which instrument is playing.

So the tone initian is responsible for most of the articulation, as well as for most of their identification. What makes it so is its irregularity, its bifurcation nature, somewhere between harmonic and noise. Musically here seems to be the interesting part without which music would be much less rich. Still, this does not mean that the steady part has no role in the sound, it has a timbre and a pitch. But indeed, the irregularity of this part most often is also not static but shows changes in blowing or bowing pressure, vibrato, noise, and many other additional features. This is why it is most often called a 'quasi' steady-state, as at least for an acoustic instrument, there is never a real steady state. The sounds are always fluctuating, which makes them interesting.

Still, again this does not mean that there is no music possible or produced with no perfectly real steady-state. In electronic music, one is perfectly able to produce such sounds. Still, also here, most sounds are modulated with effects like flangers or phasers, reverberation, delay, wha-wha, detuning, bit crushing, and an endless possible set of effects, modifying the sound out of the steady-state.

When comparing the initial transients of musical instruments throughout the book, we will see that although the scratching of a violin and the spitting of a trumpet at their respective tone onsets distinguish them clearly audibly, still also here internal similarities are present, which makes the mechanism of identification even more interesting to understand. Still, for this, we first need to discuss some more musical instruments and then turn to the music psychology and perception side.

The Saxophone Reed

The reed in this whole process is, therefore, basically a valve. It opens and closes and allows the air stream to flow more or less. This is also because of its heavy damping.

Still, it stays a small plate that can vibrate on its own. One can click it by hand and hear a very small, high-pitched sound. The lowest frequency of this sound varies from reed to reed, still, it is 3000 Hz.

Now all structures start vibrating if they are excited, if somebody knocks on them or if they are hit by some mallet or stick. So the back impulse from the saxophone tube hitting the reed will make it vibrate too. Now, of course, first of all, the reed closes the gap to the mouthpiece. Still, after this, it will vibrate a little bit longer on its own, normally about one or two more up and downs, which are indeed much smaller vibrations compared to the basic movement.

Still, this leads to a slight modulation of the air jet. This modulation is not changing the basic pattern, but it is influencing the strength of the frequency in the sound, which is at the frequency of the reed vibration. Now, this frequency in operation is normally

1500 Hz. The decrease 3000 Hz of the reed when clicking with a finger and that in playing is caused by the air stream, which is heavily influencing this vibration and is slowing it down.

Still, the frequency, and also the strength of this reed vibration, depends on the precise geometry of the reed. As mentioned above, manufacturers build reeds for special purposes: jazz, rock, classical, etc. These reeds differ in right this frequency of the reed, enhancing more or less about the frequency 1500 Hz.

Also, this reed frequency is independent of the tone played. So no matter which tone or articulation a musician uses, it will always be there. Therefore it is part of the basic character of the sound. Interestingly, when it comes again to identifying musical instruments, this region is used by listeners, and when playing pitches above this region, even with the initial transient present, one can hardly tell a saxophone from a trumpet.

More Wind Instruments

So until now, we have discussed wind instruments with a single reed, like the saxophone. The clarinet is very much alike in terms of the driving mechanism we are interested in. Basically, there are two other instrument families, the brass instruments and flutes or organs. To make our discussion not too long, we omit the brass instruments and take a closer look only at organs and flutes, as some new aspects appear here.

Still, only briefly, brass instruments are lip-driven, where lip means the player's lips. The air pressure inside the player's mouth opens the lips and releases an air jet, very much like that seen with saxophones. The main difference is that here the lips open with the pressure, while with saxophones, the reed closes. The reed, as well as the lips, are indeed nothing but valves opening and closing, and therefore allowing airflow or not. Both only vibrate very little on their own, as we have seen.

Such a valve-like behavior is not there with flue organs. Basically, there are two kinds of organ pipes, those with reeds, the reed organ, and those with a labium, the flue organ [4]. The reed organs are very much like saxophones, the flue organs are different. Here the air is taken from a wind chamber where a compressor is blowing wind in. This wind chamber can be pretty large in great church organs, and several people can easily stand inside. To make the pressure inside this wind chamber strong, there needs to be a compressor constantly blowing wind in. Formally, these were the choir boys, who were moving a big pair of bellows with their feet as long as the organist was performing. So the performance depended on them, if they were too slow, the pressure was not high enough, and the organ stopped. Nowadays, it is an electrical compressor pumping air inside the air chamber.

If an organist presses a key on the organ, on the top of the wind chamber, a small hole is opened, where air will flow out of the chamber. It flows into the foot of the flue pipe. This food is closed on its top, with only a very thin slit at the outer side. This is where the airstream or air-jet comes out. Right above the slit, at a distance of some centimeter, is the labium, a plate which is a cut edge at the beginning and

R. Bader, *How Music Works*, https://doi.org/10.1007/978-3-030-67155-6_7

increases in thickness upwards. Above this, the pipe tube is placed, where the labium is only the start of. The tube is closed at all sides, except that gap of a few centimeters between the slit and the labium. The tube itself may be open or closed at its upper end, changing the pitch of the instrument.

So when the organist presses a key and air flows out of the slit towards the labium, a tone is produced. The interesting thing is that a very similar pitch is produced when splitting off the organ tube from its foot and its labium and only letting the labium system vibrate. This was not the case with the saxophone. One can blow only in the saxophone mouthpiece with no tube attached, and after some tries, get a squeaking tone, which is hard to control. With the flue pipe, there is a regular tone appearing without the tube attached to the labium, where the labium pitch may not perfectly fit that of the whole pipe but comes pretty close to it. Still, the tone produced without the tube is much lower in volume, actually too low for a church. Therefore the labium part of the pipe would perfectly be enough to make the pipe work, still, one needs a tube to make the sound louder. But again, this is different from the saxophone, where the tube was needed to decide about the pitch at all. So with the saxophone, the pitch was determined by the tube, with the organ pipe, it is determined by the labium and the tube.

Before we see what happens when putting both systems, the organ pipe labium, and its tube together, we need to have a look at the labium system first. This is considerably different from that of a reed instrument.

When the organist presses a key, the air stream or air jet of the respective pipe is leaving the slit at the end of the organ food and starts moving towards the labium in free air. We had a similar situation with the saxophone, where starting from a small gap, a narrow air stream entered a much wider mouthpiece. There, according to the Navier-Stokes equation, is a velocity difference between the airstream itself and the air right above and below it. This produces a pressure difference, which again leads to a nonlinear change of the flow direction towards one side, the side which had only slightly less of a difference. Insofar we are close to the saxophone.

But now this airstream changing direction and becoming more and more a vortex is interacting with the labium. The labium has two sides, one is the outside of the pipe, one the inside. As the flow started its way opposite of the labium, both sides are equally likely to be chosen by the air jet. Still, as it has to move somewhere on its way to the labium, it will close the side it is already going. There it forms the vortex, right as with the saxophone, and the main air jet ends. The main vortex may slip into smaller ones and decay into noise.

Still, again this vortex produces a pressure gradient, the musical sound, which is then propagated in space and is the musical sound we hear. But now the situation is different from that of the saxophone, as there are now two sides of the labium. As the vortex is forming on one side only, the pressure gradient is mostly on this side too. Still, again for the air jet coming next from the slit towards the labium, there are two possible ways to go. The first way is to take the side of the labium the old vortex has gone. Still, there is a strong pressure gradient with heavy, turbulent damping, which looks like a wall for the new air jet arriving. The second way is the other side of the labium where nothing happened yet, and where the pressure is normal. Of course, the newly approaching air jet will take the side with normal pressure, so it takes the other side compared to the first vortex.

So the new jet will form its vortex on the other side, then also producing a new pressure gradient there, and start decaying. Meanwhile, on the other side of the labium, the pressure gradient has moved forward, and the situation became normal there. So clearly for the air jet approaching only then, the situation is vice versa, one side with a normal pressure gradient, one side with a strong one. It will decide for the lower pressure, and this game will continue, producing a regular periodicity, an oscillation, a musical pitch.

As this is a system on its own, not driven by another system and not coupled to any other, and as it is showing a periodicity, it is called a self-sustaining oscillation system. It organizes itself and therefore is again a self-organizing system where a complex geometry together with a strong nonlinearity of the Navier-Stokes equation leads to a very simple output, a stable periodicity, or oscillation. Without the nonlinearity in the Navier-Stokes equation, this would not work, as it produces the nonlinear direction change of the air jet while it is traveling from the slit to the labium. Also, without the labium, it would not work: it presents one side to be affected by the pressure gradient of the other, giving way to a new air-jet entering this labium side.

Now when attaching the tube to the organ food, the situation changes insofar, as now the flute plays with the pitch of the tube. Still, this only works if the pitch of the tube is not too far away. Ideally, both pitches are the same. Still, when using a tube with a very different pitch, either there is no sound at all, or the sound becomes inharmonic and complex, and bifurcations and multiphonics are produced. Still, there is very little research on this, as this is not the normal case, organ pipes have one fixed pitch, not like saxophones.

Indeed, the interaction between the labium and the tube is very much like that of the returning impulse with the saxophone. The returning pressure impulse from the tube interacts with the air jet by applying its pressure to one side of the air jet. This determines that the air jet is moving to the outer side of the labium. The next vortex is, therefore, to the inside of the tube, producing a new pressure gradient, which is moving through the tube, is reflected at its end, and returns to the air jet again. The speed of sound in air is about 343 m/s, the air jet speed is around 3–10 m/s, so about ten times slower. Therefore the traveling sound impulse can travel longer distances in the same amount of time. Therefore, the time it takes for an impulse once through the tube is about the same time as it takes a new air jet to move from the slit to the labium, forming a new vortex. Still, this is not perfectly the same, and so if both processes happened independently, we would have two pitches at the same time. Still, the tube and the labium part are coupled, where the returning impulse is moving the air jet and forcing it into its periodicity.

The boost of amplitude this interaction results in, which is the reason for attaching a tube at all, is then caused by the returning impulse of the tube, which is then radiated into space as musical sound.

Two organ pipes can also synchronize when standing next to each other [87]. Then the two impulses of the two instruments interact through the air between them.

When the two pitches are slightly different, they synchronize to one single pitch. This is known by organ builders and used when neighboring organ pipes need to be in tune. This might be a problem because the temperature and humidity in churches can vary strongly in winter and summertime. As temperature and humidity change the speed sound, the pipes might go out of tune. Still, when placing them next to each other, they can synchronize and still have a common pitch.

Recorder and Transverse Flute

A recorder flute also has such a labium as does have the flute. Still, there is also a tube attached, and we would not be able to get it to reasonable pitches without this tube. This is also true for many flutes around the world. The Balinese *suling*, the Kachin *sum pyi*, the Japanese *shakuhachi* are all flutes with a labium, may they be end-blown with a wind channel like the Western recorder or like the *suling*, may they be transverse flutes like the *sum pyi* or may they be end-blown without an air channel, like the *shakuhachi*. Unlike flue organ pipes, when detaching the tube and playing the instrument, often only a very high-pitched tone can be heard. The fundamental process of tone production is the same, of course, still, the labium system on its own does not produce a pitch similar to that of the tube.

This is not surprising, as these flutes need to produce different pitches according to the fingering the player chooses. When he or she opens or closes different holes, the flute will play different notes, of course, otherwise, we could not play melodies with it. This is different from the organ, where each pipe is only playing one note. If the player chooses to play a different note, he needs to use another pipe. This is why the organ has so many pipes.

So the recorder and the transverse flute need to work slightly differently. If a transverse flute had a labium region which is tuned to one of the pitches, the instrument is actually playing with regular fingering, playing other notes than that would not be easy, bifurcations would occur, and strange sounds would be heard. In this case, it is unavoidable to tune the labium to a pitch that is far from any of the pitches the instrument is normally playing.

Still, the mechanism is still working, as again, the impulse returning from the tube is interacting with the flow from the player's lips and will therefore determine a regular pitch, the one the player chooses. We could expect that it would be easier for a musician to play a note on a flute when the labium would be tuned to the pitch he wants to play. Still, we do not know this for sure, as only very few research went into this problem so far.

Still, the mechanism of producing a new pressure impulse at the labium region needs to work, as also with these instruments, the impulse traveling along the tube is heavily damped, and it is distorted while traveling through the tube. The heavy damping is clear when the player stops playing. Like with the organ pipe, the tone is nearly immediately gone. This is very different from, e.g., a guitar tone, where after plucking a string, this string sounds for a long time on. When a flute player stops playing, the tone is gone nearly immediately.

Still, it is interesting to go into more detail here with the transverse flute. Indeed, stop playing will bring the tone to an end immediately. But when investigating the energy of the system, only about 2–3% of the sound energy in the transverse flute is radiated into the room so that we can hear it. 97–98% of the energy is not radiated and stays in the tube of the instrument. The situation is getting even worse when considering the energy the player needs to drive the instrument to play a tone. Only again, 2–3% of the energy of blowing will be transferred into sound energy inside the tube [29]. Therefore only about 0.03–0.04% of the energy the player needs to play the instrument will become the sound we hear from the flute. In terms of today's energy policy, this is very inefficient indeed!

But now, if nearly all sound energy in the flute stays in the flute, we would expect the flute to still play on for some time after the player has stopped blowing in the instrument. This is straightforward, as all energy which is not radiated into space is still in the instrument, and if there is a sound wave inside the instrument, it will progress on, constantly radiating for quite some time.

Still, as we know, this is not happening. The reason, therefore, clearly is that the energy in the flute is also strongly damped, dissipated, is sucked, and gone inside the tube. The mechanism for this damping cannot be the damping of sound in air. Sound in air is nearly not damped at all, as there is no friction within air. Indeed, the only reason why a sound wave is getting weaker in air is a thermal or temperature process. A sound wave, as we have seen above, is that particles in air are getting closer together or drifting further apart, compared to the case of no sound wave present. But if one is pressing air particles closer together, their temperature will rise. It is because temperature, internal energy, is the very fast movements of air particles on a microlevel. When forcing air particles closer together, they will bump into one another much more and, therefore, their movement will become stronger. As this movement is temperature, the temperature will rise.

If the temperature rises when particles are getting closer, it will fall if they drift further apart. A sound wave, therefore, is also a difference in temperature between regions with higher and adjacent regions with lower temperature. But if one region has a higher temperature than a neighboring region, there will be a flow of energy from the region with a higher temperature to the one with lower. But this will lead to a relaxation of particles in the higher temperature region and a speeding-up in the lower region, the particles will move further apart or come closer together. So the difference between the regions becomes less.

As this difference is the strength of the sound wave, the sound wave loses strength, it decays, it is damped. This is the only reason why sound is decaying in air. If one is thinking of building a new concert hall and wants to make it sound good, one main aspect of doing so is to decide how much damping the different frequencies should have. A room acoustic designer can do this mainly by deciding about the material at the walls of the concert hall, here the damping occurs. In room acoustic simulations and calculations, therefore, the damping in air is neglected, as it is so small compared to that at the walls that we cannot hear it.

Coming back to the flute, we now have the situation that the sound wave is in the instrument, and only very little energy is leaving the tube and is radiated into the surrounding air. So we would expect the instrument to play on for some time after the player has stopped playing. This is not the case, and therefore the energy in the tube needs to decay fast. As this is not happening by damping in air, the only candidate left to damp the sound in a flute is the walls of the tube.

And, indeed, flute builders know this by heart if a flute has rough walls inside or if the air channel of the flute is too narrow, it is very hard to play the flute at all, as internally it is too heavily damped. We all know this from blowing through a straw. The thinner the straw tube, the harder it is to blow through it.

Now we might wonder if both cases, blowing through a tube and a sound wave traveling through it, are really the same. Indeed, they are because a sound is nothing but a flow of particles. This flow is much faster than blowing, and it is traveling back and forth. But the basic process is the same, it is a flow of particles. And indeed, also a sound wave can be modeled with a Navier-Stokes equation perfectly, not only the blowing of wind of a player. Still, again, as discussed above, the model needs to be compressible, otherwise, we do not have a sound at all. But then a sound wave is perfectly there as a flow of particles through space, traveling with the speed of sound and behaving like a sound wave in any way.

So there are two reasons for damping of a sound wave in a tube, its width or diameter and the roughness of its walls. We have already discussed the first one above when considering turbulence. If flow through a tube is getting too fast, it becomes turbulent, and the flow nearly instantly stops. So for very high blowing pressures, we would expect the system to fail completely, as also a sound wave could not travel through the tube at all.

Still, the more prominent part are the wall losses, where rough walls inside a wooden recorder flute make it very hard for a player to produce a tone at all. The reason for these wall losses is not perfectly clear, still, it seems that it is caused again by turbulence. The sound wave is flowing along the surface of the tube and is disturbed by the wall. This is like small obstacles, which stand in the way of the wave. Of course, the wave will be disturbed then, leading to small vortexes and turbulence. Again as discussed above, such turbulence is heavily damped, and this will explain the heavy damping at the walls.

Again thinking of the basic process of tone production in the flute, it is necessary that there is heavy damping, otherwise, the impulse would bounce back and forth the tube, leading to an inharmonic spectrum of the sound. Only if it is heavily damped, it is mainly traveling once through the tube and is mainly sucked by the internal damping and its interaction with the blowing flow from the player. Only then the new impulse produced by the turbulent process at the labium looks perfectly the same as the previous one, enforcing a perfectly harmonic overtone spectrum, and therefore a musical tone.

Another property of flues is that they change the content of their harmonics from lower to higher notes. Lower notes do not only sound lower in pitch but are also quite dark in their tone color. Indeed, low notes do only have about three loud partials, the fundamental and two overtones. This is very different when playing higher notes on the instrument. Then the timbre is getting much sharper, and more overtones are present. This is caused by the shape of the impulse produced at the labium. The flow in the instrument with lower notes is slower as the blowing pressure of the musician is smaller. Therefore the impulse is rising only slowly up and down and then looks quite similar to a sine wave, not perfectly as there are about two additional overtones. Still, the rise and fall of this impulse are right the overtone structure of the sound, and therefore lower notes have only a few overtones.

When the player wants to play higher notes, he needs to increase the blowing pressure. Then the raise of the impulse is faster, turbulence inside the tube is raising fast too, and therefore the impulse decays fast accordingly. Then the impulse is sharper and therefore contains many more overtones.

This is necessary to understand the bore-profile in flutes. Flutes have a so-called bore, a change of the diameter throughout the length of the flute. Often an instrument maker has a special bore-profile, and it is possible to associate bore-profiles to instrument builders, which is of some importance with historic flutes [174]. The bore-profile is known to shape the amplitudes of the notes played on the instrument. Renaissance instruments have bore-profiles enhancing the second partials, the octave above the fundamental, while Baroque flutes have the third partial, the fifth above the octave stronger. The Renaissance sound is, therefore, a bit simpler when playing the flute alone as with the Baroque flute. This is necessary as the Renaissance flute is mostly played in ensembles. The repertoire of these Renaissance ensembles is mainly polyphonic music, where there is not only melody and an accompanying harmony like with Baroque music. Polyphonic music of this age is meant as the interaction of voices, which are all equal, bass, tenor, alt, and soprano, all playing their melodies that fit into each other to form a harmonic whole. With such a musical ideal, it is helpful to have a flute sound that is not prominent but able to fuse with the other flutes. Therefore a Renaissance flute ensemble is like an additive synthesizer known from electronic music today. Contrary, in Baroque times, the ideal of a flute as a solo instrument developed. In such a case, it is necessary for the flute to have a distinguishable sound. Here the enhancement of the third partial helps a lot.

Now whether the second or the third part of a flute is stronger is determined by the bore-profile. When the impulse of a low note is traveling along with the flute, it looks very much like that of a sine wave, as the impulse rise and fall is so slow. Then the back coming wave and the next wave produced at the labium meet, and a kind of a standing wave is present in the flute. Such quasi standing waves are very well known and have points where they enhance themselves and those where they cancel each other out. The reason is simple. If one wave wants to travel to the right and the other to the left with the same strength, the resulting motion will be null. Such a point is called a nodal point of the wave. Other regions where both waves act in the same direction are called anti-nodal points.

Now the two mentioned partials, the second and the third, have different shapes, and, therefore, their anti-nodal points are at different places. If a bore-profile is narrow at a place of an anti-node, this wave cannot stretch out too much compared to the case where the bore is wide at the place of such an anti-node. If a wave cannot stretch out too much, it cannot be too strong, and therefore its amplitude is weaker. With such a system in mind, an instrument maker can shape the amplitudes of a flute by changing the bore-profile.

Still, as on a flute, many notes are played, and each of these notes has different places where the anti-nodes of the second and third partial are, a single bore-profile will not work that way with all notes. Also, as impulse shapes are getting steeper with higher notes, the impulses are getting sharper and, therefore, no real standing waves are present in the tube. Therefore such a bore-profile is only working with lower notes, and its influence on higher ones is barely predictable. Still, it will change the higher note timbre too, and therefore a builder with a decisive bore-profile will have a distinguished sound on his instrument.

Free Reed Instruments

When traveling through Northern Thailand or in Yunnan, Southwest China, in Laos or some eastern parts of Myanmar (formally Burma), one can experience the fascinating sound of mouth organs, the *khaen*,[58] the *hulusheng*,[59] the *sheng*, as it is called in mainland China, or the *sho* in Japan. These instruments have multiple pipes that lay next to each other, with their lower ends attached to a gourd or other kinds of small wind chambers. Within each pipe, a small reed can freely vibrate up and down. It is driven by the air pressure in the wind chamber, which again is caused by the player blowing into the instrument. Each pipe has one finger hole at its side where the player places fingers on.

One fascinating thing is that if one is blowing into the wind chamber with all finger holes open, there is no sound at all. Only if one is closing a hole this pipe will start sounding in its pitch. This is convenient as if all pipes would always sound at the same time if driven by the player. Of course, the player wants to play melodies and therefore needs to be able to choose which pitches to play.

The pipe is not sounding because the open finger hole at its side is placed such that it disturbs the impulse caused by the reed, traveling down the pipe so strongly that the pipe cannot sound at all. The reason for this is not perfectly clear and up to investigation [13].

Still, another thing is even more interesting with this instrument. If the blowing pressure of the player is low, the pipes are indeed sounding, no matter if the finger hole at the side is open or not! The pitch of each pipe then is different from the pitch when the pipes are played with regular strong blowing pressure. So each pipe has two pitches, one with high and one with low blowing pressure.

The reason for this behavior becomes more clear, as we find that the pitch with strong blowing (and closed finger hole) is that of the pipe, caused by the length of the pipe, while the pitch with low blowing pressure is caused by the pitch the reed has when vibrating on its own without a pipe attached. Therefore with high blowing pressure, the reed is forced to vibrate with the pitch of the pipe length, and with low blowing pressure, the reed can vibrate on its own and forces the pipe to vibrate with the pitch of the reed.

So again, we have a nice example of a self-organizing instrument, where one vibrating system is forcing the other into its frequency. We also have two options of who is slaving whom, depending on the blowing pressure. And again, we have a sudden change between both cases, at a certain blowing pressure, the situation changes suddenly, and the slaving is turned around. We also have the hysteresis loop again, where one needs more blowing pressure to turn from the reed vibration into that of the pipe than one needs to go the other way, decreasing the blowing pressure to return to the reed vibration again.

Again the precise mechanism for this is still not perfectly clear and under investigation. Still, the phenomenon is there and most likely explainable by the fluid dynamics already found with saxophones and flutes above. The reed is driven by the blowing pressure and is therefore displaced. When it bears down a bit, the air can flow around it, causing turbulence below the reed. This releases the pressure on the reed, as the air can now flow around it, and the reed starts moving back. This again disrupts the airflow around the reed, and a new pressure is put on top of the reed, driving the mechanism anew.

Again the sound of the instrument needs to be produced by the turbulence, the first large vortex of the air jet flowing around the reed. This can be estimated for sure because of the sharp sound the instrument produces. This sharp sound has many overtones that are all contained in the impulse, which is caused by the first large vortex behind the reed. Still, the reed itself is vibrating in a nearly perfect sinusoidal way up and down, so only in the fundamental frequency of the sound with no overtones at all. So as the resulting sound has many overtones and the reed has none, the sound cannot come from the motion of the reed, it is produced by the first large vortex of the turbulence behind the reed.

So it is no surprise that the mouth organ shows such self-organizing behavior like a sudden change of the forced and the forcing system. It has a highly nonlinear driving mechanism, where turbulence flow plays a crucial role. Still, we hope to get more insight into this in the future.

Indeed, there are more free-reed instruments in the world, the blues harp is such an instrument which again has a slightly different driving mechanism and sound behavior. The harmonium is another example which a very soft and low sound used mainly in churches or for other more spiritual music. Indeed, the harmonium is the end-point of a development of musical instruments based on friction processes, much like the bowing of a violin, with instruments like the *glasharmonica*, the *clavicylinder* or the *terpodion*, starting mid-18th century. Although a different driving mechanism, the sounds are very similar. We have discussed the history of these instruments above and will be concerned with the acoustics, the reason for such a sound in a section below.

Shawms, Oboes, Double-Reed Instruments

Shawms are double-reed wind instruments where in the West, the oboe might be the most prominent one. Still, many more examples are used all over the world, like the bagpipe or the *zurna*. The latter has been used in the Turkish military orchestra and spread all around Turk countries, down to India, and in many countries in Southeast Asia in many forms, like the Nepalese *sahanai*, the *horanewa* 17 of Sri Lanka or *hne* or *dum da* 17 of Myanmar, the *surna* in Southern China and so on.

The instrument has one advantage, it is mainly loud! The Heavy Metal music in Middle Europe up to the 20th century was bagpipes and hurdy-gurdy, another loud instrument, working with friction. This music was played mainly by peasants but also for military use. It is outdoor music for parties, festivities, marches, and war, where one would need loud instruments to be heard at all.

That works even better with drums, and the *Janizaree* band of the Turkish military orchestra is a very influential example of such a drum-and-shawm band, playing the Turkish *zurna* together with drums [236]. This ensemble became very popular and spread mainly by Muslim traders. The Nepalese *panjan baja* (lit. five instruments) consists of a *saranghi* and drums [250], the music played at Singhalese Buddhist *pujas* (rituals) each day all over southern Sri Lanka consists of the *horanewa*, a shawm and drums. Even the very sophisticated *hwain wain* orchestra of the Bama, the main ethnic group of Myanmar, has the *hne* playing together with drums of several kinds, the most fascinating is the *pat wain*, a drum circle of pitched drums unique around the world.

Still, of course, there are other beautiful features with zurna-like instruments next to loudness. The Scottish bagpipe might also have been used as a talking instrument. Musicians raised with a Gaelic language as their mother tongue sometimes claim to be able to speak with a bagpipe in the Gaelic language, where each pitch shall correspond to a vowel. They claim that this was possible before the bagpipe became part of the military orchestra, where playing techniques changed. Then the pitches might have been heard as vowels, and the melismas, small and fast tone patterns as consonants. If this is true or not is not perfectly clear. Indeed, *piobaireachd*, playing the Gaelic bagpipe, has a notation called *Canntaireachd*, a solmization like the European Do-Re-Mi...

In terms of tone production, oboes are not very different from other reed instruments like the saxophone or clarinet. Still, instead of a single reed, they use two reeds attached next to each other, therefore, they are also called double-reed instruments [2]. The reason why they are so much louder is that the two reads cause a much shaper impulse than the single reed instruments can do. Such a sharp impulse contains much more overtones than a softer impulse, and therefore there is much more energy in the sound, it is heard as much more prominent.

The reason for this sharper impulse is that the two reeds do not end in a mouthpiece immediately. There is a small tube where the air first needs to pass through until it can enter the mouthpiece. Remember that the tone of reed instruments is mainly produced by the first large vortex or the turbulent flow in the mouthpiece. As long as the air is traveling inside the very small tube behind the double-reed, it cannot build up a vortex. Still, it will do so right after this tube. As the speed of the air jet is very fast in the small tube, simply because the tube is so small. Then entering the mouthpiece suddenly leads to a very abrupt and strong vortex right behind the small tube. This leads to a very sharp impulse, which then has many overtones. So, in the end, it is the abrupt change between the small tube and the mouthpiece, together with the high speed of the flow, which causes the sharp impulse. The saxophone has a fast-flowing air-jet but with a smooth transition of the diameter of the tube, leading to a much smoother sound. The organ pipe has not such a fast air-jet and is therefore also much smoother.

Still, the basic mechanism is not different from that of the saxophone, and so we again find all phenomena of a self-organized system, there needs to be a certain blowing pressure to get a sound at all, there is a hysteresis loop of on- and offset of sound, and there is this very precise synchronization of higher partials of the sound caused by the interplay of returning impulse and reed displacement.

Brass Instruments

Trombones, trumpets, horns, or the dijeridoo are operated by making a funny noise with the mouth, pressing the lips together, and blowing air through them to result in a periodic sound from the lips. When pressing the lips at the mouthpiece of a trumpet and doing this again, a tone is heard. This has nothing to do with brass, the dijeridoo is made of wood, and there are trumpets made of plastic and are 3D printed, which work the same way. Still, as most instruments working like this in the West are made of brass, they are called brass instruments.

The situation here is quite similar to that of reed instruments. The air pressure inside the player's mouth opens the lips, an air jet enters the mouthpiece, produces a vortex, which produces a pressure impulse. This impulse travels through the tube, is reflected at the tube's end, and returns to the mouthpiece. There it interacts with the lips and triggers another impulse to be produced. The vibration of the lips follows the length of the tube, therefore, the frequency of the lips are forced into that of the tube. Also, the harmonic overtone spectrum is nearly perfect, which is not to be expected when considering the complex tube system of the instrument, and therefore, again, a synchronization of the overtones needs to happen, the one we already discussed with reed instruments like the saxophone. So this instrument is again a self-organizing one with synchronization of the overtones and the forcing of the lips vibration into those of the tube.

This also becomes obvious when locking at the lips vibration using a high-speed camera.[60] They vibrate regularly, and for a short time period, they open to release an air jet into the mouthpiece producing the sound impulse, which contains many overtones. Still, the lips themselves again vibrate in a more or less sinusoidal way, and those overtones cannot be found by the lips vibration. Therefore the sound cannot come from the lips vibration themselves but are produced by the first large vortex of the turbulent air jet in the mouthpiece.

Furthermore, the sound impulse needs to travel through the tube of the instrument in a complex way. Trumpets or trombones do not have finger holes where the tube is interrupted, and therefore the tube length does not vary because of a shortening of the tube by finger holes. Instead, the tube length is changed by adding additional length to the tube. With the trumpet, this is achieved by pressing valves that open or close additional small tubes and therefore increasing the overall length of the tube. With trombones, this is much more clear to see, there the player changes the length of the tube by sliding up and down the large tube bow.

So the impulse needs to travel through many tubes that change direction, have kinks and sharp edges, and end at a horn, which is shaped in a complex way. Therefore, as already found with the saxophone or the clarinet, the impulse caused by the first vortex in the mouthpiece will be altered in a way that the different overtones are stretched or compressed, and therefore are no longer harmonic one to another. Still, again as with the saxophone, the resulting sound of the instrument is very harmonic, which is a robust thing, it is hard to play inharmonic overtone structures within the normal range of playing. Therefore we would expect a similar mechanism of synchronization of these overtones as with the saxophone, and indeed the returning impulse will only trigger the player's lips to close and open again to release a new pressure impulse which is perfect as the one before, and therefore a harmonic overtone series is guaranteed.

The only difference between a reed and a brass instrument is that the valve opens in another direction. With reeds, the valve will close during playing, while with lips, it will open. In other words, when not applying air pressure and no sound is produced, the reed is open, and the lips are closed. When applying air pressure, the reed will close and open, while the lips will open and close, so vice versa. This is not a fundamental difference in terms of tone and sound production, still, it is the basic difference between the instruments.

Indeed, playing a trombone or trumpet is different from a saxophone or clarinet in many other ways. Trumpets players need strong lip mussels. These mussels are hard to build up, and after one or two hours of intense playing, the lips might be so tired that the player is not able to continue. Also, when not practicing the trumpet for about a week, the mussels are gone, and the player needs to start practicing the instrument right from the start as if he had never played before. This is not so important with reed instruments, and it might be one reason for inventing the more complex reed system instead of just playing instruments with lips that do not need any complex reed or even double-reed construction. Double-reed players do produce their reeds normally on their own, it is very hard to do it for others, and although one can buy double-reeds, a professional player will always build them himself. Still, a saying among musicians is that you need to build a bathtub full of double-reeds until you have one, which is actually sounding! This is different in other cultures. The *shahanai* of Nepal or the *horanewa* of Sri Lanka both have leather double-reeds. Each reed side is build of three or four layers of leather, mainly from the water buffalo. To make these reeds sound is quite some effort!

Additionally to the high nonlinearity of lip-driven instruments like trumpets or trombones, that is turbulence as the sound production, they have another interesting nonlinearity, which causes what is called 'brassiness'. This brassiness is known from Big-Bands or marching bands, but also in the classical orchestra when these instruments play loud some licks or fills. This sound is very bright, strong, and in a way, brassy. It is very much different from the sound when played with a lower volume. The brassiness, therefore, is an important part of the instrument.

Now additional brightness means more overtones. As we have seen already, more overtones are achieved when the impulse traveling in the tube is sharp. So increasing the brightness of the instrument means increasing the sharpness of the impulse. In brass instruments, this is achieved through the very high air pressure inside the instrument when playing loud. Then the inside pressure might be up to 140–150 dB. As we know, for the human ear, 120 dB is the threshold for pain, and 150 dB might damage the air instantly forever. Indeed, when this 150 dB impulse is leaving the instruments, it decreases its strength down to about 100–120 dB right at the bell, and, therefore, this is not too damaging.

But with such high air pressures, a sound wave in air is no longer traveling the same way as with lower air pressures. As we have seen already, normally, a sound wave in air is not changing its shape. Now with high air pressures, the speed of sound is increasing. This is due to the molecule's air consists of. An airwave is nothing but molecules moving a bit, and therefore driving or pushing neighboring molecules. After pushing the next molecule, the one which has pushed has done the job and will return to a rest position. This is why sound in air is different from wind: in wind, the molecules flow continuously, in sound, they only move a bit, pushing the next molecule and returning to their initial position. So an airwave is only a flow of energy, not of matter.

Now the rules of this pushing are governed by the forces molecules apply one to another. These forces are electric potentials, which are very strong when molecules are right next to each other and decrease nonlinearly to very low forces when they are a part away. With low sound pressure, the distances between molecules are quite large, and the movement of them quite low. Then the force one molecule applies to its neighbor is small, and if it moves twice as much, the force applied will also be about twice as much. Remember that this is the definition of linearity, when one unit is doubled, the other unit needs to be doubled too. If not, if the second unit is tripled or changed in any other way except of doubling, we have a nonlinearity, and strange things might happen.

Now the interaction between molecules, strictly speaking, is never really linear. Still, with low sound pressure, and therefore low interactions, it is nearly linear, and so we can assume the sound wave to propagate the normal way. Still, when the movement is getting larger, the basic nonlinearity of the force interaction becomes considerable. Then moving a bit further towards the next molecule will increase the force applied to it tremendously. We then have a nonlinear interaction, and again strange things will happen.

One of these things is brassiness [186]. At the peak of the traveling impulse, the interaction of molecules is very high, while only a bit further up, it tremendously decreases. Therefore at the peak of the impulse and slightly behind where the air pressure is also quite strong, the molecules push much stronger towards the direction the wave travels to. Still, just right before the impulse, the force is not so strong, and, therefore, the wave travels much slower. This leads to a steepening of the impulse, where the molecules from the back push strongly towards the peak, but the ones before the peak slow down the movement. The peak is getting steeper and steeper, and therefore the edge of the impulse at its front end becomes more and more a wall.

Now from the Fourier theorem, we know that such a strong change, from zero to a maximum pressure, nearly instantaneously means that there need to be very many high and strong frequencies present in the impulse. This is right what brassiness is. Therefore, the nonlinearity of the air is producing this brassiness sound of brass instruments. This is not itself a self-organizing process, but another important sound aspect caused by nonlinearity.

By the way, one knows this sound also from elephants [99]. When they blow their horn, they also use their trunk and produce very high pressures in it. The same thing happens here, the air becomes nonlinear, and the trumpeting of elephants is the same effect as the brassiness of a Big Band horn section or a classical orchestra brass sound.

Articulation and Noise with Wind Instruments

The *shakuhachi* is a Japanese bamboo flute. It is often used for meditative music, where melodies are simple, but the music has a multitude of different sounds. The *shakuhachi* is end-blown, which means that the bamboo stick is open at both ends, and at the upper end, there is a labium cut at the side of the rim. The player blows at this labium, on the rim, and covers most of the rest of the open hole with his lower lips.[61]

The sound of the instrument has a lot of noise, next to the played pitches. It is the wind noise that appears as a result of the turbulence at the labium. As before, the first large vortex is producing an impulse that travels through the bamboo stick, is reflected, comes back, and interacts with the player's air jet in such a way to produce a harmonic spectrum. Still, next to the first large vortex, there are many smaller vortices following this larger vortex, building a vortex street, where the vortices produced by the decaying largest vortex become smaller and circulate faster. These smaller vortices do not have the chance to build up impulses that travel down the tube, are reflected, and come back to interact with them, simply as they are too small, and therefore are dissipated, damped, and gone fast. Still, they all are there, and each of these smaller vortices produces a pressure gradient, a sound, which is radiated from the instrument. As very many smaller vortices appear in a chaotic manner, many of these small pressure gradients are produced, many small sound impulses. As these are chaotic, irregular in time, we hear a noise.

All other wind instruments also produce this noise, and we clearly hear it. Still, with the *shakuhachi*, this noise is particularly loud. But noise is also found with the transverse flute. The reason is that the turbulence is also happening outside the instrument, and therefore the noise is not covered up, as in the case of saxophones or trumpets, where the noise happens within a mouthpiece and first needs to travel through the tube to be radiated. Still, also these instruments can be played in a quite noisy way, which might be wanted.

This also holds for the trumpet. Miles Davis was one of the first to transform the trumpet from the military signal sound, as used in Big Bands, into an intimate sound, playing very noisy to enhance an intimate feeling [245]. And indeed, many people associate noise with intimacy in such musical cases. Here the noise is colored noise but still containing a pitched note with low volume.

Now the *shakuhachi* sound is associated with a meditative state of mind. Indeed, the instrument is often used in Zen Buddhism. Noise is one part of this meditative experience, and therefore the instrument can be played solely producing noise without a pitch for a while. This kind of noise is also found in nature, it is very close to the sound of trees with many leaves where the wind goes through. The sound production is very similar, as the wind, an air-flow, is disturbed by the leaves, and therefore go into turbulence. Each small vortex is producing a small pressure gradient, an impulse, and as there are so many of them, the resulting sound is that of colored noise, just like that of wind instruments. It is interesting to see that not only is the sound the same, but also the physical tone production is just alike.

The sound of a waterfall is another example of a sound very close to colored noise. Water is another example of a flow. Indeed, physically, for both, the flow of air and that of water, the Navier-Stokes equation holds, and the mechanisms of turbulence are right the same. Therefore most of the sounds of a waterfall are produced the same way as that of a wind instrument. Still, with a waterfall, some sound is produced by water hitting stones when fallen down, which is a different kind of tone production. So not all of the sounds of a waterfall can be found to come from turbulence, still, most of it is indeed produced by turbulence.

In acoustics, there are basically two kinds of noise. White noise contains all frequencies with the same strength, the same amplitude. All other kinds of noise are called colored, where some colors are specified. So, e.g., pink noise has amplitudes, which are $1/f$, where f is frequency. So higher frequencies have lower amplitudes in the noise sound. Such noise is heard by us in such a way that all frequencies have about the same loudness. Remember that the physical sound pressure and psychological loudness can be very different. Interestingly, such $1/f$-noise (called 'one over f noise') is produced by self-organizing systems. This is another clue that sound perception is tuned to self-organizing physical systems. The human brain activity follows a $1/f$ noise [92], loudness fluctuations in Bach's *First Brandenburg Concerto* are also $1/f$-like [258], river flows [188], bird songs, or environmental sounds often follow a $1/f$ rule [271].

A closer look at the $1/f$ noise is interesting in terms of the predictability of a sound. The $1/f$ rule more precisely reads $1/f^\alpha$. If $\alpha = 1$ we have $1/f$. If $alpha = 0$ we have $1/f^0 = 1/1 = 1$. This means that the amplitudes of the spectrum do not depend on frequency anymore, they are all the same, and therefore the sound is white noise. If $\alpha = 2$ we have $1/f^2$.

Now physically, white noise is a completely random process. What is happening at a time point is not at all determined by any previous time points of nothing which has happened in the past. On the other side, $1/f^2$ physically corresponds to Brownian motion, the motion of moving molecules, what we call heat. Their movement is very much predicted, they depend on all previous motions of all particles. Their movements look random, but mathematically they are perfectly determined! Now the 1/f noise is in between. This means that what happens at one point in time is determined by previous events to some extent, but not completely, a certain kind of randomness is still there.

Musical sounds can be complete noise, as we have discussed above. Still, such music is often perceived as unpleasant [257]. On the other hand, music showing the 1/f behavior is found to be pleasant. As the output of a self-organizing system is not perfectly random, but also not perfectly harmonic, shows bifurcations and some kind of transients, noise, etc. the sounding output of a self-organizing system seems to match the needs of our brains, of us and therefore makes us feel pleasant.

Still, it appears with environmental noise that this is not the only way to make us feel pleasant [271]. We turn to brain dynamics below and will see how pleasantness is a complex topic. What sounds pleasing to someone might sound ugly to someone else.

So noise of the 1/f kind, pink noise, is found in nature. Indeed, in Zen Buddhism, to become as natural as flowing water or as blowing wind is an ideal, and producing a similar sound might remind the player or the listener of that. Then noise has a meaning, is not perfectly random. Still, this meaning is not rationalized in terms of expressing it verbally, it might not even become clear to players or listeners. It is expected to 'work', to set the listener in a state reminding him or her of a 'natural' state of being. If this is about to work, it needs to happen inside the brain, and we will come back to this after discussing self-organization, synchronization, noise, and other aspects of brain neural networks below.

But the *shakuhachi* has some more articulatory possibilities. The player has a small distance between his lips and the labium, where the air jet leaves his mouth and then hits the labium. The player might vary this distance slightly by changing the angle he holds the instrument. When he lowers it, the distance becomes longer, when he raises it, the distance becomes less. With longer distances, the played pitch goes down, and with reduced distance it goes up. The reason is that with a longer distance, the returning impulse from the *shakuhachi* tube has more space to interact with the labium. Therefore the interaction time is raised, delaying the next impulse produced at the labium. Therefore one whole cycle of the impulse, production, traveling down, traveling up, and producing a new impulse is enlarged. A longer cycle means a longer periodicity of the pitch. The played frequency is the reciprocal of the periodicity, therefore, the pitch goes down. When the lips come closer to the labium, the opposite happens, the periodicity becomes smaller, and the pitch rises up.

Many other musical instruments have similar techniques to raise or lower pitches. With a saxophone, changing the lip pressure so that the lips press down the reed stronger or weaker will raise or lower the instrument's pitch. This is because the reed has more or less time to open or close the mouthpiece and, therefore, more or less time to produce a new pressure impulse. In all wind instruments, when in- or decreasing the blowing pressure, this will happen in similar ways. Lower pressures will produce lower pitches. Of course, the production of a new impulse is happening faster if the air jet is entering faster into the mouthpiece. Increasing the blowing pressure is doing right this, when increasing the speed of the air-jet, the pitch must rise.

Another articulation technique of the *shakuhachi* is to make pitch glides between two neighboring pitches by slowly covering or releasing a finger hole. Then the finger slides very slowly above a finger hole, and the pitch gradually de- or increases towards the pitch of the next open finger hole.

Yet another articulation is overblowing the instrument. When raising the blowing pressure considerably, at the time previously, one impulse traveled through the tube now, two do travel. This is caused by the increased speed of the air jet approaching the labium. After the vortex has built up, it is removed so quickly by new air coming from the player's mouth that the jet wants to change direction towards the other labium side much faster. If this fits an impulse coming back, a stable oscillation will appear at twice the frequency compared to the normal case when not overblowing.

These two cases, regular blowing and overblowing need not appear exclusively. With a certain blowing pressure, certain raise or decay times, and certain ways the air pressure is controlled, the player is able to produce both pitches, the fundamental and the overblown one. Then one hears two pitches, an interval of an octave. This sound, combined with noise, makes three distinguishable sound elements within one instrument.

There are many more possibilities and uses of the instrument in contemporary music, known as extended techniques, including talking into a wind instrument while playing, using a fast oscillating tongue to produce a flattering sound, the so-called flatter-tongue, playing with two wind instruments in one mouth at the same time, like the jazz legend Roland Kirk made popular next to many others. These techniques add additional variations, fluctuations in time and frequency, explosive sounds, inharmonic components, and noise in many variations.

The field of noise and experimental music nowadays has become part of mainstream pop, ambient, EDM, Trap, Dubstep, and many other musical styles. They also include soundscapes, with more and more became part of the experimental and art music scene over the last decades with recordings of cities, industrial sites, animals, nature, etc. Many consist nearly solely of noise. Noise is also a necessary part of Independent music with crunchy guitars and noisy drums, distorted vocals, or bass lines. Crash cymbals, snare drums, or shakers mainly produce noise. Noise is part of Rock, Punk, or Metal music with distorted guitars and distorted other instruments. Other instruments coming close to noise are those with very sharp sounds, which are so dense, that they come very close to noise, like bagpipes, dijeridoos, or the hurdy-gurdy, the metal guitars of ancient times and the middle ages, as discussed above.

So noise is as much part of music as are harmonic sounds. It is interesting that it became more and more part of mainstream music over the last decades. Indeed, it is often used as a reference to nature, rough or berserk behavior, basic instincts, or trance-like states rather than culture, where the latter is closer to harmonic sounds. Still, noise is also often accompanied by high intellectual reflections about music, consciousness, society, or politics. Free jazz was understood as freeing the individual from social restrictions, blowing one's mind, but also to open up political or social repressions. The intellectual art scene, as well as the soundscape community, discuss aesthetic and artistic topics, environmental issues, globalization, gentrification, social injustice, or human rights.

The use of noise to free one's mind or to free the individual from social constraints is often found. Still, there is no need to have this association, and one is free to find noise to act in a different way, use it for relaxation and chill-out, as a soundscape for envelopment, or as healing sounds. Some are just annoyed. So there is not only one noise, and there is not only one way to deal with it. For some, noise might act to free, for some, it is the opposite. Some might use it for reflection, for some, it does not mean anything.

Still, this does not mean that the origin of noise and harmonic sounds is arbitrary, as we have seen with many examples so far. Random noise does appear mainly with systems not self-organized, partly determined 1/f noise is partly deterministic and therefore produced by an organizing system. Harmonic sounds need a very complex physical system to be produced, resulting in a harmonic overtone series.

Therefore the rules of noise and harmonic sounds cannot be found in the sound or the reaction to it, it can only be found in the production process and the system of musical instrument, listener, musician, production, art, and culture as a whole.

Friction Instruments

Friction instruments are those whose mechanism of sound production includes a process of sticking and slipping one thing onto another. We have discussed some of them already below. The most popular are violin, cello, and their family members of bowed musical instruments, where the hair of a bow is gliding over a string in a process of constant sticking and slipping. Still, there are many others. The *lounuet* from New Ireland is a wooden block played by sticky fingers gliding over wooden plates, sticking and slipping over them.[62] The terpodium is a keyboard instrument where the keys press bars covered with leather onto a rotating wooden cylinder, where again the leather is sticking and slipping over the cylinder.[63] With the glass harmonica, a finger touches a rotating wet glass cylinder, and by sticking and slipping over it, it produces a musical tone. The clavicylinder uses longitudinal waves of a stick to produce sounds, where again, the fingers glide along the bar, sticking and slipping along it.[64]

All these instruments can produce harmonic overtones, although the tone production process is very complex. Still, the same process might also result in a scratchy sound, a squeaking, or some noise. Such noise might be wanted as with the *lounuet*, which is about to mimic frog calls or the call of the hornbill bird. With violins or the terpodium harmonic sounds are mostly intended, still also not always.

The process of sound production in friction instruments might be very different. The terpodion has a very short sticking time compared to the slipping time, with the violin, it is vice versa. The terpodion timbre is mostly caused by longitudinal waves of the bars, while its pitch is determined by the transversal movement of this bar. This difference between pitch and timbre also holds for the *lounuet*, where the pitch is determined by the plate the finger rubs over and a combination of air cavities below the plate, while the timbre depends on the kind of rubbing. This is similar to a violin, where the pitch and timbre are caused by different mechanisms.

© The Author(s), under exclusive license to Springer Nature Switzerland AG 2021
R. Bader, *How Music Works*, https://doi.org/10.1007/978-3-030-67155-6_8

The bowing process has mainly been studied with the violin.[65] Here the bow is covered with rosin to make it more sticky. While playing, the bow sticks to the string and displaces it more and more when the player moves the bow in one direction. This continues until the bow tears-off the string. Then the string is slipping along the string until it is more or less at its resting position again. But then it sticks or glues to the bow again, and the process starts all over, the bow is again displacing the string.

Now basically, there are two reasons why the string is teared-off the bow, changing from sticking to slipping. The main reason is that each tear-off leads to two impulses that travel along the string in both its directions, are reflected at the strings both ends, and return to the bow-point. One impulse travels to the bridge first, and one travels to the nut. The one who first travels to the bridge is back to the bow-point fast, travels through it, as the bow is only about to catch the string again, is reflected at the nut, and returns to the bow-point again. Then it has made the whole roundtrip on the string back and forth. The other impulse first traveling to the nut returns to the bridge when the bow is already displacing the string. It, therefore, returns earlier compared to the impulse first traveling to the bridge.

Indeed, it is the impulse that has traveled a whole round trip back and forth the string who tears-off the bow from the string again, as it disturbs the sticking. It is necessary that this impulse which travels a whole round trip back and forth the string tears-off the bow, and not the other one, as then the fundamental frequency of the string is that of a string with such a length. If the other impulse did the tear-off, this would not be the case, as this impulse is back much faster. Then the frequency of the played note would depend on the position of the bow on the string, which in regular bowing is not the case. The bow-string system, therefore, needs to be adjusted such that it is the impulse traveling the whole round trip to make the tear-off.

Still, this need not always to work. Another possible reason for such a tear-off is that the increasing tension the string acts on the bow due to its increasing displacement becomes so strong that the bow tears-off. If this is happening earlier than the first impulse coming back, the pitch of the string is not that of regular playing. Furthermore, the new impulse produced again tears-off the string at yet another irregular time-point, and when repeating this, the sound is not a harmonic one anymore. Still, it is also not perfectly random, and such a sound is perceived as surface-like or thin. This happens if the bow has not enough rosin on it to be able to make the sticking strong enough, all players know the sound of a bow not covered with rosin. Then it is very hard to play in a regular manner at all.

The other case is when the bow is pressed to the string with much strength. Then the returning impulse cannot tear-off the string from the bow, as this impulse is not strong enough to overcome the hard pressure of the bow. Then the bow is displaced more than usual and tears-off only when the tension caused by the large displacement is just too much. This is especially heard when playing with slow bowing speed, resulting in a scratchy sound.

So the regular motion of bowing is only possible in a medium range of bowing speed and medium pressure, the bow is pressed to the string. It also depends on the amount and kind of rosing used for the bow, which makes it more or less sticky. Other parameters are important, too, as is the internal damping of the string, which needs to be much stronger than with guitar strings. Only with much internal damping, the returning impulse is more or less gone after it has detached the bow from the string and does not disturb the process any further in the future. A guitar string is very hard to bow indeed, mainly because its internal damping is low, and, therefore, all impulses stay on the string longer. Still, bowing a guitar is also possible when performing it carefully.

So the bowing process is hard to handle and needs much practice to master. Indeed, the process is purely self-organizing, where a complex system and a strong nonlinearity leads to a harmonic motion. The nonlinearity is the sticking/slipping difference, which is suddenly turned on and off and necessary to make the process go. Then the harmonic overtones are again very precisely tuned 1:2:3:... The reason is very similar to that of wind instruments. There the old impulse was only triggering a new one, which was produced right the same way as all previous impulse, and therefore the sound results in a very harmonic spectrum. With bowing, the returning impulse is also only triggering a new one by tearing-off the string from the bow, which produces a completely new impulse. The old one is more or less gone right after the tear-off due to the large damping of the violin string.

So we have two very different physical processes which have the same principle and work the same way very much. And both lead to a very precise harmonic overtone spectrum only because of the systems being very complex and self-organizing. Also, both systems are only stable within a certain parameter range and can easily be brought into other kinds of vibrations, producing inharmonic sounds or noise.

Yet another similarity is the hysteresis behavior we have already seen with wind instruments. There it takes more blowing pressure to change from a noisy sound with low blowing pressure to a harmonic overtone structure than it takes to return from a harmonic sound into noise again. The same holds for bowing. To change from a noisy sound with low bowing pressure into a harmonic sound needs more pressure than returning from the regular motion into a noisy one.

Therefore friction instruments are self-organized systems by heart. Deviations from the regular motion lead to non-harmonic sounds, as is the case with the *lounuet*, which is about to produce animal sounds like frog or bird calls. These sounds are not perfectly harmonic, but also not perfectly random, the right job for such an instrument to produce. Indeed, animals have a very similar sound production system like humans, and we will see that voice production is a self-organizing system too.

Guitars and Plucked String Instruments

We have seen the wind instrument and the bowed and friction instrument family to be self-organizing by nature. But how about plucked string instruments like the guitar? We would not expect it to be self-organizing, as there is a string that vibrates with its frequencies on its own, and there is no way to find any nonlinearity in this. Also, all other features for a self-organizing process are missing, like the sudden onset of a pitched musical tone out of previous noise, hysteresis loops, or more complex sounds coming from the instrument. No matter how soft we pluck the string, it always sounds harmonic. If we pluck it stronger, it will start bumping to the fretboard and make some noises, still this is more a case of distortion, which is indeed nonlinear, but is more an add-on rather than part of the fundamental process of guitars tone production.

And indeed, a string on its own is not self-organizing, it is a linear system on its own. But when examining the instruments closer, we find some behavior that does not fit into such a simple system.

Still, the picture changes when we realize that the string is not along, it is attached to a soundboard, which is again glued to ribs, a back plate, and a neck and interacting with an air volume inside the instrument. A simple theory of resonance, where the plate resonates the sound of the string, assumes that the soundboard, or the top plate, shape the string sound according to the frequencies the plate itself has. We can determine these frequencies by knocking on the plate alone. Then we get a sound that contains all these frequencies with their respective strength. With guitars, we find that the lowest resonance is the so-called Helmholtz resonance of the air inside the instrument at 100 Hz. The first frequency of the plate itself is 200 Hz, roughly speaking, and the second at 300 Hz. Still, guitars have many tones played on them in between these frequencies. From this theory, we would expect them to have a very low volume, as they do not fit to the lowest frequencies of the plate. These might have some frequency region around them, which makes them resonate to some extent, and, therefore, still, some radiation will happen, but we can measure these ranges and find them to be quite small. So, strangely, the guitar vibrates strongly at frequencies where it is expected to be very weak. Indeed, such guitars would sound very strange, as low pitches would be quite low in volume.

Another experimental finding with guitars is the shape of these frequencies on the top plate, their eigenmodes. These can be determined experimentally, which has been performed to quite some extend, by driving the top plate by an artificial shaker. Still, when playing the instrument with its strings, these mode shapes look considerably different from the ones measured with artificial excitation. Even worse, they change their shape depending on the point the different strings are attached to it. This is expected to some extent, as when driving such a plate at a node of the eigenmode, the vibration will be weak, and when driving it at an anti-node it will be strong. But this is only about strength, not about its general shape. But indeed, the shape of the eigenmodes do change with changing driving points, where the strings are attached to the guitar top plate.

The difference between driving a guitar artificially by a shaker and playing the strings is shown in Fig. 1. The upper figure shows the eigenmodes of the top plate when it is driven by a shaker, a device vibrating sinusoidally, and attached to the points indicated with numbers, so for each frequency at a different point. To arrive at beautiful pictures, the top plate needs to be driven at the points where the eigenmodes are strong to be able to supply energy into the system. The lower plots on the other side show the guitar top plate vibrating when it is forced to vibrate with the strings

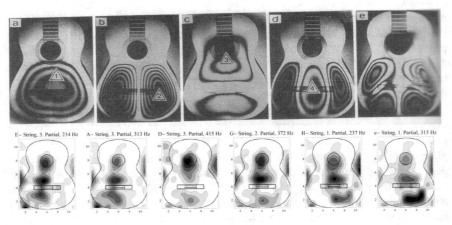

Fig. 1 Upper figure: Eigenmodes of the guitar top plate (measured with laser interferometry, From: Jansson 1971). Each eigenmode has its own frequency. Lower figure: Forced Oscillation Patterns (FOP) of a guitar top plate, which shows how the guitar vibrates when it is plucked by a string, on the very left is the low E-string is plucked, on the very right, it is the high e-string (measured with a microphone array, From: Bader 2013). All cases show only the strings partial at frequencies 300 Hz to compare the vibrations. These FOP are not similar to any of the eigenmodes shown in the upper figure, also not as a combination of two neighboring eigenmodes. Also, they show a driving-point dependency, so the strongest vibration is close to the point the string is attached to the top plate

frequencies. Then it vibrates no longer in its eigenmodes but in forced oscillation patterns (FOP). The different plots show the frequencies of string partials 300 Hz when plucking the six strings (lower E-string on the very left, high e-string on the very right). The frequencies of the string driving the top plate are a little bit different in each plot, as it needs to be, as these are frequencies the string provides and no eigenfrequencies of the top plate. Now for such a low frequency, we would expect the top plate to vibrate more or less the same in all cases, with an eigenfrequency 300 Hz. Still, the forced oscillation patterns strongly depend on the driving point, the point where the driving string is attached to the top plate. Around this point, in all cases, the vibration is strongest. Therefore forced oscillations are very much different from eigenmodes and cannot be derived from them easily [21].

To come closer to the real mechanism acting here, we now have already found a fundamental feature of tone production, the interaction between string and sound-board, and we can ask a simple question: When attaching a string to a soundboard, why do we hear the sound of the string and not that of the soundboard? This is not caused by us most often plucking the string and not knocking on the soundboard. Because if we indeed knock on the soundboard, we hear a knocking sound at first, but after the knock is gone, we hear the sound of the string, low in volume but clearly. We do not hear the string vibration itself as radiated from the string, as a string does nearly not radiate due to the acoustical short circuit discussed above. So what we hear is the vibration of the soundboard, which is taken over by the vibration of the string. Therefore, no matter which of the two we bring into vibration, the string or the soundboard, it is always the string that wins the game and forces the soundboard to vibrate with the strings frequencies.

From the above, it is clear that the energy transfer between the two is two-sided. Energy from the string enters the soundboard and vice versa. But why is the sound-board taken over by the string?

Let's do another experiment by changing the material properties of the sound-board. Wood, guitar, piano, or violin soundboards are made of, has a short sound when we knock on it. This means it has strong internal damping, and when applying energy to the plate, this energy is gone soon. This is not the case with strings, which vibrate much longer as they are much less damped. But what if the soundboard would consist of a material which does damp only very briefly, as is the case with a string? As such a guitar or piano is not that easy to build, although this is not impossible, we can still do a computer simulation.

Therefore we use the geometry of the instrument, its material properties and choose a string as well as a place where the string is struck or plucked. Then we insert all this in a computer software which knows how the structures are vibrating, we can reproduce the FOPs. This is called physical modeling, as the physics of the instrument is modeled on a computer. The most precise modeling is by taking the whole geometry of an instrument and solving the differential equations of vibration at many discrete points on the geometry.[66] These differential equations determine the displacement and the velocity of, e.g., the guitar at many points on the instrument geometry. Then the computer solves these equations, which means that for many time points, he calculates these displacements and velocities, and we can watch the instrument move in very slow motion. We can also listen to the sound such an instrument would produce, as the algorithm calculates the vibrations for many points in time. These are sampled time points, just like with a CD, which has 44100 such points per second.

There are several algorithms that can be used to do such physical modeling, each has its pros and cons, but in the end, result in the same sound. These methods are often very useful, as one can easily try different geometries, materials, couplings between instrument parts, or playing techniques. This saves much time, especially with grand pianos, which easily take about a year to be built. Trying changes in the geometry would take many years, and a method which can do this on a computer can build thousand of versions of pianos within a day.

Returning to our discussion, such methods are also useful to test new materials, and here we want to build a soundboard, which is as weakly damped, just as a string is. Now when doing so and plucking a string on a guitar or playing a note on such a piano, the sound of the instrument is very different from a regular instrument, and most people would not even be able to identify the instrument as a guitar or a piano anymore. The sound is very much reverberant and more like a sound in an ambient room than that of a regular instrument.

The reason for this strong change in sound is that the guitar or piano soundboard no longer does only vibrate with the frequencies of the string but continues to also vibrate with its own frequencies. Additionally, the string starts taking over the frequencies of the soundboard to some extend, although not perfectly.

With normal playing, the so-called eigenfrequencies of the soundboard are only present at the very beginning of a plucked or struck tone, within about the first 50 ms or less. If one is listening closely to the tone of a classical guitar, one can hear these eigenfrequencies as a knocking sound, just as one would take a finger and knock on the piano soundboard without plucking the string. Within the initial transient of the sound, this sound of the soundboard is not perceived as unwanted noise but fused in the brain with the string sound to one tone sensation. This knocking sound is an important part of the guitar or piano timbre, and if one leaves it out, which again can be done in a computer by cutting it off, the remaining sound is dull and empty. So from this point of view, the soundboard already has a strong influence on the sound.

Still, again, when making the internal damping of the soundboard small, it will continue to vibrate with its own frequencies on and on, next to the strings frequencies, of course. So in the normal case, the string wins the game and forces the soundboard to its frequencies after a small struggle in the initial transient because the soundboard is much stronger damped compared to the strings damping. With a weakly damped soundboard, the string wins this game no more, it ends drawn.

A computer model of a plain piano soundboard, one without ribs and bridge, shows this very clear. The advantage of a computer simulation is that we can apply any internal damping we want. With real pianos, this takes much more effort. Also, we do not have any distortions due to measurements, using microphones and amplifiers, everything in silico is precise. Also, a plain soundboard is shown here as we then are sure that the results are not due to any bracing or irregularities in the structure, but only due to the properties of wood itself.

In Fig. 2 such a soundboard is shown in three of its eigenmodes. Each row is one such eigenmode, the top row shows the lowest monopole mode 26 Hz, followed by a dipole 86 Hz, and a more complex one 180 Hz. These are the eigenmodes the piano soundboard vibrates in when it is vibrating on its own. Now, if we play a string, the soundboard is forced to go with the frequencies of that string. Each piano key has a different point where it is applying its vibration to the soundboard, where it is attached to the soundboard. In the computer simulation, 14 keys have been used, starting from the lowest key 1 to a very high key 74 (the piano regularly has 88 keys nowadays altogether). Although not all keys have the three eigenmodes frequencies, 26, 86, 180 Hz, we can still apply them to all keys in a computer simulation. This makes sense when we are interested in the question of how the internal damping shapes the eigenmodes when the soundboard is driven by an external string force.

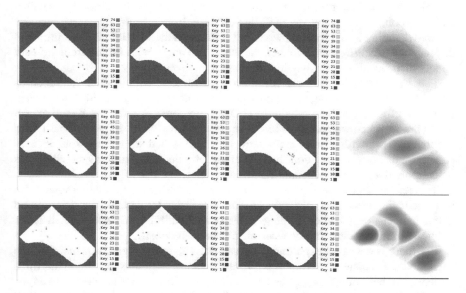

Fig. 2 A computer simulation of a piano soundboard, where each row shows a different eigenfrequency of the soundboard, and each column shows a different amount of damping. On the very right are the eigenmodes the piano vibrates in. On the other three columns, the dots show the points where the maximum amplitudes of forced oscillations are for different keys played on the piano. The eigenmode theory tells us that the mode shapes (which are those on the right column) stay the same for all keys played, so with forced oscillations, when the forcing oscillations are the same as the eigenfrequencies, only the overall amplitudes of the eigenmodes change. Therefore also the places of maximum amplitudes need to be the same. This is true in the case the piano is damped much less than a real piano is, shown in the column second to the right. Still, the next column (second to the left) shows a piano damped about the same as a real piano. Here already the places of maximum amplitudes very much depend on the driving point. The column on the very left shows a soundboard damped much more than a real piano. Here the driving point dependency is very strong. Therefore we need to consider stringed-instrument body's sound behavior more in terms of internal damping, energy distribution, and forced oscillation patterns (FOPs) rather than of eigenmodes [11]

Now again, if eigenmode theory holds the modal shape, the three shapes in the very right always look the same, no matter where the soundboard is driven, as long as it is driven by that frequency. Then, in each case of the different keys, the point of maximum amplitude with every single case needs to be the same. So for the lowest mode, it needs to be at the point where the maximum amplitude of the eigenmode is, for the higher modes, it needs to at a place of maximum amplitude, the dipole has two such maxima, the triple mode has three.

The second column on the right shows where the maxima of the forced oscillation patterns (FOPs) for the different keys are in case the soundboard has internal damping much lower than a real soundboard. There the eigenmode theory holds pretty well, nearly all maxima are about at points where the eigenmode maxima are. The column second to the left is the case of an internal soundboard damping with is realistic. There the FOPs are considerably different from eigenmodes and, therefore, a more complex distribution of the maxima for the different keys can be seen. This is called a driving point dependency, the FOPs are different from eigenmodes. If we increase internal damping even more, shown in the column on the very left, the driving point dependency is getting even stronger, the maximum amplitudes for different keys are widely distributed over the whole soundboard.

Therefore, again, we find that for musical instruments with wooden soundboards of strong internal damping, the eigenmode theory does no longer hold, and we need to take the forced oscillation patterns (FOPs) into consideration. We would expect the FOPs to be considerably different in the case we drive the soundboard with a frequency, which is not its eigenfrequency. Still, when we use the soundboards eigenfrequency, as we have done here, internal damping is determining the vibrations much more than the soundboards eigenfrequencies.

This tells us that the string indeed slaves the soundboard, a behavior known from self-organizing systems. It also tells us that there is a control parameter determining if this happens or not, the damping of the soundboard. Strictly speaking, it is the relation of internal damping between string and soundboard.

But there is a second experiment we can do. The string is one-dimensional, the soundboard is a plate, therefore, it is two-dimensional, strictly speaking, it has three dimensions, but soundboards need to be thin to still be able to vibrate strongly, therefore, we take the soundboard as two-dimensional for now. But what if also the soundboard would be one-dimensional? Then it would also be a string. What if we couple a string to a string, both with the same low internal damping.

If we do so, there is no winner of the game anymore, both strings interact and take over the frequencies of the other string while maintaining their own ones to a certain extend. This makes clear that there is a fundamental difference in the slaving behavior depending on the dimensionality of the structures. A lower-dimensional structure, here the string, is much more able to force a two-dimensional structure into its frequencies than vice versa.

So the soundboard loses because of two reasons, it is weaker damped, and it is of higher dimensions than the string. This is the reason we hear the sound of the string and not that of the soundboard. That does not mean that we only hear the sound of the string, as the soundboard is not completely damped, and the difference in dimensions is only by one (1-dimensional against 2-dimensional). Still, we can test this too, by taking our computer model again and turn the internal damping of the soundboard to very high. Then, indeed, the resulting sound, radiated by the soundboard, is perfectly that of the string, and the soundboard does not alter the strings sound in any way.

This experiment is not far away from real instruments. Antonio de Torres Jurado (Antonio Torres) (1817–1892) is maybe the most famous guitar builders ever, and at the end of the 19th century, he refreshed the sound of the guitar, which at his times was quite dull and dark and turned the instrument into one with much more brightness and loudness. There are several features that have been associated with his guitar building and which he was claimed to have invented, the 7-bar fan bracing at the bottom of the top plate or a different shape of the instrument. Still, all these features have existed before. But what all of his instruments indeed show is that the thickness of the top plate at the boundaries, where it is glued to the ribs, is extremely thin, a bit more than 1 mm. The normal thickness of guitar top plates is roughly between more than 2 mm up to 3 mm [221]. Generally, the thinner the top plate, the wider it can vibrate, and therefore the louder the instrument, which is wanted. Still, a thinner plate is about to break sooner or later, and therefore the thickness of the plates needs to be chosen carefully. To arrive at an optimum, the top plate thickness is not the same all over the plate but is carved thicker and thinner, according to the ideas of the instrument builder.

Now a plate with a thickness of slightly above 1 mm is no longer a plate. Wood, cut down that thin, can be rolled together like a carpet. Therefore it acts more like a membrane, and therefore such top plates are also called membrane plates or diaphragm plates. Those parts of the plate with such a small thickness cannot move anymore like the rest of the plate, and therefore behaves differently compared to the rest of the plate. Indeed, the velocity of the sound wave entering this thin region decreases tremendously, and such wood is even much stronger damped than normal. But this means that a sound wave entering such a region of a Torres guitar near the to plate boundaries is hardly reflected by these boundaries and is practically not moving back into the top plate middle again. Nowadays, such boundaries are also called acoustic black holes [206], as waves entering them slow down and never come back.

Torres was very much aware of this fact, as to demonstrate the building principle of his instruments he constructed a paper-mache guitar which had his 'normal' top plate but where the ribs and the backplate were built of pape-mache, which is so heavily damped and has such a low sound velocity that it can be considered as not moving at all. He played this guitar claiming that it sounded right like his regular guitars, so he wanted to say that in these guitars, only the top plate vibrates, and the other parts are not moved by the top plate.

This forms a situation very similar to that of a top plate very heavily damped. In both cases, the waves entering the plate do not come back from the sides and, therefore, cannot build up and eigenfrequencies of the plate. Theoretically, this is the same as if one would have a very large plate, theoretically infinite. From such a plate, no waves would be reflected from the sides, too. Such a plate has really no behavior at all and therefore is not able to shape the sound of the string in any way.

For Torres guitars, this means that only very few weakly returning waves from the boundary of the top plate cancel out each other with other waves on the plate. Then there are only very few standing waves, only very weak top plate eigenmodes, which could shape the strings sound, and the sound of the guitar is more string-like, more bright, and loud. Such experiments with membrane soundboards have also been performed with pianos at the end of the 19th century and found to be of an interesting timbre [117]. Contrary, soundboards made of metal were also tested, those this very low damping, and found not be suitable for pianos, right because of the reason that in the sounds, many frequencies of the metal soundboard could be heard.

This also means that different kinds of damping, as present in different kinds of wood, lead to strong changes in the sound of the piano or guitar. The geometry does play a major role, too, because it determines the possible eigenfrequencies the soundboard is shaping the strings sound with. Still, the amount and way of this shaping process are determined by internal damping. As with middle and higher frequencies, the number of eigenfrequencies of guitars and pianos is so high, and they are all so nearby, the shape of the instrument, its geometry does not play too much of a role here anymore. It is rather the internal damping determining the sound. This is different for different types of wood, and so the internal damping of wood seems to be the most important parameter for wooden instrument sounds.

But returning to the question if pianos or guitars are also self-organizing systems, we have already found a slaving process, the forcing of one part of the instrument by the other to take over its frequencies. We also found a strong nonlinearity, which is the sudden change of the dimensionality, from one-dimensional to two-dimensional. So the vibration traveling along the string can only do so in one direction, but when entering the soundboard, it can suddenly move into all directions of the plate.

Still, some features are missing, like the sudden onset of a harmonic sound coming from noise or a hysteresis like with wind instruments. Here, again, a model can help. This time we can use the IPF, which worked very well with wind and friction instruments, determining these sudden onsets, noise, harmonic parts, and hysteresis. To use it with guitars or pianos, we need to extend it due to the increased dimensionality of the instruments. So a guitar consists of a top plate which is driven by the string and gives it to the ribs, the sides of the instrument, which again makes the backplate vibrate. All make the enclosed air move as well as the neck. So we need to include more reflection points to the IPF than was the case with wind instruments. There we only had one reflection point, the bell of the instrument, which acted back to the reed or the mouth. Now we have several points acting back, which we can easily include by splitting the energy the string supports into the energies of the different parts and let them act back to the string. With pianos, we can assume that the soundboard is so wide that the different directions the waves can travel on it can be taken as additional reflection points.

But when doing so, the algorithm behaves in an unexpected way: it stays perfectly stable without any bifurcations, sudden changes in its behavior, or hysteresis. All over the possible playing range, the sound is perfectly harmonic. In other words, if we increase the dimensionality by increasing the number of reflection points, we make the slaving process perfect for all possible kinds of articulation, strong and weak plucking of the string or plucking position. No matter what we do, such a system is always stable.[67]

This does not mean that it is no longer self-organizing. Indeed, it is strongly self-organizing, but to such an extent that it cannot be driven into chaotic behavior anymore. This is the reason why stringed plucked or struck instruments do not show bifurcations or the like. Still, their sound production process is very complex indeed, and only because of this complexity, again, we have a very simple output, a harmonic spectrum. The part the struggling between string and guitar body take place is the initial transient of the sound, the very first beginning, which is complex and chaotic. It also shapes the basic character of the instrument a lot. Therefore the self-organizing nature of guitars indeed shapes the guitar, piano, or violin sound tremendously.

Practical Consequences for Guitar Acoustics

The next pages might be of interest to instrument builders and especially guitar builders. Readers who are not too much interested in the details and tricks of instrument building can skip this section without leaving the flow of the book. Still, if one is interested in the practical consequences of the principle of self-organization of musical instruments, please go on.

From the mechanism of self-organization and impulses traveling through the guitar, we can derive several rules-of-thumb connecting the building of a guitar with its sound. These rules are not heading for a 'perfect sound', as sound, of course, is a matter of taste. Still, it is possible to predict certain parameters of the sound of an instrument from building principles.

We have seen that damping is one important aspect in slaving, the one less damped wins the game. In a guitar, the wooden body is more damped than the string, therefore, the string always wins the game. Still, a guitar body is not damped completely and can vibrate on its own for a short time, the time it takes for a knock on the body to disappear. If this ability to vibrate on its own is made even shorter, the body will not only take over the frequencies of the strings but will also do so without much changing the amplitudes of these frequencies. Contrary, a body that is damped less will still try to counteract the string and does so by enhancing frequencies which it would vibrate with on its own while at the same time being not be so strong in its own frequencies.

Therefore a body that is nearly completely damped will only sound like a string. While a body not so heavily damped shapes, the strings amplitudes more and more into its own direction. Therefore the less the damping of the body, the more 'character' or 'individuality' a guitar will have. The more it is damped, the more it will sound with a simple spectrum of a string, therefore sound with more precision and brightness.

Additionally, a body which is damped heavier will most often sound louder. This seems strange as more damping means fewer vibrations. Still, with heavier damping, the complexity of the vibrations on the body becomes less, the forced oscillation patterns of the body are more simple. Complex patterns often cancel out themselves when it comes to sound radiation. If one part of the body is moving down while another is moving up, both cancel to quite some extend, and therefore the sound is getting lower in volume. Indeed, guitars also have so-called evanescent waves, waves that are strong on the body but cancel each other out only some centimeters away from the body. A recording engineer needs to know this, a guitar recorded in the near-field, right before the top plate, sounds very much different from one recorded about one meter in front of it.

A radical case is that of a bass-reflex guitar. There the top plate, back, and ribs are made of extremely thick wood, these plates might be 2.5–3 cm thik. Plates of a thickness of one centimeter or more are so rigid, they practically do not move at all. These guitars have only a very thin vibrating area around the bridge. This area acts as a membrane, basically only moving in and out, but without much complex forced oscillation patterns. Indeed, this guitar sounds much stronger in volume because the vibrations are monopoles, only moving in an out, and nothing cancels here, both in terms of eigenmodes as well as sound radiation.

Increasing the damping of a guitar body can be done in several ways. Most often, lacquer leads to strong additional damping, depending on the kind of lacquer. Also, different kinds of woods have more or less damping.

Still, there is another way of increase damping by making the boundaries of the top plate very thin, maybe only about 1–2 mm. Wood that thin is no longer stiff but can be rolled or curved, it is no longer able to vibrate reasonably. Therefore it is heavily damped, vibrations going into it become slower and are not really reflected inwards again. Then all vibrations from the strings are basically going into the top plate, moving towards its boundaries, but do not considerably come back. In such a situation, there are no eigenmodes of the guitar body anymore, and therefore, again, the top plate sounds like a string alone without much shaping of sound by the top plate. Antonio Torres used this method, as discussed above [221]. Diaphragm plates are also discussed in piano soundboards [117]. Such guitars or pianos are sometimes also called membrane guitars.

A similar behavior appears when the small wooden rims gluing the top plate to the ribs, the sides of the guitar are too thick. Then the reflection from the top plate boundaries is strong, and the guitar has much character, sounds low in volume, and also has a quite narrow sound. Building these rims thinner leads to a louder sound, more open, and free. This, like most of what has been said before about the guitar, does also hold for violins to some extend.

A guitar where the sides and back do not really move does also lack of sound details, which normally come from these parts. The ribs and back plate contribute to the middle frequency range of guitar sounds, the back more to the lower mids, the ribs more to the higher. The neck strengthens the high frequencies. The Helmholtz, of course, is the lowest frequency, normally 100 Hz. Still, these parts are only moved by the top plate in a slaving manner. It takes a while at the beginning of a played note for them to come in. It sounds as if the guitar would react slower to a played note. For guitarists, this is an important aspect of playing, the reaction time to a note. This parameter is purely psychological as physically, the guitar starts vibrating right after the pluck. Still, some sounding guitar components, the air inside, the back, and the ribs, only come in later compared to the top plate. This sounds like a slow attack, often making it hard to play the instrument. The top plate itself always starts right at the beginning of the tone, and therefore, the instrument sounds with much more attack if this top plate is moving without back and ribs.

Good tonewood is supposed to be hard. This is a common saying in instrument building. How hard some wood is, determines the frequencies it vibrates in. But as we have seen, the guitar body is mainly slaved to vibrate with the frequencies of the string, it can only shape the amplitudes more or less, depending on the amount of damping. The reason why hardwood is preferred is that it is not breaking so easily compared to softwood when it is under stress. The strings of a guitar act on the top plate with about 60 kg, quite some load. If wood is too soft, the top plate simply breaks. Hardwood is more able to withstand these tensions.

Softwood could also stand this tension when it would be thicker. If a top plate of a guitar would be about 4–5 mm in thickness, the softwood might do too. But then the plate is so thick that it has a hard time to vibrate. It still does, but with only very small displacements. But these displacements do move the air above the top plate, and therefore lead to the sound we hear. If the displacements are low, then the sound is low too, and the instrument is too low in volume to be used.

So the idea is simple, use a wood as hard as possible, which allows the plates to be as thin as possible. This makes the instrument as loud as possible. Indeed, over the last about two centuries in guitar building, there was a loudness war, the louder, the better. So instrument builders tend to make the top plate as thin as possible before it breaks.

Another aspect is stress in the guitar body. Back plates of guitars, but often also top plates, are glued to the body such that it is curved a bit. The plate itself is flat but made a bit larger than the ribs. When gluing both together, the plates bend out to fit into these ribs. Then the plates are under tension. This tension is known to increase the brightness of the sound.

Still, wood is flowing. We know this from cupboards there plates loaded with books over the years might start to bend down. When removing the books, the plate keeps this bowed shape, they have flown. Basically, everything is flowing, constantly changing its shape, which is called viscoelasticity. Stones flow, although very slow and little, wood flows, oil or water flows, all at different speeds, but all the same way, molecules shift and reconfigure according to forces like gravity or, with guitars, string tension. So when a top plate is glued to the ribs of a guitar under tension, the plate will get accustomed to this shape and flow in such a way that the original tension is gone, while the shape remains the same. Still, these instruments are known to sound more bright.

The reason for this is that plates do not only vibrate up and down, they also vibrate in their plane direction. This is called a longitudinal vibration. If one is knocking on the side of a plate, it is vibrating in its plane. The sound from this vibration is not too loud, as it is only radiated at the plate sides, which are very small. So this kind of vibration does not contribute to the sound of a guitar. But when the plate is bent, when it is curved, this vibration is triggering the up and down vibration of the plate little by little. So the sound of the in-plane vibration is transferred to the up and down vibration, which has strong radiation, and we can therefore hear it. As the speed of sound in the plane is much higher than that in the up and down direction, the in-plane sound is much brighter. When transferred to the up and down direction, it enhances the brightness of this kind of vibration. Therefore the overall sound becomes brighter. But this is not caused by the tension of the plate but by its curved shape.

Inhomogeneities in the thickness distribution of guitar top plate also contribute to sound. Guitar builders take much care in terms of which part of the plate should have which thickness. These often range between about 2.5 and 3 mm. The thickness distribution might also be asymmetric so that the distribution on one side of the guitar is different from that of the other side. It is hard to measure the differences in the sound when performing such changes in general. The reason is that the wood of each guitar is different, and therefore guitars with the same shape and geometry can easily sound different. Still, when modeling guitars on the computer, we can keep all wood parameters perfectly the same and find out about the changes.

Indeed, these thickness changes, as well as different placing of the bracing, the ribs glued to the top and back plate to stabilize them, cause much fewer changes in the sound than expected. One needs to listen very carefully to hear a change at all. Only dramatic changes lead to a considerable difference. What does sound different is when the overall shape of the instrument is changed. Then instruments sound change their timbre a lot. Still, within the subtle sound changes, some rules-of-thumb can be formulated too. Changes around the bridge will affect all frequencies, while changes around the soundhole and above it will affect only the mid- and high frequencies. This is because the forced oscillation patterns on the top plate caused by the string are nearly not present around the soundhole but concentrate on the region around the bridge. Also, asymmetries of these changes can lead to an increase in how 'stereo' a sound is, how large the instrument appears to us, still, they cause a slight decrease in loudness.

Yet a feature most known to guitars is the Helmholtz resonance. This is the air flowing in and out of the soundhole. When it flows in, it compresses the air inside the instrument, which pushes the air around the soundhole out again. But then inside the guitar, there is an underpressure that pulls the air around the soundhole in again. This motion was first described by Hermann von Helmholtz in 1863 and caused loud sounds.

In guitars, the Helmholtz vibration is determined by the air volume in the guitar and the size of the soundhole. Normally it is tuned to 100 Hz and is responsible for the bass of the instrument. Without it, guitars would not have much bass, as the instrument body is too small to vibrate in such low frequencies. Classical guitars tune the Helmholtz just beneath the frequency of the low open A-string, while with Western guitars, it is around the low F♯, the second fret on the low E-string. The lower the Helmholtz, the heavier the guitar sounds. This is because of a psychoacoustic effect of sound, low frequencies sound large, high frequencies sound small. This Helmholtz is very well known and a standard tool in instrument building.

So altogether, it seems that using the right wood is more important to the sound of an instrument than its shape or small changes in the thickness of the top plate, or even the fan bracing. To put it the other way around, a guitar builder who finds a good piece of wood and builds a guitar from it needs to deal with its properties, and therefore it might be reasonable to make fine-tunings afterward.

It is getting harder and harder to get good tonewood. The more people live on earth, the more wood is needed, and trees are grown fast and cut early. Also, climate change will produce a change in the grown woods, at least in Europe. The use of spruce in instrument building is mainly caused by a decision in the Middle Ages when commercial wood farmers started to plant fast-growing woods. With the changing climate, spruce or maple is not growing fast enough anymore, and at the moment, the search for alternative woods in Europe is on.

So instrument builders for years try to use different materials replacing wood or using other kinds of wood. As damping seems to be the crucial parameter for tonewood, it is a new challenge to find hybrid materials, combining wood with carbon fibers or other polymers, or laminar or plywood, which is more stable and can therefore be made thinner and louder. Top-ranked electric hollow-body guitars have always been built of plywood. There are already examples of these methods which sound very close to wood, sometimes even better.

The challenge here will be the design of material where the damping can also be tuned depending on frequency. Generally, higher frequencies are damped much stronger than lower ones, and, theoretically, these rules follow an exponential decay curve. When measuring the damping of wood, this general rule indeed appears, still, there are many exceptions all over the frequency range, where suddenly some frequencies are damped less or more than expected. This leads to a very lively sound of musical instruments made of wood, which is often lacking when artificial material is used, which follows the general rule much more exactly. A plastic guitar sounds like plastic. Still, it is not a basic problem to build a material which deviates from the theoretical exponential decay rule and design a desired damping spectrum. We will see how this develops in the future.

The Human Voice

We found so far several musical instruments mimicking singing, animal songs, or screams, expressive articulation. So to close the section about musical instruments and then turn to the human brain and perception of music, we need to have a look at the human voice. We would expect it to be a self-organizing system too, and indeed it is. It shows all aspects of self-organization, a complex physical system that can produce perfectly harmonic sounds but can also sound rough or like noise. The transitions between harmonic sounds and noise are again happening suddenly, and there is hysteresis. One can try this immediately by singing a note with low volume and continuously decreasing the lung pressure. All of a sudden, the tone breaks down, and we hear a whisper, a wind coming through our mouth. Now, if we increase the lung pressure again, the tone will start again, but we need more pressure to make it start than we have used to make it break down (this is hard to experience, still found in measurements.[68] The same thing happens when going into falsetto, the high-pitched voice above our breast voice, suddenly appearing with increasing lung pressure and with different on- and offset pressures. Just try!

So we need to have a look at the production side of the voice to decide whether this is caused by a complex, nonlinear system or if there might be a different reason.[69] Physiologically the situation is complex but well known. Muscles press our lungs together, leading to a flow of air from inside the lungs through the trachea and our larynx and finally through our mouth. The larynx consists of two vocal folds, two tissues at the left and right side of the trachea. These folds might be relaxed when not using them and stretch when we produce a tone for speaking or singing.

The stretching of the vocal folds is caused by muscles attached to them. The stiffer the vocal folds, the higher the pitch produced. When singing or speaking normally, these vocal folds will open and close periodically while at the same time being moved up and down. So the folds are going through a circular motion, upwards and sideways when opening, and downwards and joining together when closing. This movement is crucial for tone production. Each time the folds open, an air stream is going through the open slit, move into the area just above the folds, and produce a vortex there, just like with the saxophone, trumpet, flute, or organ pipe. This vortex produces a pressure gradient, which is the transformation of the flow into sound. This pressure gradient is moving upwards, is shaped by our mouth to form the vowels, and leaves our moth, we speak or sing.

Still, if the pressure from the lungs is not strong enough, the folds do not move periodically, and we hear a noise, maybe a whispering voice. If the folds are not perfectly the same, left and right, because of damage, illness, or the like, then the perfect periodicity is no longer working, and we hear a rough voice.

Still, why is this system working that way? Indeed, there is no conclusive answer accepted in the research community until now, but several suggestions. First, we would ask why the folds move at all. Suppose they are closed at first without pressure from the lungs. Now when lung pressure starts, they would open to release this pressure. But this could be it. The lips are open to some extent, the air can flow through, and this stable state could go on until the lung pressure changes again. There would be no need for the lips to close again. Maybe if the lung pressure onset would be very sharp, they could bounce open and close a couple of times, still, then they would stay at rest. Indeed, the tissues the folds are made of are heavily damped, and, therefore, there would be not too many bounces.

So there needs to be a force that brings the lips back in again. Of course, there is a force that closed them in the first place, the muscles pressing them together. But this is counteracted by the lung pressure, so this cannot be the cause.

Historically the first idea was that the mechanism closing the lips again is the same, which brings a plane into the air, the Bernoulli force. The wings of a plane split the air into two parts, one is moving above and one below the wing. Now wings are constructed such that the way above it is longer than that below. So the air moving above it is stretched, while that moving below it is compressed. But compressed air wants to expend, while stretched air is sucking surrounding air into its region. So the wing is once pressed upwards by the air below it and pulled upwards by the air above, the plane moves up.

This mechanism is supposed to work with the human voice too. Therefore we need to remember the circular movement of the lips, not only opening and closing but moving up and down. The reason is that the lips are not too flat, they are rather quite long in the direction of the trachea. Therefore a lower part of the lips might move in, while the upper part might move out at the same time, causing that circular motion. But then the lips, when not perfectly closed, would look like a tube which once will become thinner at the top and wider at the bottom, only then to become thinner at the bottom and wider at the top.

Now a jet moving through the folds when they are nearer at the lower end and wider at the upper will flow fast through the small entrance and will slow down when moving upwards. But such a lower speed will produce a pulling of the structures around it inwards, which could be the force bringing the upper lips part closer again. Then the lower lips would move apart, as forced by the lung pressure. After this change, the situation is vice versa, the lower part of the lips move together again. This will then look like a rotation of the lips while opening and closing. All this is then caused by the Bernoulli force.

This idea is interesting but fails to explain why there is a sudden change between whispering and normal singing and normal singing and falsetto. There is no reason why a perfectly harmonic motion should not take place over the whole range of lung pressure. So although Bernoulli definitely holds, it cannot be the whole story.

To improve the situation, in the 70th, the first computer simulations of the vocal folds could be performed. This was done with right the idea from above, each fold has two parts, an upper and a lower one. So we can model each fold with two masses. Therefore, it is called two-mass model. They are pushed one to another by the muscles acting on them, and they are pushed apart by the airflow. This model does not use the Bernoulli idea at all. Indeed, it works very well, when there is a medium lung pressure, the system moves periodically, a musical tone or a speech sound is produced. When reducing the pressure, suddenly, this motion breaks down, and noise appears. And indeed, with high pressures, a falsetto voice comes into place. It also shows the hysteresis effect, so we might say we are done.

The problem with this model is that we do not finally understand what is happening. We can make it move inside the computer and observe its movement. But what is the basic principle behind it?

To get closer to such a principle, another model in between the two-mass model and the Bernoulli effect was suggested, using a Van-der-Pol oscillator. This is again a differential equation very much like that of the vibrating mass point we have discussed above. The mass is the vocal folds which vibrate when driven by the air jet from below again. But the Van-der-Pol equation has a special feature coming with the damping. This damping is assumed to depend on the displacement of the vocal folds. When they are moving in a small oscillation, with low amplitude, the damping is as usual. Still, when they move wider out, the damping turns into its opposite. It becomes an additional energy supply, it is driving the lips even further.

This seems unusual at first, and indeed this is impossible when we would simply have one normal moving mass. But the vocal folds are more complex than that as we have seen. The Bernoulli idea, the two-mass idea, they both try to understand why the lips are moving at all with a constant stream. The Van-der Pol model is just another approach to this.

So with low displacements, the system is normally damped. Then the lips cannot move very much, the energy supplied to them by the air jet is damped out very fast. So the lips do move, but with very low in amplitude.

Still, from a certain threshold of airflow on the damping becomes driving, and the folds start moving. This strong upraising of movement is counteracted by the restoring force, of course, that is, the lips torn back by the muscle they are attached to, and due to the fact that they are attached to the trachea, the air column. So they move back and forth in a regular motion then.

This model now explains the sudden onset of the voice at a certain threshold. The Bernoulli model could not do this. The two-mass model could do this, still, we could not solve the two-mass model to arrive at another equation, which shows why this behavior is like it is. We can only ask a computer to solve the two-mass model for time point after time point, iteratively. Then we get a series of these time steps, the time series, but we do not know the basic behavior.

This is different with the Van-der Pol model as this can be solved like an equation, it can be solved analytically. Then we can look up at which lung pressure the oscillation starts and other such aspects.

So we might find we have finished yet. But also the Van-der Pol model fails to explain several things, falsetto or the rough voice. As the model fails to explain these, it cannot be the end of the story.

There are other explanations for the vocal fold motion. One is a model that is very attractive, as the differential equation it uses is only of first order with respect to time. Such an equation does not vibrate, it can only decay. So when the vocal folds are displaced by the air jet, when they only would move back when the jet is diminished. But this model is more complex by including a nonlinearity to the displacement. So when the displacement is small, the nonlinearity is very much as if the equation would be linear. But when the displacement is getting larger, the nonlinear model results in much larger values than the linear one does. This could correspond to the behavior of the vocal folds as a larger system that has a lower and an upper edge and moving in a circular motion.

This model does about the same as the Van-der Pol model. When the lung pressure is low, the nonlinearity does not play a role, and the folds do not move in a regular vibration. But from a certain threshold on the nonlinearity suddenly becomes relevant, and the folds move in a regular oscillation producing a pitched sound. Technically this is called a bifurcation, as from a sudden threshold on the folds behave in a fundamentally different way, from decaying to oscillating. Such a bifurcation is called a Hopf-bifurcation. This bifurcation is right the transition from whispering to the singing voice. The model also shows the hysteresis loop when moving in and out, from whispering to singing and back. So everything seems fine at first.

The model has the advantage that it can easily deal with small differences in the lung pressure, a thing the Van-der Pol model has problems with. In a Hopf model, when the lung pressure is slightly in- or decreased, as is always the case with normal singing, the oscillation is disturbed a bit but will then return to a regular motion very soon.

But there is one problem. When the lung pressure is getting even bigger, from a certain point on this returning to a regular motion is no longer happening, and the model predicts that the voice would become louder and louder and even louder without stopping doing so. This can't be true! So again, this model has a drawback and is not the whole story.

Another idea is that the vocal folds behave in a similar way than does the saxophone or brass instruments. This idea comes from the question of whether the sound of singing stems from the vocal fold vibrations. As we have seen, the folds themselves move in a regular way, which looks very much like a sine wave. But the sound released from them is very much impulse-like. One can test this by using an older shaver, which makes a buzzing noise. When holding this shaver to the neck, the vibration of the shaver is transferred into the trachea and replaces the sound the vocal folds do. By moving the mouth and lips, one can form normal speech without using the lungs and the movement of the vocal folds.

So the sound the folds produce is not at all a sinusoidal but is a snaring, rattling sound. We know this behavior from the saxophone, the trumpet, or from reed instruments. There the saxophone reed, the player's lips, or the free-reed all move only up and down and act like a valve. These parts are not moving in a way the sound sounds. There it is the first vortex of the turbulence, which produces a pressure gradient, which is very sharp, and this is the sound. The valve only determines the fundamental periodicity, the sound is produced by the pressure gradient of the first vortex.

This needs to be true for the vocal folds, too, and indeed when modeling the flow through the folds a first, a strong vortex is formed out, producing the sound. So for sure, the vocal folds work the same way as reed or brass instruments do.

So one might assume that then this pressure gradient is also acting back to the vocal fold, in a similar way the returning impulse from the tube of a saxophone or trumpet does. But things are not that simple here. The saxophone reed is not changing its tension as the vocal folds do. The pitch of the voice is determined by this tension, that of the saxophone or trumpet by the tube length.

Still, this is not true with reed instruments, like the harmonica, as there is no tube attached. So in both systems, we could assume that a constant air jet is displacing the folds (or the reed), and it would stay open forever if not a strong pressure gradient behind the folds (or the reed) would act back to it, forcing them to return. Such a system would need to have a transition from whispering to singing voice, as turbulence is strongly nonlinear, and a weak air jet would not be able to produce a good first vortex with a strong pressure gradient. Also, hysteresis would be there for sure, as when the system is moving already, one could take some energy out without making it break down again. Falsetto could also occur when the lung pressure is so strong that the lips do not return to their normal position, and the whole system can oscillate with a smaller circle of the folds motion.

Still, this is only a guess by now, and further research needs to be done to get deeper into the influence of such turbulence on the fold motion.

For sure, the folds act like saxophones or trumpets in terms of sound production. They are also highly nonlinear and a self-organizing system, which all models assume, and which is confirmed by the sudden on- and offsets from whispering to singing and the hysteresis.

So the analogies between wind instruments and the singing voice are not only those of similar sounds. Indeed, wind instruments are mimicking the singing voice on a very physical level. As the real mechanisms of the singing voice are still not perfectly understood until now, this cannot have happened by instrument builders looking into the throats of people and copying the mechanism. Why they are similar, we cannot know for sure. Still, we have seen that such systems are very robust and produce a perfect harmonic spectrum and therefore are superior to other systems. So once found, they were preserved and improved.

Although much more could be said about musical instruments and the singing voice, what we have found until now is enough to understand the basic mechanisms. Nearly all musical instruments are self-organizing systems that show a simple output, the harmonic overtone spectrum robustly, only because they are very complex, nonlinear, and consist of several parts that interact in a complex manner. So self-organization is the very core of musical tone production. An exception may be percussion instruments, although they are much more complex than one might expect and also often very nonlinear. Still, they differ as they do very seldom have a harmonic spectrum.

The next step is to have a look at the perception side of music. This happens in the ears and the brain. Again in what follows, we will not at all cover the whole range of music perception. But this is not our aim. The aim is to understand the basic mechanisms which work to produce hearing, the sensation of sound, tones, timbres, rhythms, of music. We will see that many of these mechanisms are very much the same as those of music tone production. This will lead to the question of the egg and the chicken, the evolutionary development of the hearing system, and that of musical instruments. But in the end, this will not need to be decided as it is more of a historical interest. What is more important is the question if our brain is basically a musical instrument and musical instruments are a brain. But let us develop this in the next chapters.

Neurophysiology of Music

Music perception and production happen in the brain. Although a tremendous amount of research has been devoted to the brain when listening to or producing music, the mechanisms are not at all understood in full. There are different approaches trying to localize tasks like rhythm identification [248], where pitch is heard [10, 18, 59], where musical syntax and logic is processed [157], or where timbre is perceived [58]. The production of music in terms of movements and motor actions, as well as possible interactions between music and language, vision, or other cognitive tasks, were investigated and often located in the brain.

Still, the brain consists of neurons that are highly interconnected. On average, each neuron connects to about ten thousand other neurons, often far away, even between the left and right brain hemispheres. So the brain is a network, a neural network of highly nonlinear nature showing synchronization, self-organization, oscillations, and many other similar effects. Generally, the trend to localize single cognitive tasks with single brain regions is more and more replaced by the network idea.

Indeed, neural networks have been investigated for many decades, and simplified versions are used in many computer applications, from internet search engines or automatic machine control to classification systems of information of any kind.[70] Neural networks are so universal that they basically can be applied to any tasks which organize, sort, or identify something. This is no wonder as they are derived from mechanisms working in the brain, which again is able to perform all these tasks.

Still, this does not mean that all brain regions can perform anything. In music therapy, an astonishing healing success is the improvement of restricted movement abilities of Parkinson patients, which often have tremors, shaking hands or legs. When these people listen to a steady rhythm and perform simple movement tasks, the auditory part of the brain is able to take over some tasks the motor brain area normally does but now fails to do [248]. After some training, the movements indeed improve, which is often a great relief. Still, this is one of the very few examples where such a success could be made, and most often, such interactions fail.

So this is indeed more stressing the localization idea in the brain, and it is indeed realistic to assume that certain neurons have specialized to certain tasks over the development of a human. Physiologically, when listening to music, the brain does not have much choice at first, as sound does enter through the ear, and all neural nuclei which come right after will deal with this auditory input. These nuclei are the same for all humans and are very similar to those of apes, cats, mice, or rats, next to others. So there are universals in the organization of the brain brought along by evolution, which seems necessary to perform about the same tasks everyone has in common.

Such tasks are often very simple for us as we are so used to them. When someone plays a guitar string, everybody hears a pitch, and by playing different fingerings on the guitar, we hear a melody. Although in rare cases, people are not able to hear such melodies, called amusia, most hear the melody completely automatically. Still, when trying to make the computer replicate this task, it is a highly complex thing to do, and until now, no algorithm has been invented, which performs as good as humans do. This holds for most tasks in music that computers fail where humans have not a problem with, although quite some research has been put into finding such algorithms [12].

So we are in the middle of nature and nurture, localization and network. Still, the situation is even worse in terms of research as the brain is constantly changing. We learn permanently, which happens in the brain by changing neurotransmitter interaction strength between neurons, the way neurons interact with one another. Some changes are temporal, some are permanent. Still, this plasticity of the brain means that a brain is the way it is only at one time-point and different at right the next, changing within a few milliseconds. But how to investigate a thing that is constantly changing?

Self-organizing Models of the Brain

All models which deal with the brain as a whole are self-organizing ones. Walter Freeman suggested the global interaction of neural interaction in the brain at different levels [92, 162]. The smallest level has about 10 000 neurons, the largest in the whole brain. Such units show oscillations for short time intervals of a few milliseconds. Then the amplitude of the oscillation decreases, and a short, chaotic period follows. Very much like musical instruments work, steady-state time intervals where a clear pitch can be heard is followed by a new tone, which has an initial transient at its beginning that is chaotic.

The rhythms in the brain have historically been sorted into regions called alpha, gamma, or theta wave bands.[71] When measuring the spectrum of the whole brain, a scaling law appears, as already discussed above, in terms of noise. When plotting the brain rhythms in a log-log plot over a wide range, a steady slope appears. The slope tells us that the noise the brain makes is not Brownian, not perfectly random, but also not short-ranged, not perfectly deterministic. It is in between, which means

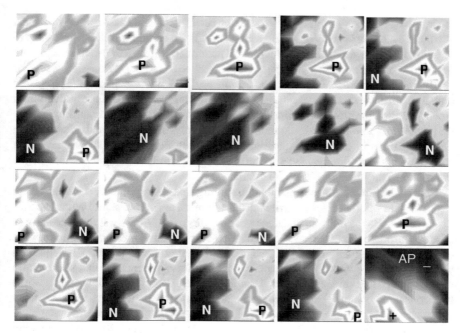

Fig. 1 A spatio-temporal pattern recorded from the brain of a rabbit, here reacting on a visual stimulus, a flashlight, using an electron array of 8 × 8 electrodes. Each electrode had a distance of 0.79 mm to its neighbor. Therefore, the whole field is only a plane of 5.53 mm × 5.53 mm. The figures show the strength of the neurons of a frequency 20–25 Hz. Each frame is a time point, starting with the very left top frame, following the top row to the right, continuing with the second row, etc. A vortex can be seen, which would have its center outside below the frame. The vortex lasts for the first 13 frames, so until the 4rth in the third row. Then the pattern repeats. One cycle takes about 10 ms. The vortex figure remains for about 200 ms [163]

that there is a dependency on what happens now on what has happened before. This holds over the whole brain. This scaling law tells us that all brain areas interact one with another.

To investigate single regions of the brain Freeman and his colleagues used a sensor array, a two-dimensional field of sensors on a micrometer scale, placed inside the brain of animals. Such sensors do only record the action potential, the firing of a few neurons. With EEG measurements, where the sensors are placed outside the brain on the skull, very many neurons are measured at the same time. On the other side, one can record the action of single neurons inside the brain. Here the idea was to measure a neural field in space and time, where one sensor covers several neurons (Fig. 1).

Nearly all animals scan their environment on a molecular basis we call smell or olfactory senses. As already discussed, human hearing is also evolutionarily derived from another sense of touch, where the hair cells called the line element in fish transformed into the cochlear. As olfactory sensing seems to be the first sense to emerge, evolutionary Freeman was focussing on the brain area associated with smell and placed the sensor array there. He found that when a certain smell was presented to the animal's sense, a certain spatio-temporal pattern appeared in the brain area. A kind of circular activation of neurons in the field could be found during the time a regular oscillation happened. When this oscillation broke down, the pattern disappeared only to reappear with the next regular oscillation after the chaotic phase. As different smells resulted in different spatio-temporal patterns, it is straightforward to assume that smell is a certain spatio-temporal pattern. As a smell is a qualia, something appearing to us in our consciousness, it is again straightforward to assume that a conscious content is a spatio-temporal pattern of the electromagnetic field in the brain during the appearance of that qualia.

This makes much more sense than assuming that a single neuron would fire in case a certain smell is detected. All neurons are more or less the same. What happens inside of them is the built-up and decay of electric energy. So why should the activity of one neuron be the experience of one smell and the activity of another neuron the experience of another smell? Still, when a different complex electromagnetic field changing in space and time appears, the conscious experience of that sense could easily be that field. Indeed, as there is nothing else except this field, we can assume that conscious content, experiences are spatio-temporal patterns of some or many neurons.

Spatio-temporal patterns have also been found with hearing [203]. When placing 18 electrodes on the primary auditory cortex, called A1, of Mongolian gerbils, such patterns appeared after the animals had learned that they were expected to move through a door when a sound of rising pitch was played to them and that they should stand still in case of the sound of a falling pitch. The auditory cortex is small in these animals, therefore, the electrodes were placed only 0.6 mm apart. The primary auditory cortex (A1) gets the input of the ears through the auditory pathway. The last stage before the cortex is the thalamus, where a region called the medial geniculate nucleus is connected with the A1. Up to A1, the auditory nerve fibers are tonotopic, which means that each fiber is expected to represent exactly one frequency. This is not perfectly true, as we will see below. Still, definitely at the A1, this tonotopy is given up, and when the nervous spikes go beyond A1 into the rest of the auditory cortex and beyond, nerve fibers are no longer associated with certain frequencies. So A1 is an interesting place to put such an array of electrons on.

Ohl and colleagues doing this study found that the animals at first were able to discriminate against the rising and falling sounds. As these sounds were varied, always starting at 2 kHz and raising to 2kHz, 6kHz, 8kHz, etc. the animal needed to learn that there are basically two kinds of sound, those rising and those falling, although there were several sounds rising and several others falling. They called this categorization different from Kant's idea of categories being machines extracting quantity, quality, relation, and modality. Still, the idea is clear, and we could call them categories, schemata, boxes, or whatever. The point is that from a certain training on the gerbils had an 'Aha' effect, they understood what the game was all about, learned that raising sounds were meant to go, no matter what kind of rising it was, and that falling sounds meant to stand still, again no matter how the sound did fall.

Now in the case when the categorization was established in the animals, they showed a clear, repeating pattern of in their A1, different for each case, a spatio-temporal field. This was not the case before the understanding when the animals were able to discriminate different sounds but did not get the clue. Then there were repeating structures in the A1, too. Still, they corresponded to neighboring electrons. Remember that the A1 changes from tonotopy in the input to patterns at its output. So although there also were repeating spatio-temporal fields, they were very different in case the animals did just hear different sounds and when they were able to categorize them into raising and falling sounds.

Another model trying to explain the behavior of the whole brain is that of Karl Friston [95]. He uses an analogy of thermodynamics. Brains, like all living beings, exist far off the thermal equilibrium. They are highly organized to maintain their differentiation and therefore allow action, perception, consciousness, they act living. This physical fact is used in analogy to brain perception and action. If the brain is in a state of knowledge, it is expecting a certain input from the world outside. When we hear something, this enters our brain and might not be what we did expect, we are surprised. Using the mathematics known in thermodynamics, how physical devices act in such cases, he can show that perception of a sound, that of a bird chirping, is adapted by the brain to result in a brain state which corresponds to the state the bird's chirp was produced by the bird.

This brain model is especially interesting as it is not trying to model the brain by using all single neurons, but by using a physical law, we are sure it needs to hold for all physical devices, therefore also for the brain. As it is able to explain perception as well as action by the brain trying to adapt while at the same time maintaining its state of being far away from thermal equilibrium, the brain as a whole can indeed be understood in this way, a self-organizing one.

The model has another epistemological part. The bird's chirp detected in the brain leads to state-space variables, which are the same as those variables used to build the chirp in the bird. Remember that the human voice is a self-organized system, and therefore, the parameters needed to produce a sound are only a few able to produce all kinds of articulation. The Friston model shows that these parameters of sound production are reproduced in the brain as just these parameters. In that case, the brain acts just like the bird in terms of sound production. This is a very different kind of perceptual process compared to the idea of the brain taking the input from the world and, by doing signal processing, trying to make sense of it. This signal processing idea is close to the idea of Kant for whom the thing itself (Ding-an-sich) is principally not perceivable. We can only make estimations about it. The Friston model, on the other side, means that we can understand the outside world as a parameter space, therefore how it works, its very nature. We can do so because both sound production and perception are self-organized actions that behave with the same physical laws.

The epistemological idea of Kant has also been challenged by interpretations of the Grothendieck lemma, a mathematical finding. This finding is purely mathematical at first. Still, it was the basis of object-oriented programming language and was also used in art theory. It might get a bit formalistic now, still, this is what math is all about.

A category in mathematics is a set, where all elements can be transformed one into another by rules. We all remember set theory from school, the set of all positive numbers, of all square roots of two, etc. A category is a set, but all elements can be transformed one into another by rules. So if a set would consist of the numbers 1, 2, and 3 a rule to convert 1 into 2 would be 'plus 1': 1 (element 1) + 1 = 2 (element 2). If we can invert the rule, here using -1, now start from element 2, and we get back to 1, this transformation is called isomorphic.

Now the idea of the Grothendieck lemma is to take two categories. One is a normal category, the second is a special one, a pure set, where there are no rules connecting the elements. In terms of an object-oriented programming language, the first category is equivalent to writing the code, where variables are connected by rules. The second category then is the compiled code only consisting of pointers to memory addresses, where there are no rules left anymore, as is the case in the code itself. Compiling the code means to take category one and transfer it into category two, the category which is a pure set, the pointers of machine language code.

Now the question is simple: If we only have the second category, the pure set, the pointers, is there an isomorphism which transfers these pointers back to the original category one, the programming code precisely, without errors or left-outs? This is a very important question in terms of large software packages with thousands of classes and millions of code lines. Will it work in all cases, no matter how complex the software code is. The mathematical answer is yes, there is an isomorphism from the pure set category back to the original category of programming language.

This lemma has been transferred to art theory [191]. Is there a way to perfectly understand a piece of art? When we listen to a musical piece, can we derive all elements of it from just listening to it? When listening, we collect information, therefore, we fill category two, the pure set, with information. We do not have the rules connecting the single observations. Still, we try hard to understand the melody, the musical instruments playing, the genre, the musicians, etc.

Using modern signal processing methods, it is not possible to derive the single tracks of a mixed musical piece from only the mix. The reason is very simple, when a guitar and a bass play the same note, their overtones are the same. How would one know how much amplitude a certain overtone belongs to the guitar and how much to the bass? There is no way. Still, we clearly hear the instruments in our perception. Therefore we are indeed able to derive the piece from the single observations. Still, it is better to know about the instruments, the musicians, the genre, and use all this information. But how if we would try to reconstruct the musical piece in our minds just like the musical instruments play? Such a perception is called perception-by-production. Indeed, we perceive multi-pitched tones differently for instruments we know compared to instruments we do not know [10]. When we know the instrument, our auditory pathway uses the information of timbre, a wide field of neurons in the auditory cortex. We use the sound itself to judge if an interval is a fifth or a major second. When we do not know the instruments, we go the abstraction way, try to reduce the sound down to simple periodicities, and detect the interval through these simple periodicities. This is pointing to an activity of the brain in terms of the amount of surprise and a strategy of periodicity reproduction in cases of surprise, and a strategy of timbre reconstruction in terms of familiarity.

Still, until now, there is no proof that a perception-by-production really happens. Still, the Grothendieck lemma shows that it is principally possible, and the Friston model that it is physically plausible.

Hermann Haken has a purely mathematical approach to brain dynamics, formulating the brain as a synergetic system [112]. With highly virtuosic math, he deals with brain dynamics analytically, meaning with mathematical equations on a piece of paper only, without the need for a computer to model the system. Most brain models assume that one can describe the differential equations which hold for a single neuron and that one can add the input from other neurons to this equation. Then the math is set up, but the equations also need to be solved. When there are thousands of neurons or even a handful, these models do no longer try to find an analytical solution to this large set of equations. As an example, an analytical solution to a differential equation is the sinusoidal as the solution of the wave equation. Rather such large systems use a computer to solve these differential equations by programming so-called iterative methods, where the next time step is calculated from the previous time. This gives a lot of numbers that later need to be post-processed to make meaning of them. An analytical solution, on the other side, has the great advantage that one has the rules written down right away in a simple form, and one can understand the behavior of a system on a piece of paper. Still, very easily differential equation systems become too large to be able to find such a solution. Haken is one of the very few achieving an analytical solution for a large set of neurons, virtuosic indeed.

He uses his synergetic computer model to show that such brain dynamics are able to identify patterns. The synergetic computer consists of many template patterns. Each template pattern is a spatio-temporal field, mathematically a vector changing in time, just like the spatio-temporal neural fields discussed above. The brain dynamics need to identify which pattern matches an input, something we hear or see. Therefore the brain is changing according to the input as long as the correct pattern is found. This works very well with a simplified neural model of neurons firing when an input threshold is reached.

The model is also able to deal with Gestalt perception. There the input is not the full pattern but lacks certain parts, like a melody where only the first few notes are played or a picture where some part is missing. We have already discussed this in the introduction chapter of music psychology, where binocular visual fusion, combining the input of the two eyes into one picture, was found with the cat to be a self-organizing process in the brain [94]. This can analytically be shown using the synergetic computer to work with simplified neural networks mathematically in an analytical way. Here again, it is a spatio-temporal field in the brain, which is the Gestalt, the conscious content.

Yet another example of a whole-brain model is that of György Buzsáki, again a self-organizing one [57]. He goes more into detail about how brain rhythms come into place and how they are built up in the brain. He is mainly concerned with the question of how we know where we are in space. If we walk through a building, we do not see the whole building at each moment. Still, we have a clear idea on which floor and on which side of the building we are. Buzsáki, like others, finds this spatial representation in rats and mice in the theta-band brain synchronization in their hippocampus. The hippocampus, or archicortex, has a quite regular array of excitatory as well as inhibitory neurons, the latter also called interneurons. Local neurons are locally connected, forming feedback-loops. Only very few connections to neurons far away exist. In nonlinear dynamics, such an array is called a small-world network and known to synchronize very distant areas of a network by only those few far-reaching connections. Mathematically one can show that such small-world networks are able to synchronize completely as a whole. It only needs such very few far-reaching connections to achieve that all elements oscillate with the same phase. Phase is meant here the same way as with sinusoidals discussed above. With neurons, it means different neurons at very different parts of the brain fire at the same time. We call also this a phase-synchronization.

Now the hippocampus has just such an arrangement, all neurons are connected to neurons right next to them in a very regular manner. Still, only a few have connections to far-reaching areas. Then a phase synchronization happens, which appears in the theta-band of oscillation. In experiments which mice, it is known for many decades now that this synchronized behavior is associated with the conscious experience of being at a certain place in space. Different spaces are represented by different neurons in the hippocampus being active, therefore producing a complex spatio-temporal field representing the conscious content, just like with the Friston model.

Practically all models of the whole brain are self-organizing models. They find all kinds of typical behavior of self-organization in the brain, synchronization, sudden phase-transitions, hysteresis, etc. Indeed, the brain is highly nonlinear and highly connected, therefore there can be no doubt that the brain is a self-organizing device.

In terms of perception and action the spatio-temporal fields building up for some milliseconds seem to represent the conscious content, the qualia. This makes sense in many ways, as it is not clear how a single-neuron activity physically would be different from any other single-neuron activity. A conscious content needs to be that differentiated than as different content exist, which is easily possible with spatio-temporal electromagnetic fields. Still, as consciousness is widely debated, we should have a closer look at some arguments and findings next.

Music and Consciousness

How consciousness comes into place is not really known and very much under debate until now. We distinguish consciousness from unconsciousness, the state where we sleep dreamless or where we are knocked out because of illness, a hard stroke on the head, by anesthesia, or medical treatment. Then there are states in between, where consciousness is fading away, like when going to sleep. There are states of mediation or other kinds of so-called altered states-of-mind. Psychoactive drugs are able to extend our consciousness and make things blur, flow, or very colorful.[72]

All these states of consciousness are of a different kind, and we also might find that all senses have a different kind of consciousness, where hearing is different from seeing, smelling, touching, or thinking. We can argue that what we call consciousness is a content of our consciousness, a sound we hear, a flower we see, or a thought we have. So then we can find all altered states of consciousness caused by meditation, drugs, sports also conscious content. When consciousness is fading, it might get darker around us (a visual cue) until we would reach such darkness that we call it unconscious. Strictly speaking, this could also be a kind of content, one of void, darkness, absence. Unconscious is also used when things in us are supposed to exist or happen without us knowing. They are not in our focus or our intention, speaking in phenomenological terms.[73] Again there is no content and, therefore, no consciousness. So from this point of view, consciousness is not a big deal.[74]

Still, consciousness is something we still find fascinating as we would not expect all things to have consciousness. We know that humans have it, and we assume that animals also feel and think to some extend. What about trees and flowers? Only a few people would find stones or a table having consciousness, and those who do are called esoteric. So there is something special about consciousness in contrast to unconsciousness.

But why do we need to discuss consciousness in terms of music at all? The problem is that of the difference between the brainstem, which historically is expected not to have consciousness, and the neocortex, where consciousness was traditionally located by most researchers. Humans, apes, mice, rats, and higher animals have a neocortex. In humans, it is located right under the skull as a layer covering most of the brain. Below this layer, additional important parts of the brain are located, the hippocampus, the thalamus, the midbrain, or the brainstem, next to others. Lower animals, fish, spiders, or flies do not have a neocortex, not a brain as we have.

In biology, there is a saying 'no brain, no pain', which means that animals with no neocortex do not have consciousness and therefore do not feel pain. This is especially important in fishing or when cooking crabs alife. Modern ways of fishing with very large nets, where the fish are squeezed to death even underwater, would be highly brutal when fish would feel pain. Indeed, recent research finds crabs to indeed react to electroshocks which they are exposed to, indicating a painful experience, meaning that they indeed feel pain, and, therefore, also have consciousness [76].

Returning to music, distinguishing a brainstem from a neocortex is a problem too. As we will see below, the human auditory system is part of the brainstem and the midbrain until the neural signals reach the primary auditory cortex, so the auditory part of the neocortex, called A1, through the thalamus. But until reaching A1, much information processing is happening, much detection of, e.g., sound localization is performed, and filtering happens. Not all information is actually reaching the A1. Still, we actually perceive and hear that filtered information, which indicates that also the lower parts of the auditory system are part of our listening experience and therefore are conscious.

Indeed, we cannot be sure about what consciousness is, but in the light of such an important issue in terms of music and listening, we need to go a bit deeper into the problem of consciousness and what it might be.

First of all, we associate consciousness in the brain. The brain is a neural network. But such networks also exist in the solar plexus and, as discussed above, in subcortical areas like in the auditory pathway, which ranges from the ear to the neocortex. A neural network consists of neurons that have electrical charges and are connected to other neurons via axons and dendrites, small channels that transport electrical charges. These electrical charges are ions, molecules that have more or fewer electrons than they have protons. Remember that in nuclear physics, protons have a positive elementary charge, and electrons have right the same amount of such a negative charge. Protons, glued to neurons, form the core of atoms and are surrounded by electrons. When an atom has the same amount of protons and electrons, the charges cancel, and the atom seen from outside has zero electrical charge. The most simple atom, therefore, is hydrogen with one proton and one electron. Atoms important for the charge within neurons are sodium $_{11}Na$ with 11 protons and electrons or potassium $_{19}K$ with 19 such protons, next to others. But both also exist with one electron missing, which makes them Na^+ and K^+. To build up an electrical charge in a cell, the so-called sodium–potassium pump. This pump works in the cell membrane and is a very complex molecule called sodium-potassium-ATPase. It transports three Na^+ atoms out and two K^+ atoms inside the cell. Therefore the amount of positive charge in the cell decreases, and the cell becomes negatively charged. There are many other kinds of such transport through cell membranes, still, this is the most important one.[75]

In the end, neurons have an internal negative charge of about 70–100 mV, quite much for such a small cell. When a cell is triggered by other cells sending their electrical impulses to it, the cell will discharge and send its charge to other cells it is connected to via its axions. This sudden discharge from about 70 mV to around 0 mV is called a spike. It is the activity of the neuron triggering other neurons in the neural network.

As all neurons in the brain work like this, produce spikes after being triggered by other cells and again trigger other cells with its spikes, the whole system forms a large network where every cell is connected to all other cells directly or indirectly. The amount of influence of a cell onto the cells it is connected is very different. It might be large or small, it might be positive or negative. When it is positive, a target cell will send its spikes faster when more incoming spikes from other cells arrive. But this may also work the other way around. A cell might be inhibited to send out spikes the stronger it is triggered by connected neurons. These two processes are crucial for a working network. The excitatory process enhances the activity, while the inhibitory process leads to a selection of information and of information reduction.

The amount of how much one cell triggers the others might change too, which is called the plasticity of the cells. Cells are able to change the amount of how much they are influenced within minutes, seconds, or milliseconds, which seems to be crucial for learning. If one learns a melody or a rhythm, we expect some neurons to change their plasticity and therefore learn. Plasticity might also be the formation of new connections between neurons, which also often happens, especially if we learn new things. Still, this process is much slower than changing the influences between existing connections [153].

So far, the traditional picture of the brain and how it basically works as an interaction of neurons with spikes and changing plasticity. Below we will discuss advanced models, which go into the details of neurons and are based on quantum mechanics. Still, for the moment, we will stay on the neuron level to follow our discussion about consciousness. If this discussion needs to be altered when including quantum mechanics in the picture is not clear yet.

Before we get deeper into details of how the brain processes music, we need to go back to our consciousness problem to address the problem if also lower brain areas, those from the ear to the neocortex, might also have consciousness. To do so, we need to take the physical picture of the brain.

In physics, there are four known basic forces all things are made of, the electromagnetic force, the strong and weak forces, and gravitation. Modern physics also discusses dark energy and dark matter, as with the existing forces, many experimental findings in cosmology, like the speed galaxies rotate or the speed galaxies, far away from us, travel, cannot be explained with the gravitational laws we know today. Still, these problems only appear on very–very–large scales, stars, and galaxies, and are not really expected to be relevant for such tiny things as the earth or even our brains.

To determine the cause of consciousness, it seems to be enough to scan the existing four forces. Gravitation may be excluded, as also astronauts only subject to very small gravitational forces have consciousness too. The strong force is the one gluing protons or neurons together and is only relevant in nuclear fusion or fission, a process found in the sun or nuclear power plants but not really in the brain. The same holds for the weak force, which also does only appear with β-radioactivity.

So from this point of view, the electromagnetic force is the only left to form consciousness. Indeed, when we measure brain activity, e.g., with an electroencephalography (EEG) measurement, all we measure is electricity the brain is producing. There are also ways to measure the activity of single neurons, which is done by electrodes put right next to single neurons, they also measure electricity. This is not surprising, as electricity is the way neurons work, building up electrical charges of about 70 mV and spiking, so transferring these charges to other cells.

But we can go even deeper into all kinds of chemistry happening in the brain, which all is exclusively built on electrical changes. Atoms and molecules do only interact with each other due to such forces. Two molecules charged the same way dispel and move into opposite directions. These movements are possible because the molecules have mass, and therefore have inertia, so they can be accelerated or slowed down. Indeed, inertia (a mechanical feature) and heavy mass (a feature of gravitation) are the same. Gravitational force means that bodies are attracted only because of their masses like the earth is forcing all things towards it. But these forces are so weak that they do not play a role when two molecules are close together. The electrical force is much, much larger than the gravitational one.

So, in the end, to explain consciousness, we are left with the electromagnetic force alone. But does this mean consciousness is electricity? If so, everything would have consciousness. This sounds esoteric and no longer scientific. Still, if we assume consciousness not to be electricity, then we are forced to find another reason for it, something which would then be beyond physics. But if we follow this line, we are indeed esoteric! We might hope that there is another physical force not yet found, still, as we are able to find so many conscious contents, like seeing, hearing, thinking, etc. as activity of the brain, measuring electricity, and as consciousness is present to us every day and all day, so prominent that we would expect it to be strong and therefore detectable with our physical means, it is very unlikely that there is a physical force not yet found explaining consciousness.

So if we stay on scientific grounds, it is reasonable to at first follow what is already there, the electric ground for consciousness, and see how far we get.

One main idea of how consciousness comes into place in a brain is that the brain is in a medium state of synchronization. Interestingly, when measuring the activity of neurons all over the brain at the same time, we find that many neurons spike at the same time. They are synchronized. Furthermore, many neurons spike regularly, often around 40–50 times per second, which is at 40–50 Hz. In brain research, this is called the gamma band associated with brain activity, thinking, and perceiving. Also, long-range music perception seems to correspond to enhanced gamma-band oscillation [38]. Other bands are the alpha band from about 1–8 Hz and the beta band between 8 and 40 Hz. So in all bands, there is a regular spiking or oscillation. Still, the gamma band is the one found when the brain is active and conscious.

Practically all regions of the brain fire more or less in all bands all the time. But when we are thinking, the gamma band is most active. Still, also this activity happens more or less in different brain regions. But what is even more interesting is the phase in which these different brain regions oscillate or spike. It is like with a musical rhythm pattern. If neurons in the auditory cortex, the one located at both sides above the ear, are spiking with 40 Hz, and the visual cortex, mainly located at the back of the head, is spiking with 40 Hz too, this does not mean that all neurons fire at the same time. It may be that the neurons in the auditory brain fire right in between those of the visual cortex. Then we say they are in anti-phase. Here we use the same terminology as with acoustic waves, where a cycle is one oscillation, here the time interval between two spikes. Then within this time interval, we can be right at the start, in the middle of it, or at any other time point within this time period, so at a certain phase. Phases are again taken to be a part of the circumference of a circle with radius one, so they are 2π. So when the neurons of the auditory cortex and the visual cortex are in anti-phase, the auditory neurons fire at the phase π of those of the visual cortex, right in the middle. They could also spike at $\pi/2$, or $3/2\pi$, etc.

This phase relation between different brain regions can be smaller or larger. If many neurons in the brain spike with the same phase, at about the same time, we have the state of epilepsy, also called hypersynchronization.[76] In such a state, the normal brain functions no longer work, and people lose their consciousness. They start to shake, as also the motor cortex, the brain region triggering the movements of the body, located at the top of the head, is perfectly synchronized to all other brain regions, and as all cycle regularly, the motor neurons send out the signals to the arms and legs to move in this oscillation.

So epilepsy is the state of nearly perfect brain synchronization—and a state of no consciousness. Still, there might be an oversynchronization of certain brain areas, e.g., between the thalamus and the cortex, called thalamocortical dysrhythmia, or hyperactivity of the limbic system. Then people are still conscious but show strong depression. Depressed people often have a reduced field of consciousness, of what they remember of the past, visualize for the future, and see as their social and personal space in the present. They do not think about possibilities for future planning or remember good things in their past. They forget possible relations with friends or relatives and do not take possible actions into consideration, which they might do to improve their situation. This seems to be reflected in an over-synchronization of the different brain parts, which all oscillate too simple, not much differentiated. Of course, this relation is only an analogy and not a scientific proof.

Of course, there needs to be more evidence about the role of global or local neural synchronization in the state of consciousness. But synchronization between different brain regions has also been found when perceiving music. Techno is a musical style where the DJ is building up a texture of sounds and beats that become denser and more intense, building up musical tension. Dancers expect such tension raise, which happens dozens of times during a Techno night, and react on it happily awaiting the climax of this tension raise, the moment where the Bass Drum comes in, and the four-to-the-floor beat starts. In an EEG experiment where listeners were played such a song of Electronic Dance Music (EDM) measuring the gamma-band oscillations of the brain during these tension build-ups, a strong synchronization increase of different brain regions happened until the climax was reached. After the climax, the brain regions de-synchronized again only to build up again with the next tension raise. Of course, the synchronization was never complete, otherwise, people would fall in epilepsy [120].

That external visual effects can cause such epilepsy is known from music videos nowadays often found in Hip-Hop or in Heavy Metal music, where a stroboscopic light effect is present all through the song.[77] Before the video, a warning is added for people of potential epileptic tendencies to warn them watching the video. Here it is a visual effect regulating the amount of synchronization in the brain.

Global brain synchronization was found important for attention, expectation, memory, or certain tasks. Baars describes this as a 'theater of consciousness' in this Global Workspace Theory. Conscious content like imagined episodes of life, fused senses of auditory or visual input when visiting a music performance of watching a movie are found to be caused by synchronization of many brain areas [9]. The visual binding of the input of the two eyes into one picture is found to take place when neurons in the visual cortex synchronize [94]. Towards time points of expectation, like waiting for a traffic light to become green, several brain areas increase their synchronization [56]. Also, synchronization of cortical areas in case someone hears music or any other sound was found [166]. Long-term and working memory seem to interact through neural synchronization [84]. The list could be continued.

There seems to be a relation between the oscillations in the brain and conscious states, and even to consciousness itself, as this is gone with full synchronization. As the spiking is nothing but a changing electromagnetic field alternating in space and time, one idea of consciousness is that it simply this electromagnetic field. We measure this field with EEG or other techniques, we can alter it by listening to music, watching a video, or applying an artificial electromagnetic field to the brain, and we find states of consciousness, tension, attention, expectation, or memory to be represented by it. So why not simply accept that our normal states of everyday-life consciousness is a medium synchronization of all those neurons, and the endless amount of different sounds we hear, things we see, what we might think, hope, or feel are just different kinds of synchronization and oscillations in the brain.

This view is indeed attractive, as we see that total synchronization means a loss of consciousness, we find dreams to be states of mind which are different from daytime states, and indeed the oscillations and synchronizations are different in both kinds. We also can understand that to obtain consciousness, we need really many neurons doing the right things, and therefore smaller neural networks cannot have such conscious states. And we need not assume that other electromagnetic fields present in tables or stones might be conscious too because they would simply be too primitive to have such conscious states humans have. We do not need to assume that computers have consciousness, something sounding strange to all using them, as our present computers use a single processing chip operating serially one operation after the other, an electromagnetic field much too primitive to be conscious. Then the task would be to go into more detail and find out what electromagnetic field is which conscious state.

But what we can do is to return to our discussion of how music works, and especially to the question if subcortical areas, all the brain regions starting from the ear up to the auditory cortex in the neocortex, are conscious or not. We can argue that we would not expect them to be able to think about complex problems or to plan future actions, but taking their complexity and synchronization into considerations, as well as their many connections from left to right hemisphere on many levels, as well as the bottom-up and top-down process, conscious content, indeed sound color or pitch perception, can easily happen within these neurons. We will get into more details about this below and see if this can be justified even more.

There are other theories about consciousness based on quantum mechanics, which assume that consciousness is a quantum effect that appears when a certain amount of quantum fluctuations have happened.[78] As the brain is very active with a lot of neurons, these theories calculate that all 20 ms a conscious state will appear. This corresponds to a frequency of 50 Hz, which is a center frequency of the gamma band of brain activity we have already found to be present during many psychological phenomena. Still, the 50 Hz can also be explained by the physiology of connected neurons and the chemical processes happening.[79] Some claim that the time a neuron needs to refresh after it has spiked is the reason for the 50 Hz, so the time it needs for the chemical processes to build up the 70 mV in the cell after spiking, the sodium-potassium pump. Some assume that it is the number of connections that leads to this 50 Hz. So there need not necessarily be a connection between this frequency and a quantum phenomenon.

Still, it is interesting to see that many molecules within a neuron act as quantum objects rather than just like classical objects [228, 229]. One such quantum phenomenon is that a molecule cannot be in arbitrary states but rather has so-called eigenstates, states with certain frequencies. Mathematically this is the same as what we know from Acoustics, e.g., with strings of guitars or violins. A string is vibrating with a certain pitch, which again consists of a harmonic overtone spectrum, as discussed above. The fundamental pitch of a low A-string of a guitar is 110 Hz, its first overtone is 220 Hz, its second overtone is 330 Hz, and so on. These are the frequencies in which the string vibrates. It does not vibrate with 150 or 200 Hz. It only does so with its own frequencies.

Now a similar thing happens with molecules on a quantum mechanical level. They have certain eigenstates with eigenfrequencies they are in. The analogy to the vibration of a string is the vibration of an electron. Classically, an electron is a particle that has a position in space at a time point and moves through space. Quantum-mechanically, an electron is a vibration that is present around the kernel of an atom and extends over an area. Indeed, theoretically, it has no boundaries at all, and each electron is distributed over the whole universe. Now the electron can only vibrate in the frequencies the atomic kernel allows it to. As the kernel has many eigenstates, an electron can be in one of them and can jump from one to another. The higher the state, the higher the energy the electron is in. If an electron jumps from a lower state to a higher one, it loses energy, which is radiated through a photon, a light particle away from the atom. This is the light we see emerging from bodies. So the quantum world is different from the classical, where quantum objects, like electrons or atomic kernels, are not particles but objects 'vibrating' in eigenfrequencies and eigenstates.

Every-day objects are indeed particle-like, we exactly know where they are. They do not appear to us like vibrating clouds that stretch endlessly into the universe. Indeed, quantum mechanics only appears at very small scales, so, e.g., a hydrogen atom as a radius of about 1 A, which is 10^{-10} m, a tenth of a nanometer. 1000 nm is $1,1000\mu$ is 1 mm, so 1 A is quite small indeed. Still, quantum theory does not give a boundary for the transition from the quantum world to our classical world. Indeed, it does also give no reason for this transition, one of the unsolved problems of quantum mechanics. And although traditionally physicists do expect molecules, consisting of about a dozen atoms, no longer behave like quantum objects, recently, more and more cases are found where quantum phenomena are reported on larger objects. One example is Rydberg atoms, which have such high eigenfrequencies of the electrons that these electron clouds are still strong at distances of nearly 1 mm [96]. Then the atom is about one micrometer big and still behaves quantum mechanically. So there is evidence that also larger objects are still in the quantum world. Another example is fullerene balls, molecules consisting of 60 carbon atoms, which also show quantum mechanical behavior, despite their large size [6].

Neurons consist of many different molecules, among them are microtubules built out of different amino acids arranged in a double-helix. They are part of the so-called skeleton of a cell, the cytoskeleton (from cyto = cell), so the cell skeleton. They stabilize the cell, but they also do transport other molecules through the cell, which is important for the electrical abilities of neurons. There are certain motor proteins, proteins that move, which are attached to the microtubules and move them through the cell.

Each microtubule consists of about 300 atoms. Molecules so large are not expected to act quantum mechanically as a whole but indeed microtubules do [33]. They can be brought into higher eigenstates by applying certain specific voltages to them. If a microtubule is in such a state, it stays in it even when applying different voltages to it. Then it acts as an amplifier with a certain amount of amplification. This microtubule can be brought into its ground state again by applying a certain specific voltage, again with opposite polarization. Then with another specific voltage, it can be brought in a higher eigenstate, where it again stays until it is released from this state again. So these molecules have a memory, and it is under debate if they might act as part of our memory system.

Microtubules act as amplifiers, as we have seen. When applying a voltage to one end of it, at its other end, a higher voltage is present. They manage this by using charged ions attached to them from the sides, using the voltage of these ions [267]. Such behavior is known from electronics, a transistor acts like this. Electronic circuits have three basic elements, transistors, capacities, and resistors. Another part of neurons is acetin, which is also part of the cytoskeleton and is a stiff structure, not moving like the microtubules. Electrically acetin acts like a wire with resistance, it is a resistor. Neighboring wires of acetin then act as capacitors. So within each neuron, all elements necessary for an electrical circuit are present, again there are millions of them in each cell. So it is possible, although not proven yet, that each neuron is an electrical circuit on its own.

Furthermore, the cytoskeleton is not fixed at all times. Microtubule-associated proteins (MAP) are molecules sticking microtubules together, fixing the cytoskeleton. But these MAPs may disappear in a cell and therefore make the microtubules move, which is necessary for the basic operations of the cell. Then they reappear again, sticking the microtubules together, but then in a different configuration. If cells also act as electrical circuits storing memory, then a rearrangement of the microtubules would mean a change in the memory content. Indeed, in many cases, such rearrangements have been found. One case is reported where the neurons in the auditory cortex act like this when the task is to learn that a certain frequency is good to have in mind [268]. In the cells, the MAP proteins did disappear only to reappear after a certain time when the learning process was over.

Investigating the electrical and quantum mechanical microstructure of cells is a recent scientific branch called Nanoneuroscience. It is open where it will lead to, still, it is a promising field. When each neuron is an electrical circuit on its own, then the capacity of the brain is heavily enlarged to an extent not to be estimated yet.

To again come to our discussion on consciousness, if consciousness were a quantum phenomenon, it would work with photons and electronic states, which is right what the electromagnetic force is we have discussed above. The differences would only be on the precise mechanism, but it would still be the same physical realm of just these electromagnetic interactions. So nearly all research of today dealing with consciousness is using the idea of electromagnetism as the basis of consciousness.

As we now have found that subcortical brain regions are also expected to have consciousness, it makes sense to now discuss the physiology of the ear up to the auditory cortex in terms of many musical parameters like pitch, timbre, rhythm, spatial listening, and many others, expecting that the perception of these parameters would be here. Of course, the neocortex plays an important role in music perception too, and we will discuss this in the following. But there, we will see that the connections become more and more associative and no longer too hard-wired. The abstraction of auditory events in the neocortex might help in music theory, still, the reason we hear music for most people is the sensation of timbre, rhythm, melodies, etc. Without the phenomenon, the analyzed musical structure is indeed comparable to a bus schedule, telling us when and where each event in music happens. This will lead us to fundamental discussions about meaning and emotions in music and how they are related to cultural, social, or political aspects of music. Still, to understand this, we need to discuss the ear first.

Reconstructing Impulses—The Ear and the Auditory Pathway

The way the acoustic sound wave, entering the ear, is processed until it reaches the neocortex in the primary auditory cortex is indeed very complex and not yet understood in full. Basically, the waves travel through the outer and middle ear, enter the inner ear through the tympani, and is transferred into electrical neural spikes in the inner ear, the cochlear. Then it goes through several neural nuclei, regions of neural cells, like the cochlear nucleus right behind the inner ear, the trapezoid nucleus, the medial geniculate, and several others.

These neural nuclei are seen as closed elements from a physiological standpoint. If one investigates the auditory pathway, the way from the inner ear to the neocortex physiologically, one finds these nuclei, which are connected by axons going from one nucleus to the other. This is the physiological reason for identifying several stages in the auditory pathway. But there also is a neurological one. Most investigations on hearing are done with animals that are supposed to have very similar auditory systems compared to humans, like mice, rats, cats, or owls. Still, György von Békésy (1899–1972), the first to have correctly understood the mechanism of traveling waves in the inner ear, at the end of the 19th century, where in 1961 he got the Nobel Price for, was making his investigations at the human cadaver [39]. Therefore he needed to investigate the inner ear right after the death of a person because only after about an hour after death, the inner ear does not work the way anymore when people are still alive.

When investigating neural connectivity, there are several methods to visualize the activity of neurons and their axons, so the connection from one neuron to another in the brain.[80] Next to the physiological findings, this neurological connectivity is another source of information if and how one neural nucleus is really interacting with another nucleus. It might be that there is a physiological connection, but it is not really used to change the sensual input neurologically. This can be caused by neurons acting in an inhibitory way, as discussed above.

Indeed, nearly all parts of the auditory pathway are connected one to another, and they are interacting in both ways, up and down, from the inner ear up to higher neural nuclei, and from the neocortex down to lower ones. These bottom-up and top-down processes are happening at the same time, forming loops. These loops can trigger other nuclei and form more loops. In the auditory cortex, several of these loops have been identified with different operations [233]. Indeed, to go from the inner ear right up to the neocortex can be done by only one or two neurons in between. This also holds the other way around, the neocortex sends neural signals down to the inner ear in several ways, the fastest is also by two neurons within this top-down process.

As these connections are on nearly all levels of the auditory pathway, and as many loops have been identified, more and more scientists do no longer use the terms bottom-up or top-down but speak of a complex system, which is self-organizing when an auditory input is present. Neurons spike and trigger other neurons, which again spike to third neurons, which might be connected to the initial spiking neuron. Then the loop is closed, and the process is self-sustaining. After a while, a pattern appears, which is a balance within the neural network, a self-organization, and maintains itself for some time. It might change or break down because of internal or external reasons. But within some time, it is able to maintain a certain activity. Therefore it produces a complex temporal and spatial electromagnetic field, a pattern. From the discussion above about consciousness and conscious content, it can be expected that this self-organized electromagnetic field is the conscious content, the sensation of sound, pitch, rhythm, or other musical elements.

This means that an input of acoustic waves in the ear will not only be present when it is there, but it will also prolong a bit longer, a thing we call memory. Of course, when a piece of music ends, we clearly hear that it is over. But we also hear some melodies, some timbres, or rhythms. These memories might not be as clear as the original, immediate sensations, but they are clearly there. So the brain needs to maintain the sounds, which need to be prolonged spiking of neurons in neural networks. The self-organization of these neural networks allows such memory.

This is also called short-time memory, lasting about 3–5 s. As the cause of this memory seems to be the echo of the auditory input in the brain, it is also called echoic memory [238]. Such a sensation will fade away after a while, and some seconds later, we might remember that we have heard a musical piece, but it becomes harder to still remember details of timber or melodies. In a phenomenological view, short-term memory is also discussed as now-points (Jetzt-Punkte) [103].

Long-term memory needs neural plasticity, a lasting change in the brain. As far as we know today, long-term memory is caused in two ways, by the change of the coupling strength between neurons as well as due to the building of new connections between neurons. Such a change can be caused by altering the interaction strength between neurons. This happens within milliseconds. It can also mean the built-up of a new axon. This needs at least three presentations of the same stimulus [153]. We will come to long-term memory later.

For now, it is interesting to see that we hear not simply the input sound from outside but that we take this sound and act on it, the neural network is going into self-sustained states.

These states are also actively working on the input, according to our wishes and needs. When we listen to a piece of music, we can concentrate on the guitar line, the vocal text, the bass line, the drums, the timber, or many other aspects of the music. We focus our attention on one or the other parameter. Although it is not clear yet what attention is in terms of neural activity, we know that the brain can actively suppress information to enhance others. Theoretically, the neocortex, deciding to want to listen to the bass line of a musical piece, could act from the neocortex right to the inner ear and tune it in a way to suppress higher frequencies and enhance lower ones. This action could take place not only neuronally but indeed acoustically by stretching the basilar membrane in the inner ear. We will come to this below. Indeed, until now, there is no experimental evidence that this is happening just because one cannot make measurements in the inner ear of humans while listening. Still, from a neuronal standpoint, this is indeed possible.

Until now, we only did discuss the auditory pathway up to the primary auditory cortex, called A1 in the neocortex, where traditionally consciousness is meant to be. Of course, there are many other parts in the neocortex where certain tasks associated with music happens, the Broca region responsible for musical syntax [157], the cerebellum found to be active in the detection of musical rhythm [248], the front brain when it comes to texture and rhythm identification or the supplementary motor cortex when motion action is proposed [207], like dancing or playing an instrument. Still, it is also debated if these and other parts are really to the core of music. So if we extract a musical texture, a score with melodies and rhythms in time and space, all that we are dealing with are events happening in time and space. There is a saying in musicology that this also holds for a bus schedule, events in time and space. Indeed, these events are supposed to cause a musical piece and show its organization. But a score itself is nothing but a piece of paper with dots and lines on it. We would never be interested so much in such a sore if it would not be put into sound.

Nevertheless, these elements are part of music, and we will come to them below. But we should start at the part where music perception begins, the ear. There is very much to be said about the ear and the auditory pathway, and there is even more still unknown today. But it is fascinating to discuss some processes very well understood.

One central neural process of much importance in the whole brain is neural bursts. Very seldom, only one neuron spikes along. Most often, it triggers many neurons around it, which spike at about the same time, producing what is called a neural burst, many neurons spiking synchronized at about the same time point. These bursts repeat in time, maybe once every second, maybe 50 times per second. When measuring the brain with an EEG, we cannot measure single neurons, but we measure the electrical field of many neurons on average. So when neurons burst with a certain frequency, say 50 times per second, which is 50 Hz, then in the EEG, we find an oscillation at 50 Hz. This is the gamma band of neural activity discussed above.

We have already discussed the synchronization of different brain regions. This synchronization means that the time points two or more bursts at different regions of the brain spike are the same. So every single burst at, say, the auditory cortex might happen at exactly the same time point as every single burst in the forebrain. Then both regions are synchronized.

The synchronization we talk about now is found deeper. It is the synchronization of all neurons within one single burst. So this synchronization is producing the burst at all. If each neuron fired at a different time, there would not be a burst but a continuous spiking of neurons. Only when all neurons spike at the same time such a burst appears.

Indeed, these bursts are found all over the brain, and they are also found at the output of the inner ear, where the acoustic wave is transferred to the electrical spike. So we need to get closer to the mechanism of the inner ear to see why the ear produces such bursts. We will indeed find that most music is enhancing such bursts, but also that the ear is indeed built to even enhance the degree of synchronization, making the bursts more precise. This is called coincidence detection.

Transfer of Acoustic Electricity into Neural Electricity: The Inner Ear

The acoustic wave traveling from a musical instrument to the ear is a changing pressure in the air. It changes in time, and it is different at different places in space. When it enters the ear, it is modified when it travels through the outer and middle ear. The outer ear is collecting the sound using the ear conch or pinna. Then the sound travels through a channel until it reaches the eardrum. The channel is a tube, like a very short flute, which would vibrate on its own around 3000 Hz. This means that frequencies in the incoming sound around 3000 Hz can travel through this channel much better than the other frequencies. So around 3000 Hz much more energy arrives at the eardrum and is much louder to us. So for these frequencies, we are much more sensitive than for others.[81]

Another aspect of the outer ear is that women have a considerably smaller pinna than men. This was already found by Justinus Kerner, we already had to do with. In his Ph.D. of 1805, he was investigating the hearing of humans and also of animals and was the first to find a difference there [108]. Indeed, in loudness investigations using music as input (rather than sinusoidals or noise, used in most experiments in loudness) it is striking to see that women find music of the same sound pressure level, the same physics, considerably louder than men [223].

The sound then hits the eardrum and moves it back and forth. The eardrum is connected to three bones, which transfer the sound to the oval window of the cochlear. These bones are connected to a series of muscles that relax when the incoming sound is loud. By relaxing, they transfer energy from the eardrum to the inner ear become less, and therefore we are protected against noise too loud. Still, this relaxation of the muscles is not too fast, and therefore a sudden rise of the pressure will still damage the ear physiologically.

Now through the oval window, we enter the most astonishing part of the ear, the inner ear, or the cochlear. It is a spiral of about three rounds becoming smaller towards its top. Altogether it is about 3, 5 cm long. This is determined by the hole in the surrounding bone it is embedded in. The cochlear consists of two main channels which are both nearly round and part of the bone structure, an upper and a lower channel. Between both channels, there is a small gap of about 1 mm. Within this gap lies the so-called basilar membrane. As this basilar membrane wounds up through all three spiral rounds, it also is about 3, 5 cm long. But it is only about 1 mm wide and therefore more a rod than a membrane.

On this basilar membrane are the hair cells, called stereocilia, which transfer the mechanical energy of the incoming sound wave into electrical signals. Therefore they are also called sensory cells. There are two kinds of sensory cells, the inner hair cells, which are mainly transferring information from the cochlear to the brain, and the outer hair cells, which react to information coming from the brain. Indeed, already here, we have the bottom-up and top-down processes discussed above. In a minute, we will see how they are forming a nervous loop.

One might expect that the process of transferring acoustic waves into neural spikes would be one of transferring mechanical energy into electricity. Still, mechanics is nothing else but electricity too. Sound waves have regions where the particles of air are closer one to another and other areas where they are further apart. The forces which decompress the areas with more dense particles are electrical forces pushing the molecules apart. Therefore the transfer in the cochlear is not a change in the fundamental force. It is a change from molecular forces to ionic ones, as the transfer of electricity in the brain works with ions and not with an electrical charge like in electrical wires. But even more important, it is also a transfer of the impulse shape, as we will see now.

The inner and outer hair cells are called this way because of their place on the basilar membrane. Still, they are considerably different. The inner hair cells, placed on the basilar membrane, are also connected on their top to another membrane, the tectorial membrane, which is above the basilar membrane. The outer hair cells are free on their top. This leads to several differences. Among them is the reaction of the cells to very low frequencies below the hearing threshold, so below 20 Hz, called infrasound. Such sounds are produced by wind farms, which transmit sound at frequencies between 1 and 20 Hz mainly. The inner hair cells do not move with such frequencies, as they are connected twice to membranes, which both are moving about the same with very low frequencies. Still, the outer hair cells are only connected to one membrane and therefore move. This might be the reason why people around wind farms often feel seasick. The outer hair cells tell them that they move slowly while their eyes tell them they stand still.

Now the cochlear has a very special feature concerning the movement of the acoustic waves which travel through it. The pressure entering into the cochlear is increasing the pressure in the liquid (perilymph and endolymph) inside the cochlear. Lymph, like water, is very hard to compress, it is a hard medium. The harder the medium, no matter if it is matter or a liquid, the higher the speed of sound. The speed of sound in air is about 343 m per second. This is not too much and definitely much slower than the speed of light. Therefore, when we see a person standing far away making noise, it takes a while until this noise is heard, although we might see immediately that this person is making that noise. The slow speed of sound in air also makes echoes possible, and we are able to estimate how far a lightning was away by counting the amount of seconds it takes between the lightning and the thunder we hear. When it took about five seconds, the lightning was away five times 343, which is about 1, 5 km.

Now the speed of sound in water or lymph is about 1500 m per second, five times more than that in air. As the cochlear is only 3, 5 cm long, the sound takes only about 0.023 ms to travel through the cochlear from bottom to top. Therefore in the lymph of the cochlear, the pressure is the same all through. Of course, it changes in time, but at each time point, the pressure acting on the basilar membrane is the same throughout its length.

In contrast to the very fast speed of sound in the lymph, the speed of sound on the basilar membrane is very slow and not constant. The basilar membrane is like a soft tissue, quite like some meat. It is not moving much, and when hitting it, the movement is very fast gone. The speed of this movement depends again on how hard, or here we better might say soft, the membrane is. The trick of the ear now is that this membrane is soft at its beginning at the oval window where the sound comes in, and it is getting even softer when spiraling up to its end. Then at its beginning, the speed of sound on it is about 100 m per second, while at its end, it is only about 10 m per second. So the waves traveling on it are slow and even do slow down. Furthermore, this membrane is heavily damped, as we know from meat or flesh. When it is brought into movement, this movement very shortly dies out. And indeed, when the lymph stops acting on the membrane, making it move, it stops moving nearly immediately. So it does not act as a string of a guitar which moves a long time after it has been plucked, no, its movement is very fast gone.

All these components are necessary to make it possible that this system is sorting the sound coming in, splitting it into its single frequency components. The slowing down of waves on it leads to a steepening of a wave with a certain frequency up to a point on the membrane where it is most steep. Beyond this point, the movement suddenly breaks down (Fig. 1). This point on the membrane with the largest displacement is different for different frequencies. Lower frequencies have their highest peak towards the top of the spiral, higher frequencies have it towards its beginning. At the middle point of the membrane, a frequency around 2000 HZ has its place, its best-frequency as it is called. With this mechanism, the inner ear, with a purely mechanical system, is sorting the single frequency components of a sound by placing them at different positions along the membrane.

The mechanism of steepening of waves towards a place on the membrane is the result of the three features, the slowing down of the speed of waves along the membrane, the heavy damping of waves on it, and the movement of the membrane by the lymph, all along its way at one time with the same strength. With only one feature missing, it would not work.

The transfer of this mechanical (electrical) energy into neural (electrical) energy, the spiking of the nervous cells on the basilar membrane, the stereocilia, is a consequence of the membrane movement. The inner hair cells appear in bundles on the membrane, which move the same way. They are not standing perfectly straight on the membrane but at a certain angle. On their top, they are connected by so-called

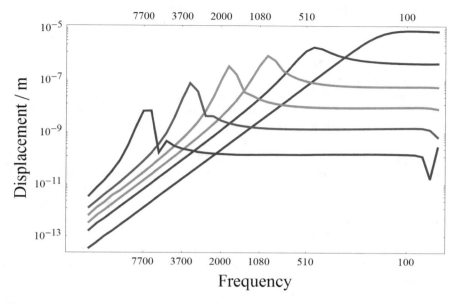

Fig. 1 A sound wave with a certain frequency is making the cochlear move such that for each frequency, a maximum vibration appears at different positions on the cochlear. In the figure, the cochlear is unwind from its spiral into a linear rod. The different curves show how much the cochlear vibrates over its length when it is driven by different frequencies. Each frequency has an amplitude maximum at a position called best-frequency (from [20]))

tip-links, very small molecule threads where we are now on the scale level of only several micrometers, near the nanometer scale. Now when the membrane moves, the hair cells next to each other move left and right in parallel. When they move in the direction of standing up, the tip-link is relaxed. But when they move in the other direction, this tip-link is a thread too short for the distance between the tips of the stereocilia, as these hair cells are sheared now against each other. So this small thread tears on the cell wall of the stereocilia, opening a very small gap. In this gap, ions from around the stereocilia flow into them. As we have seen above, this changes the electrical charge of the cells, and the cells depolarize. They spike and send this spike at their lower end through an axon to the next nervous cells outside the cochlear. Then the transition between acoustics and nervous energy has happened.

How Do We Hear a Pitch

So the continuous pressure change of a sound wave is transferred into a discrete signal of single spikes. When a sinusoidal of say 100 Hz is entering the ear, the resulting spike signal would be one spike at each cycle, where the basilar membrane is steep in one direction causing this spike. Of course, there are so many inner hair cells around each place where the membrane is most steep, and all of them will spike, resulting in a nervous burst. We have found such bursts happening all around in the brain, and so are not too surprised to find them here again. So a frequency of 100 Hz will lead to 100 bursts each second.

The issue becomes more complex when a musical tone of a piano, guitar, or violin enters the ear. These sounds have many frequencies, which are all harmonic one to another. There is a fundamental frequency of, say, again 100 Hz, which comes along with its higher partials of 200 and 300 Hz, etc. So the 200 Hz partial will act on another place on the basilar membrane, causing spikes to leave from there. The 300 Hz partial again has another best-frequency place and so on. The 200 Hz partial is twice as fast as the 100 Hz one and will therefore produce twice as many bursts as the 100 Hz sinusoidal. With most musical instruments, the partials start at the same time point, have a phase relation of zero, and therefore the pattern of the 100 and 200 Hz partials together will be such that first, both sinusoidals spike together, then only the 200 Hz spikes on its own, then both spike together again, etc. The more partials there are, the more bursts will be produced, still, as all partials of a musical sound are harmonic, they all will spike at one time-point altogether. So there is one very large burst of all partials followed by many smaller ones. The first burst is distributed over a large range on the basilar membrane, and therefore over a large range of different nerve cells, the others are only present here and there.

Therefore a harmonic overtone structure produces one large burst at the fundamental periodicity of the played pitch, 100 Hz in our example.

This is not happening with inharmonic sounds or noise. Again there might be very many frequencies present in the sound, causing different parts of the basilar membrane to spike, but these spikes are irregular in time, causing no clear burst at one time-point. Indeed, with such sounds, we do not hear a pitch, there is no fusion of all partials into one pitch sensation. If there is an inharmonic sound with only a few partials, we perceive all of them solely, so no fusion into one single pitch happens. Of course, if the partials are too many, like in a sound of flowing water, there are too many partials to be able to distinguish single frequencies. Still, there is also no pitch sensation with such sounds.

So we might say that when we hear a pitch, this happens with a regular strong bursting of spikes in time. But there is one problem. What if not all partials of a harmonic musical pitched sound start at the same time. What if there is a phase shift between them. Indeed, when we produce such sounds artificially, we do not hear a difference between them when they are synthesized to sounds that do not move. When the sound changes in time, we hear such phase differences very well. Still, we cannot hear them when the sounds are not changing.[82]

It is interesting to see that indeed this is not a problem at all, as the ear, as well as several nervous nuclei right behind the ear, synchronize partials with different phase in time, so that again, one strong burst is present. Within the ear, this process is automatic and part of the transition process from the acoustic wave into the nervous spike. When two frequencies are incoming, which are harmonic but out of phase, the traveling wave moving over the basilar membrane makes one slow down while the

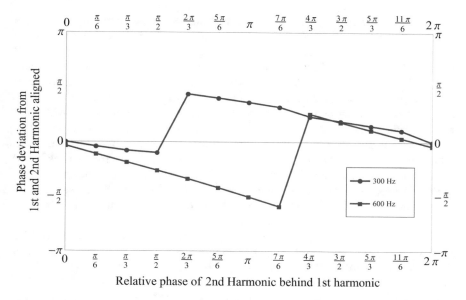

Fig. 2 Here is an example of two sinusoidals synchronizing in phase during the transition of a sound wave into nervous spikes in the cochlear. In a computer simulation, a cochlear was built as a physical model. Then two harmonic frequencies, 300 and 600 Hz were used as input to the cochlear. At first, both were in phase. Then over the range of π, the higher phase was shifted with respect to the lower phase. The plot shows the output phase of the two spikes leaving the cochlear for the different phase relations between the 300 and the 600 Hz sinusoidals. One would expect the 600 Hz phase to go through the π phase, as it does, and the 300 Hz phase to stay at zero (on the horizontal line in the middle). Still, the 300 Hz phase tries to closely align with the 600 Hz phase, it synchronizes to it. This is the first mechanism of coincidence detection, where phases in random order are aligned to form one strong impulse, one strong spike burst again

other speeds up to synchronize such that they nearly spike at the same time point. This is due to the interaction process between the waves and the spiking, which is highly nonlinear. The nonlinearity here is caused by the sudden spiking at one time-point which, is not a continuous process but rather a very sudden change. Then the spikes leaving the cochlear are again nearly perfectly in phase again one to another (Fig. 2).

The rest of the desynchronization of this burst is synchronized in later nervous nuclei. The next nucleus after the cochlear is the cochlear nucleus, followed by the trapezoid body. In both nuclei, further synchronization happens to end in about the same burst as would have been there when the sound would have entered with all partials in phase one to another.[83]

So synchronization is a major part of the system favoring harmonic overtone structure and concentrating them into precise bursts.

There are other theories of how a musical pitch is perceived. The discussion on pitch, started with Helmholtz, had shown that each sound consists of multiple overtones. Only then the question arose how it can be that we do not hear a multiple of tones, but rather one pitch. The problem became even more evident as when a sound is inharmonic, we indeed can hear the single frequencies. But when the sound has harmonic overtones, they fuse to one pitch sensation, to a single musical note.

Carl Stumpf was the first to consider this problem on a music psychological level [242], and found that there needs to be a physiological reason, although he was not able to tell which. During the 20th century, several other ideas arouse, which are mainly variations of two basic ideas.[84] The first idea is that of a residual sound. When there is a harmonic overtone structure in a sound with frequencies of, e.g., 100, 200, 300 and 400 Hz, etc., then the distances between these frequencies is 100 Hz. And indeed, if in a sound like this the 100 Hz fundamental frequency is missing, but the overtone structure is the same, then with 200, 300 and 400 Hz, etc. we still hear a fundamental pitch at 100 Hz. The timbre of the sound is different, but the pitch is the same. This is called a residual pitch.

It is used with organ pipes. Very low organ pipes of frequencies below 100 Hz need to be very long to produce such low tones and are therefore expensive to build. So if one needs a very low frequency of 27.5 Hz, the lowest A, a combination of two pipes, one of 55 Hz and one of 82.5 Hz will do. The difference between 82.5 and 55 Hz is again 27.5 Hz, so these three frequencies are a harmonic series. When the 82.5 and the 55 Hz pipes are played at the same time, the resulting pitch sensation is 27.5 Hz, although there is no pipe of such a frequency playing.

So one pitch theory is that there is a mechanism in the brain to detect the harmonic relations between frequencies and if there is a common difference between them, to take this difference as pitch. Although this theory sounds good at first, until now, in the brain, no such mechanism has been found.

The other idea, how pitch comes into place is to have time-delayed loops in the brain between neurons. When a sound repeats with a periodicity of, say, again our 100 Hz example, which is 10 ms, and when this sound is going through a loop of neurons which takes exactly this 10 ms for one round, then the returning neural spikes of this loop would coincidence with the new spikes produced by the sound, and the loop would become stronger and stronger. Then we would know that there is a 100 Hz periodicity. Again until now, there has not been found any such loop system working this way. Although there are loops in the brain we discuss below, this idea would mean that there are as many loops as there are possible pitches. Such a system would need to be quite large and would most probably be found yet. It also would be quite inefficient, with a loop for each possible pitch.

Yet a third idea was discussed but found not to be sufficient for pitch perception. With the cochlear, we have seen that each frequency has a place on the basilar membrane where it has its best-frequency. The nerves leaving the cochlear therefore transport a pitch. This is called tonotopic representation, from topi, places for the tones. This tonotopic representation of frequencies continuous until the entrance to the neocortex, the primary auditory cortex A1 we have already mentioned. There the nerves for different pitches arrive next to each other along with the nerve bundle.

Still, this system is not very precise. The places on the basilar membrane for frequencies are quite blurred, and one single frequency will not only lead nerve cells at exactly one point on the membrane to spike, but it makes a certain range around the best-frequency place to become active and fire. There is a mechanism to suppress neurons surrounding a neuron with the highest activation on the membrane to suppress their activity. Still, this is not enough to make the activation of neurons precise enough that it would be as precise as our ability to detect pitch differences. And also, the sorting or ordering of neurons along with pitch at the entrance of A1 is not very precise, as fas as measurements have shown.

Therefore the auditory system needs the temporal development of nervous spikes to detect pitch, the bursting we have already discussed. Another evidence is pointing in this direction. Neurons in the human body normally spike 3–10 times each second. This is not very much, but for most senses, like touch or smell, this is perfectly enough. With hearing, the spiking needs to be much faster. When we hear a 1000 Hz tone, one neuron would need to spike 1000 times per second. This is physiologically impossible. We have seen that a spike means that charged molecules need to flow through the cell wall into the neuron to bring it up to about 70 mV in charge. Then it is able to spike again. But then new molecules need to enter the cell again to make it ready for the next spike. This process of refreshment takes some time, and, therefore, cells cannot spike too often each second.

Indeed, the neurons in the cochlear are the ones that are among the fastest. They can spike 200–300 times per second. But this would mean that only bursts at about 300 Hz could be detected, and therefore our pitch perception would end there. Still, there are many neurons next to each other along the basilar membrane, and so if one neuron is still recharging, one of its neighbors is again ready to take over. This interlocking works until about 4000 Hz. So indeed bursts with frequencies up to 4000 Hz are produced by the cochlear.

And we are only able to clearly hear pitches up to 3500–4000 Hz. Above this, we only hear a high sound, but without a clear pitch perception. This is represented in musical instruments. The piano, for example, has its highest pitches around 3600 Hz. Playing in this region still produces a pitch sensation, but one senses that it becomes more and more difficult to hear real pitches. It fades into a sensation of a frequency increase, but an uncertainty of pitch. So it would not make sense to produce keys beyond 4000 Hz.

In the end, there are two mechanisms working at the same time, the place on the basilar membrane as well as the bursting periodicity. But when the bursting period does no longer work, we still hear a sound, beyond 4000 Hz, but no longer a pitch. So pitch seems to depend more on the bursting rather than on the place on the basilar membrane, which is activated by the incoming sound.

A Spatio-Temporal Pitch Theory

Using all that said above, it makes sense to look closer into the spikes that leave
the cochlear. As we have seen, each frequency has a place on the basilar membrane,
where it has its best-frequency place, and the nerve leaving this place is representing
this frequency in a tonotopic manner.

But the picture changes when we make the computer simulation already used
above, but now inserting sounds in the model, at first a harmonic overtone spectrum
with a fundamental pitch at 400 Hz. In Fig. 3 such a case is shown. The cochlear is

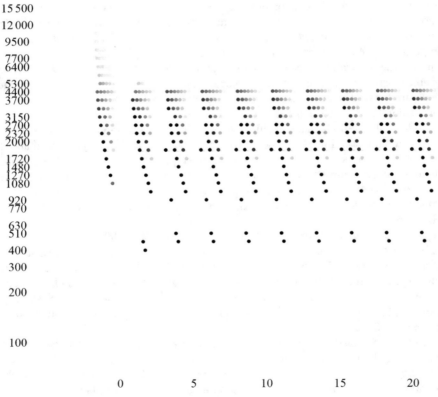

Fig. 3 Spikes leaving the cochlear, when we insert a harmonic overtone spectrum with a fundamental pitch of 400 Hz. On the horizontal axis is time, the cochlear is now shown on the vertical
axis. The fundamental frequency is represented with one spike at each new period, the overtones
have two, three, four, etc. spikes within each period, as expected. Still, as not all overtones have
equally strong energy over the whole periodicity of the fundamental pitch, *drop-outs* appear. They
form subharmonics, undertones of the respective overtones, with a lowest undertone at the pitch of
the fundamental

now on the up-axis, where each frequency has its best-frequency, the place on the basilar membrane where the wave has its maximum, where the spike is released into the brain. The horizontal axis is time [18].

The spikes output beautifully along the basilar membrane. The fundamental frequency spikes once each pitch period, so once each 1/400 s (the fundamental is 400 Hz). The second partial at 800 Hz is spiking twice within this time, the third partial at 1200 Hz is spiking three times within the fundamental periodicity, and so on.

Still, higher frequencies do not show all spikes equally strong. Of course, a spike is spiking, or is not spiking, but does not so with half amplitude. As there are about 30,000 afferent nerve fibers leaving the cochlear, the computer model does not necessarily consider all of them but assume that neighboring fibers spike at the same time point in an interlocking way, as discussed above. When doing so, depending on the amount of energy the incoming sound wave has, the amount of afferent nerves spiking will be more or less. This is indicated by the dots becoming light gray in the figure. The energy of the waveform is not able to support strong energy over the whole periodicity of the sound, therefore, the spikes become less.

But if spikes are getting sparse, gaps appear where spikes are expected, there are drop-outs. These drop-outs can have one or more spikes missing. Then the time interval between two spikes can no longer be the periodicity of the best-frequency at this point on the cochlear. If one spike is missing, the frequency will be half the best-frequency, if two are missing, it will be one-third of this best-frequency.

So we have subharmonics, undertones, appearing in the spiking pattern of the nervous spikes. Of course, the lowest possible periodicity will be that of the fundamental pitch of the whole sound, as then the waveform repeats.

This is shown in Fig. 4 when a guitar tone is used as input to the cochlear model. Again the basilar membrane is shown on the up-axis, but now the horizontal axis is no longer time. In this figure, we have summed up all periodicities between adjacent spikes, those we have just seen before in Fig. 3. Each time interval between two spikes is not shown as a frequency on the horizontal axis.

Now, something astonishing appears. We know that there are best frequencies on the basilar membrane, where the respective frequencies leave the cochlear. But now we see a strong, long vertical line at the fundamental frequency of the guitar around 220 Hz. This line is a bit thicker at the best-frequency place, the vertical axis, at right the frequency of 220 Hz. Still, it goes on through nearly the whole cochlear.

This means that the fundamental pitch of the guitar tone is not only present at one nervous fiber, where its best-frequency is, but on nearly *all* nervous fibers of the auditory system. This produces a spatio-temporal field, spatial as it is distributed in nearly all fibers up to the auditory cortex, temporal as it is the fundamental pitch periodicity which is present there.

Referring again to our idea of consciousness as a spatio-temporal pattern of neural activity, it is clear that such a wide distribution of a frequency need to prominently been perceived, compared to the timbre, which is also there but spread all over the nervous system. This explains the salience of pitch over timbre we discuss below in more detail.

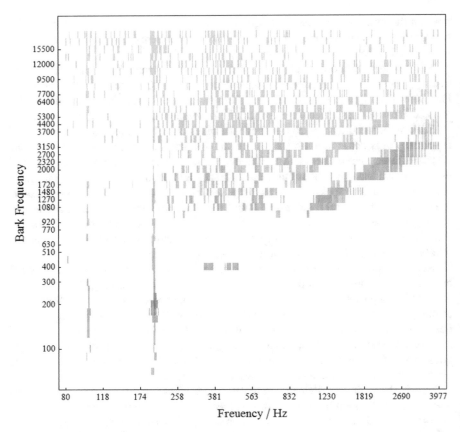

Fig. 4 Some statistics of neural spikes of a guitar sound leaving the cochlear. The basilar membrane is shown on the up-axis here. The horizontal axis is frequency. The figure shows why we hear pitch and why pitch is much more prominent than timbre for us. The fundamental pitch of the guitar at about 220 Hz appears as a long vertical line. This means that this periodicity, this frequency, is transmitted from the cochlear into the brain at all nerve fibers leaving the cochlear, not only at the position of the best-frequency of this pitch on the basilar membrane. Therefore the spatio-temporal pattern of this frequency is so widely distributed in the nervous system that pitch is much more prominent than timbre

This also points to the very nature of music, an aesthetic art, where we do not at first identify pitches, but simply perceive them, have them as sensations in our consciousness. This makes music very much different from other identification tasks happening in the neocortex. Of course, we can later refer to music theory and identify intervals, write down melodies, etc., after some music training. Still, pitch is perceived by everybody immediately, without training, and its aesthetic sensation is what gives us such joy and pleasure.

This, of course, also implies that the subcortex has consciousness too. When arrived at the auditory cortex, the sound is so much reduced, and from there on no longer tonotopic, that we would not be able to have a sensation of pitch up to 4 kHz, no neurons are still firing at this rate in the neocortex. Still, even above 4 kHz we have a sound sensation, which is not simply an identification, but a real enjoyment, a real conscious content.

How Do Tonal Systems Come into Place

It is clear that when there is no harmonic spectrum, the fusion of the partials into one pitch is no longer working, and we hear more than one musical tone. This is where we arrive at musical intervals like a fifth, a major third, or any other interval we know, e.g., in Western music. We prefer intervals of simple harmonic ratios like 3:2 for the fifth, 4:3 for the fourth, or 5:4 for the major third (Fig. 5). The simpler the ratio, the more the activity in the neurons in and behind the cochlear. This needs to be so because of the transition from the sound wave into the spike pattern. Also, the more complex an interval, the smaller the ratios like 9:8 or 10:9, the rougher they sound. Indeed, the neural patterns leaving the cochlear are getting more and more complex with smaller ratios.

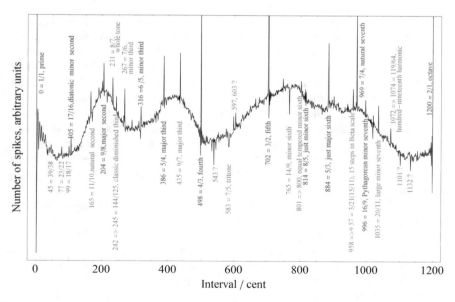

Fig. 5 Using the cochlear physical model from above, inputting two sinusoidals with all possible relations between unisono (1:1) to octave (1:2), one can compute the amount of spikes leaving the cochlear. Clearly at the perfectly harmonic intervals, sharp bursts of many neurons appear, while at irrational ratios much fewer spikes leave the cochlear

Helmholtz, in 1863 [128], was therefore proposing to build our music theory on this roughness. Although he did not know the spike patterns, we know today, physically, the overtones of two pitches sounding together in an interval might be close one to another and therefore start to beat strongly, which rough sounds. The more overtones come closer together, the rougher the sound. Therefore he was calculating for many possible musical ratios how rough they might be, not only for the simple ones we had before but also for many in between. It appears that the intervals we use are indeed least rough compared to all other possible intervals. Therefore he based our tonal system of the intervals fifth, fourth, thirds, etc. on a physiological basis.

Still, there are other cultures around the world using different intervals than those of the Western world. Indonesian *gamelan* music is such a case. It is a large orchestra built of many bronze instruments, which are arranged on a scale, like xylophones, only made of bronze. The bronze plates these instruments consist of have not a harmonic overtone spectrum but are highly inharmonic. Therefore playing one of them does not only produce one pitch, but many of them. Indeed, the lowest frequency is most often amplified by bamboo tubes tuned to this frequency and therefore are the loudest. Still, the other frequencies can be heard separately quite well, giving the orchestra a very special sound. This ensemble does not have intervals of simple harmonic ratios like 3:2 or 4:3, but something in between. It still has an octave (all musics in the world know the octave). But the other intervals are different. And they are also different from ensemble to ensemble.

The reason why *gamelan* is tuned the way it is is still not known. Many attempts have been made to explain them [234]. One explanation is that the tuning corresponds to the pitches of the inharmonic overtones. This works in some cases, in others not. When sampling the single *gamelan* instruments and detuning them to produce *lancerans*, musical pieces for the ensemble with all kinds of tunings, Western major or minor tuning, pure tone, etc. it appears that the original tuning of the *gamelan* is the roughest of all. Still, musicians of *gamelan* music perceive it as least rough and the only one really sounding like *gamelan* [260]. This cannot be explained by a 'typical' *gamelan* tuning, as each ensemble is tuned differently, and listeners were asked from all around the world, playing in different ensembles. Therefore it needs to have to do with a relation between the inharmonic overtones and the tuning, still, we do not know yet.

Many other examples could be mentioned, like *amadinda*, xylophone music of Uganda [168], and many other xylophone instruments often found in Africa, steel-drum ensembles of the Carabean [222], the *pat wain*, a drum circle in the *hsain wain* ensemble of Burmese music [19], and so on. We do hear pitches with these instruments, but along with other inharmonic or nearly harmonic partials. This is what makes these instruments so unique and fascinating.

From the standpoint of roughness, one might argue that it is no wonder that these ensembles do not have a Western tuning, as the overtone structures are not harmonic, and therefore intervals with simple ratios do not sound less rough than others. We might expect then other intervals to be preferred, which are least rough with such an inharmonic spectrum. Indeed, such suggestions have been made, but they do not work in practical contexts.

But why are these instruments tuned so differently? Asking musicians or instrument builders is mostly of no use, as they most often never thought about this problem and only work by their ears. But we might get closer to this problem when we go back to Western tuning systems.

Western tunings are nearly never really based on the simple harmonic intervals of 3:2 or 4:3. The reason is that when tuning, e.g., a piano to these intervals, we would need to do so for one key, say C-major. But then playing in another key, say D-major, B-Major, or E♯ minor, means using other intervals the piano is not tuned to. These keys will sound awfully wrong, and we cannot play in these keys. The only way we can play in different keys is to detune the original keys a bit. Then the notes do not sound perfect in the first key anymore, but they are still acceptable. They also sound not perfect in the other key, but acceptable here too. Then one has a musical temperament, where one is tempering, making compromises between keys, to be able to go through all keys in a quite acceptable way.

In Renaissance times, a so-called mean tuning was preferred, which had perfect third intervals, and one was able to use about four keys, F, C, G, and D Major. Only until Werkmeister was inventing his tuning one was able to go through all keys, although in some keys, there are still intervals which sound unacceptable [232]. Bach was among the first to try this new temperament, composing his famous Well-tempered Clavier.

In the history of Western music, dozens of temperaments have been proposed, not only by detuning existing tones but also by using more than 12 tones per octave, like 31[85] or 43,[86] only to make more and more keys sound better.[87] Still, the most common temperament today is the 12-tone equidistant tuning, where all semitones are made equally large, every 100 cents. Then all keys sound the same. In former times this was different, and therefore composing in different keys made sense, as the intervals were different, giving each key a special sound.

Now, if the temperaments of Western music are compromises, why should this not also hold for other musics of the world. In Cambodia, musicians playing traditional instruments often also play with Western musicians. Therefore they need to have an instrument that fits in both scales, the Western and the traditional scale of their music. When measuring these instruments, one finds a tuning which is right in between these two tunings [12]. This new tuning is strange and not very systematic at first sight. But this is because it is a compromise, a temperament between these two tunings.

Another example is a tuning system in Myanmar, which historically needed a change after the Bama had conquered the Thai in Ayuddha in 1763 [263]. Thai music was mainly performing ancient Hindu epics, the Ramayana and Mahabaratha. The Thai have a special tuning system which is equidistant, like the Western today, but with only seven tones per octave. There is a special sound with this tuning, and when after the invasion, Thai musicians were brought to the Bama court in Mandalay, they played in this tuning. As then the epics became popular in Burma too, Burmese musicians needed to tune their instruments to this new Thai scale. But as they were still performing their old music, they needed to retune their instruments quite often. To ease this situation, they simply invented a new tuning in between the traditional Burmese and the new Thai tuning, in which they could play both musical styles without the need to retune their instruments. So they made a compromise, invented a temperament, a new scale. This new scale is not according to simple ratios, and without knowing how it came into place, one might wonder why it has right these intervals. But with the historical knowledge, it becomes clear that it is a temperament.

This is a good example to show how music works in terms of musical tuning. There are physiological constraints, like the fusion of harmonic overtone spectra into single pitches and simple harmonic ratios leading to least rough sounds with simple and strong neural activity. But there are also social, political, and practical constraints, leading to deviations from the purely physiological tunings. This does not mean that the tunings are arbitrary and that one could choose them randomly. There is an aim to where we want to go. The temperaments of Western musical scales are still not perfect, as they are compromises rather than meeting the idle of perfect intervals, and many attempts have been made to correct this. But they are chosen because they have other advantages, namely that one can play in different keys. And often, it is right the deviations from the idle case, which make music interesting and fascinating.

After this small excursion, we should go back to the auditory system and see which constraints and findings there are beyond the cochlear.

Loops in the Auditory Pathway

The auditory system has several loops, consisting of neurons where spikes are sent around in this loop. Several loops are known, and more and more are found [204, 210]. One loop which has been investigated to a great extend is the loop starting from the cochlear, going through the cochlear nucleus to the trapezoid body, and from there back to the cochlear. The cochlear nucleus is the next nucleus of nerve cells after the cochlear. From here, the spikes are distributed to several other neural nuclei, among them the trapezoid body, called so because of its physiological shape.

We have already found that the cochlear has two kinds of stereocilia, or sensory cells, the inner and the outer cells, again called that way because they lie at the inner or outer side on the basilar membrane. The inner cells are sending out spikes to the cochlear nucleus, while the outer cells receive input from other neurons, mostly from the neurons in the trapezoid body. Therefore this neural loop starts at the inner hair cells, spiking into the cochlear nucleus, which sends the information into the trapezoid body, which again spikes into the outer hair cells on the basilar membrane.

The inner hair cells, therefore, are called afferent nerves, while the outer cells are efferents [225]. This is a basic principle of the nervous system. The senses send spikes to the brain and are therefore affecting us, while the brain is sending spikes to the body to produce an effect, for example, the movement of arms or feet. So all spikes coming from senses to the brain are called afferent, and all neurons coming from the brain to the body are called efferent.

The main job of the cochlear afferent/efferent loop is to improve the perception of single frequencies in a noisy background. This is part of the so-called cocktail party effect, where one wants to listen to the person right in front, but around are so many other people talking that it is hard to follow.

The effect is that these outer hair cells will become shorter. This is caused by motor proteins, prestin and myosin in the cells, which contract when the cell is driven by an efferent spike. Such motor activity is fundamental to body movement. If a muscle contracts, it becomes shorter. This is also achieved by cilia, cells that contract when they are affected by spikes from the brain. Although the stereocilia in the cochlear are not perfectly those of muscles, still the basic behavior is working here too, the outer hair cells contract just like muscles do.

As we have seen above, the outer hair cells are not only stuck in the basilar membrane, they are also attached to another membrane above the basilar membrane, the tectorial membrane. Therefore when the outer hair cells contract, they come under tension, which makes the basilar membrane stiffer right at the point of the contracting cell. Then this very small point on the basilar membrane is not able to move so freely as it can normally do. But if it is moving less, the inner hair cells at this point will not be affected as much as normal, and therefore they will spike less than normal too.

Now the effect is inversely strong depending on the level of incoming sound. With low sounds, the suppression is strong, while it is nearly zero with loud sounds. So when there is a noisy background, like at a party together with a louder person right in front of us, the effect is stronger on those frequencies of the lower volume background sound and is nearly not present with the louder sound we are interested in. Therefore the louder one wins, and the lower ones are tuned down.

There is also a temporal aspect coming along with this effect. As it takes about 100 ms for the loop to start working, to go at least one cycle through the trapezoid body, short attacks of sound are not suppressed at first. This is useful in cases where there is continuous background noise, like water flow or noise of a motorway. When someone talks to us, most of the sound is transient in nature, suddenly there and gone fast, like consonants are changes in pitch in speech. Then these transient sounds are not suppressed so much as the continuous background noise is, and we can concentrate on them better without being too much disturbed by the background noise.

There are many other such processes in the cochlear, many of which are still not really understood. All of them are improving our perception and tune the sound in many ways, extract information, and concentrate on aspects of the incoming sound. Another important and very well studied effect is the cochlear amplifier [189]. It is again a loop, but this time not between different neurons but within the outer hair cells themselves. It again is based on their motion, caused by the motor proteins prestin and myosin, and again starts working when the outer hair cells are displaced, at the tip link between two hairs stretch, opening a small gap in the membrane, where molecules flow in which are positively charged.

One of these molecules is calcium, which binds to the channel itself only a few nanometers after its inflow. This binding is very fast and causes the channel to close. So although it just has been opened mechanically, the channel closes again, because of electrical reasons. But then the inflow of new molecules is stopped, and therefore this binding of calcium is released too, again very fast. But then there is no electrical reason for the closure of the channel anymore, and the tension of the tip link is again opening the channel, allowing a new inflow of molecules into the cell.

When molecules flow in, the prestin is contracting. When it stops flowing in, the prestin is relaxing again. As the opening and closing of channels is so fast, the cell contracts and relaxes fast too, leading the outer hair cell to vibrate. This is an active process of the cochlear, adding movement to the basilar membrane. As the inner hair cells sense this movement, these inner hair cells become activated more than they would have been when driven by the sound coming in alone. Therefore the cochlear motion has been amplified. This process is again strong for sound low in volume and is nearly gone for loud sounds. It, therefore, helps to hear very low sounds, and if this process is lost, people often need a hearing aid.

This amplification of sound also makes the ear a loudspeaker, and with very sensitive microphones placed at the ear, one can record this sound coming out of the ear. These so-called otoacoustic emissions (OAE) are used to detect if someone is deaf. This is especially useful with infants, which often do not react, no matter how intense one might speak to them. There might be several reasons for this behavior, but one might be that the children are indeed deaf. When no OAE is leaving the ear, very likely, the children have a considerable hearing loss.

Although the cochlear has been investigated for many decades with great intensity, many processes are still not understood. There are not only efferent nerves from the trapezoid body to the outer hair cells, but also to the inner ones. They mainly also seem to inhibit on a broad range, so no matter which frequencies come in, the whole basilar membrane show suppression. Still, these efferents are attached to the axons of the inner hair cells leaving the cochlear, rather than to the cell bodies directly, like with the outer hair cells. Also, there are many more smaller effects the efferent nerves produce on the ear. But it seems that the basic mechanisms are the cochlear amplification and the efferents acting on the outer hair cells.

Other Nerve Loops in the Auditory Pathway

The auditory pathway from the cochlear to the primary auditory cortex A1 in the neocortex has basically seven main nuclei that are highly connected. We already know the cochlear, the cochlear nucleus, the trapezoid body, which is part of the superior olivary complex, and the end of this pathway, the auditory cortex. In between are the olivary complex are the nuclei of the lateral lemniscus, the inferior colliculus, and the medial geniculate. By now, it is not at all clear what these nuclei do and why they are so differentiated, although some findings have been made.

The lateral lemniscus shows a good resolution of frequency and amplitude resolution of sounds. It also seems to be part of the startle reflex, protection of the brain against very loud sounds. It is part of the brainstem and leads to the inferior colliculus, which is a midbrain nucleus. The colliculus is known to react to amplitude-modulated sounds. It is also reacting to spatial information of interaural time and level differences. If a sound comes from the right, this sound will first enter the right ear, and only a bit later arrive at the left ear. Therefore there is a time difference between the ears (interaural time difference ITD). Also, the sound level at the right ear will be louder, as it is nearer to the sound source, and as the head has an acoustic shadow, reducing the sound at the left ear. Therefore there is also a sound level difference between the ears (interaural level difference ILD). In the inferior colliculus, there are neurons that detect these differences and react to them. The medial geniculate is the last nucleus before the auditory cortex and is also connecting the auditory pathway to other parts of the brain, e.g., the amygdala, known mostly for negative emotions.

These parts of the auditory pathway are interconnected in many ways. The cochlear nucleus, right after the cochlear, is sending spikes to the next three nuclei above him, the olivary complex, the lateral lemniscus, and the inferior colliculus. But it does so not only to these nuclei on its own side of the ear but also to all three nuclei on the other side of the head, in the auditory pathway of the other ear. On the other way round, the auditory cortex sends its spikes to four nuclei below, the medial geniculate, the colliculus, the olivary complex, and even down to the cochlear nucleus. As the trapezoid body is part of the olivary complex, and this sends its efferent nerves right to the outer hair cells, there is only one station in between the neocortex and the cochlea.

There are loops between the medial geniculate and the auditory cortex forth and right back, between the colliculus and the auditory cortex, the same way. There are loops which the cochlear nucleus has directly with two other nuclei, the olivary complex and the inferior colliculus. And there are very complex loops between the auditory complex on one brain hemisphere with four of these nuclei on both hemispheres each, known today.

The role of all these loops and connections is mostly now known. Until now, we only know that these nuclei are connected and send spikes back and forth. So we have a complex neural network, where it is better to speak of a self-organized system rather than a bottom-up or top-down architecture. Still, there are so many discoveries to be made that we have only very few ideas about how the system works. Therefore research over the last decades has also concentrated on building more abstract models on how music is perceived and produced. So maybe it is good to only briefly discuss which models have been used, only then to come back to the music again.

Models of Music Perception

Science is defined by its methodologies. When we deal with music, we can listen to it or make music on our own. But this is not scientific treatment. Therefore science starts when we deal with music, or with any other thing, by choosing methods which are scientific. Basically, there are two such methods called induction and deduction. The words come from Latin *inducere*, meaning going in, and *deducere*, meaning going out.

Induction, therefore, means that we start with many experiences and findings and try to find rules or systems which may be the reason for all these findings. This is basically experimental work. One makes listening tests, EEG or fMRI studies of the brain, and from the data found, one hopefully gets an idea of what is going on. Deduction, on the other side, goes the other way. It starts from a general model, system, or idea. When such a model exists, it predicts that it will have such and such outcome. When we know how the weather works with temperature, wind, clouds, etc. we are able to predict the weather in the future. When we know how music is processed in the brain, we can predict what we will perceive when a certain piece of music is played to us.

Science uses both methods, it builds models that predict an output and then do experiments to see if these predictions were right. This process is going on and on, where a false prediction found through experiment leads to a change in the model to make the prediction better, followed by a new experiment and so on. Therefore the correctness of a prediction is a good measure for a model to be correct.

Good models are those who are most simple and still able to predict very many kinds of perception. But this is no easy task. The problem is that we invent models which we think might be correct, which indeed are correct to a certain degree. We often measure with statistics. So we do not expect the model to fit perfectly but only to a high degree.

This causes a problem. What is the reason for the model not fitting the experiment perfectly? Is it because we can never do perfect experiments in a laboratory, is it because people are different with certain taste and skills, or is it because the model itself is not good enough, and gives good results for the special task, but would fail when we ask for another task, and we would need to have a much better and much different model instead.

When investigating if music makes more intelligent, what was proposed by listening to music of Mozart, looking deeper into the data shows that this also holds when listening to Heavy-Metal, electronic, or Schlager music [3]. And when getting closer to what intelligence means, it becomes clear that it is often very narrowly defined. When investigating if listening to background music improves learning of vocabulary or other such tasks, it shows that this improves for some people while it makes it worse for others.

When we deal with political, social, historical, or ethnic aspects of music, the situation most often becomes even more individualized, and we are mainly able to describe what is going on in a musical style at a certain time in a certain place and do not really have hope to derive any general rules about music and its relations to these things in general.

When we investigate how good people are able to tell from which direction a sound comes in the context of Gaming using a 3D camera as visual input and a headphone on, it appears that it is very hard for people to tell if the sound comes from up or down. Still, it also appears that this is a problem in normal hearing situations too, and therefore part of our restricted ability.

So models predicting tasks associated with music, to improve life, or to trigger certain emotions, it appears that they need to fail, as there is no general rule for those things. They are highly individual. But if we deal with musical tasks alone, like pitches, timbre, textures, instruments, etc. then perception of people has much more in common, and we are able to build models that fit the experimental findings often very good, still also not one hundred percent.

The examples above are mainly experimental findings. Building a model, so approaching music from the other side, means that one needs to decide about the nature of the model. When we talk about hearing, many models use the bottom-up approach. A sound enters the ear, is processed by neural nuclei above it, and enters the neocortex with all its nuclei. There it is identified in terms of timbre, style, syntax, rhythm, and so on. There may be some interactions and some top-down features coming in, but basically, music perception and production are meant to be a process leading from one part of the brain to another, where each part performs a certain task.

A fundamentally different model is that of neural networks, synchronization, and self-organization. There the brain is dealing with the sound as a complex, self-organizing whole, leading to patterns and order, which then is our perception or our creative production of music. So in this model, there is a constant interaction between many parts of the brain, which results in a perception.

Now obviously, the second model comes closer to what is physiologically present in the brain. There are billions of neurons that are highly interconnected, as discussed above. As we have seen, there is only one neuron in between the way from the cochlear to the neocortex, although there are also many other ways, and the same thing down from the neocortex back to the cochlear. At nearly all levels of the auditory pathway, there are direct connections between both hemispheres of the brain. Each neuron has, on average, about 10,000 connections to other neurons. The neocortex is, therefore, highly interconnected, and driving neurons in opposite parts of the brain is happening fast and direct.

So why not choose a model which simply takes all neurons in the brain as they are into a computer and let the whole thing go. One could enter a sound at the ear and then simply watch to which pattern it leads in the brain. That indeed would be great, still, we are not that far. Indeed, there is at least one model which puts all about three billion neurons with all interconnections in a computer, still only from the neocortex and with random connections between neurons.[88] To arrive at a real model, we would need to include subcortical areas, the brainstem, and the midbrain, and know much more precisely which neurons are connected to which other neurons. This is not feasible today, we know not enough, and computer power is only about to come close to the abilities of a human brain.

So there is no perfect solution yet, still many models have been proposed to give insight into who the brain processes music. Some are neural networks, some are statistical models using probabilities, some are methods of signal processing, as are used in synthesizers or sound Plug-Ins, some are only very rough, discussing basic functionalities.

In the follow, when we discuss certain features of music and perception, we will discuss such models. It is important to see how far one gets with witch kinds of models and where they fail.

Timbre

Until now, no generally accepted music theory of timbre exists. This is in strong contrast to pitch, where we have several music theories, like the functional harmonics (Funktionsharmonik) of Hugo Riemann [220] with terms like tonic, dominant, subdominant, and the like mainly used in Europe, or like Schenkerism [78], deriving skeleton melodies and chords from a musical piece, more based in the USA. Although there also have been several suggestions to understanding timbre, there is no generally accepted theory here. So timbre is often defined as the 'other' of pitch. When we hear a musical note, we hear a pitch and a timbre. We all agree about the pitch and can associate it with a note, a small dot in a musical score. But we would not know how to notate the timbre of this sound in a way a musician would be able to perfectly reproduce it.

This does not mean that composers have no way to influence timbre. Of course, the timbre of a musical note is strongly influenced by the instrument it plays, is it a violin, a guitar, or a clarinet. And how about synthesizer timbres for different synthesizing methods, effects, and sound manipulations. If we talk about the violin, there are more dark and more bright sounds, and all can be played with complex articulations, changing timbre tremendously. Such articulations might also be part of the score. A violin bowed right after the bridge, *sul ponticello* sound different from one bowed more towards the middle of the string *sul tasto*. Also, timbre changes with the tone volume played, where louder volume is most often not only louder but also more bright. Combining instruments to a score also creates different timbres. The composer Berlioz developed a simple system of estimating round about how to achieve which timbre when combining which instruments to certain degrees [37]. But this is only a very rough estimation.

One might also wonder if a musical piece is the same when played with different timbres. Pop music or classical orchestra works are often published in scores to be played on the piano. These piano scores are arranged by a musical arranger in a way

R. Bader, *How Music Works*, https://doi.org/10.1007/978-3-030-67155-6_14

to fit on the piano so that the piece is preserved, although most of the timbre of the original piece is gone. This works when music leans on pitches, chords, and rhythms, which we can represent in such a musical score. Indeed, most copyright acts, also the German *Urheberrecht* is only looking for pitches, for melodies, not for timbre, or sound. So there is a clear understanding that a musical piece is the same when the same pitches are played, even though the timbre used has changed completely.

This does not work with electronic music, Techno, Dancefloor, Electronic Dance Music, Hip-Hop, and all musical styles which do not even have a score, like Free Jazz, Free Improvisation, experimental music, soundscapes, or the like. These musical styles rely mainly on timbre, and changing timbre here makes it another music. This also holds for much of so-called Arabian [75], Persian [81] or Indian classical music,[89] which is of extensive use of ornamentation, the playing around with the melody in nearly endless variations, fast grace notes, or small melody phrases. These *melismas* are not notated in a score. Most often, these musicians do not learn music by a score. Indeed, one would not easily be able to notate these ornamentations, as many of them are too fast to really be called a pitch. As these melismas are not only add-ons, making a basic melody more interesting, but as they *are* the melody, the flow of notes, timbre plays an important part here, and it is not really possible to make notations of them.[90]

Other musical styles play pitches but are more interested in changing articulations. The Japanese *shakuhachi* bamboo flute is played in a meditative way, where pitch changes are scarce, and the music is played by using pitch glides, kinds of overblowing, or the extensive use of wind noise from the player's mouth. Other kinds of Japanese music, like *gagaku*, knows many pitch glides from one note to another. These glides are part of the notation of a score, but they are performed in a different way than it might be expected when examining the score. Although a pitch glide is a changing pitch, it is only really one if it is performed as such. Often it is replaced by a timbre, a glottal stop when sung, or a popping sound so that the sound is not a pitch but a timbre without clear pitch information. As these elements are crucial parts of the pieces, timbre is not an add-on to a score but to the very nature of the musical piece.

The role of timbre in Western music is also often debated when it comes to money. As the copyright laws recognize the score alone, sound engineers, who work hard to produce a certain timbre for a piece, often find their work much more important for the success of such a piece than the score underneath it. In Techno music, the score is nearly always the same, a four-to-the-floor beat and some rhythmic elements on and off the beat. The buildup of tension, with instruments playing more and more dense up to the point where the beat starts, is also quite similar in most pieces in terms of the event onsets. So the pieces differ in terms of the timbre used for these events, and this is what tells a good Techno composer from a bad one. In Dub Step music, where the bass is most prominent, there are producers who have set up a recording studio only concentrating on bass-lines and only producing such. These studios are built for basses with extensive use of subwoofers, or bass Plug-Ins, not found in other recording studios. The bass lines coming from such studios are not too different in terms of pitch, it is all about timbre.

But such recording engineers are paid on a daily basis and do not participate from the amount of sales a record might have, as does the composer of the pitches. An interesting case was the legal case of Gary Moore, an internationally famous rock guitarist and singer, which was taken to court by a Blues musician only known locally, on a piece called 'Still, got the Blues', which was a hit record of Gary Moore. The Blues guitarist claimed that Moore had stolen the piece from him. Indeed, when comparing the two recordings, that of Moore and that of the Blues guitarist, they are nearly perfectly the same, even when it comes to small melismas in the played and sung melody lines. But it was the piece of Moore, recorded later, which became famous, the other did not. As Moore was a famous musician at this stage already, it is likely to assume that this helped a lot. Still, when listening to two pieces, many people subjectively find that the piece of Moore was just fantastic and great, while the other piece was a good Blues song, but no more. The difference is indeed only the different timbre, like the sound of Moore's guitar. It was also this sound that helped Moore to become famous in the first place, as often mentioned by music critics. So although the pieces are nearly perfectly the same in terms of melody, small differences in timbre might decide between a musician becoming world-famous or not.

Multidimensionality of Timbre

As mentioned before, there is no generally accepted theory of timbre. But there are several approaches. They differ by the scientific methodology used to approach timbre, as we have discussed above. One idea is to see timbre as a virtual space, where each timbre is located at a certain point in this space, and when walking through this space, one gradually changes timbre. There is a method called Multidimensional Scaling Method (MDS), which is able to derive such a space from what people hear. As this method was used extensively over the last decades with interesting results, it is worth having a look at it first.[91]

The advantage of MDS is that it is not about estimating in front of how timbre might be heard. It only plays sounds to listeners and asks them to give a judgment on how far away the sounds are one from another. Is a violin closer to a cello or to a guitar, is this guitar closer to a trumpet or to a piano? Only after having asked all combinations of sounds, an algorithm tries to fit these distances in the virtual timbre space. It really appears that this leads to clusters of instruments, where all bowed instruments like violins or cellos are close one to another, or where all wind instruments are clustered together. So the method seems to work as expected here.

What is more interesting is that nearly all studies which have been performed here ended up in a three-dimensional space. All timbres are at one point in this space. As a three-dimensional space has three axes, left-right, bottom-top, front-back, each timbre is at one point at all three axes. So a violin might be at point five at axis bottom-top, at point three at axis left-right, and at point seven at axis front-back. All other timbres are at different points, all on these axes. So we can also see how these instruments are ordered along every single axis.

These axes are what people had perceived when they judged how similar the timbres are one to another. But timbres are also physical objects and can be analyzed using Musical Signal Processing, calculating their frequency spectrum, fluctuations in time and frequency, harmonicity or inharmonicity, and so on. So the next step is to see if the instruments along one of the perceptual axis order along a parameter derived from these physical properties.

This works indeed, and is not too surprising, as listeners can only hear in a timbre what is physically there. Therefore the similarities people perceived need to be somewhere in the physics of the sound. The first finding often made in MDS investigations is that when the timbres presented differed in pitch, nobody listened to timbre anymore. As people were only asked to judge if sounds are similar or not, nobody told them to listen to timbre, or pitch, or anything. So when people are free to compare timbre in terms of what they want, they nearly always only judge sounds according to pitch. This is a strong indicator that there is a fundamental difference between pitch and timbre and that tonal systems based solely on pitch and not on timbre are a good idea, as this is perceptually strong. As we have seen before, it is also physiologically hard-wired in the neural network, and therefore we indeed separate pitch and timbre strongly.

So when interested in timbre, we need to leave all pitches the same in the sounds. When doing so, there are several parameters that people listen to the most prominent. The first one is brightness. Physically this means much energy in higher frequencies, which can be calculated from the sounds easily. A violin has many more overtones compared to a flute and sounds more bright. An oboe is brighter than a classical guitar, and a crash cymbal more bright than a double bass. This is obvious, and indeed people like brightness. One reason is that it makes sound subjectively louder, which is part of the loudness war in modern electronic music, and in the classical music of the 19th century. All sound engineers know that if they playback a track to a musician that just has recorded this track and add some higher frequencies, the musician will immediately like the track much better. But this is a short-living effect, and in the long run, people find such sounds too bright. So it is wise not to playback such a track with too much brightness, as it normally takes quite some time to convince the musician that less brightness serves his music better in the end.

A second parameter is fluctuations in a sound. Wind instrument players cannot perform with perfect stable lung pressure, lip tension, or lip distance to a labium, and therefore their timbre changes slightly all the time. Pianos in the middle range have three strings which are detuned one to another slightly—at least in the West, not too much in Japan. This detuning leads to amplitude fluctuations, a beating, which is heard as fluctuations. Pop musicians always try to add many fluctuations to sounds, which is achieved by sound effects like flanger, phaser, or chorus, making the sounds more lively. Indeed, the chorus effect is named after its origin, the chorus of a choir, where many singers do have slightly different pitches, leading again to fast and uncontrolled beatings. So many sounds fluctuate, and the amount of these fluctuations is used by listeners when comparing sounds. Indeed, fluctuation is a good physical marker when it comes to detecting note onsets, using Musical Signal Processing [1]. When a new tone starts, there are many fluctuations left. Still, this does not work all the time.

Yet another aspect of timbre with sounds having a harmonic overtone spectrum, so no percussion instruments, are inharmonic components at the beginning of the tone. Nearly all instruments have such inharmonic components, which are very important when it comes to identifying musical instruments as pianos, violins, saxophones, guitars, or the like. Flue organ pipes came in with a sound called 'chiff' or 'hiss' by organ builders [4]. As it takes a while until a regular motion of the air jet hitting the labium takes place, this air jet is struggling with the system in the beginning. Still, it produces sound there too, which is not regular, but noisy. As the air jet is only getting stronger at the start, starting from zero pressure to maximum, the process is very much like that of a 'd' or 't' consonant in speech. Therefore this sound is also called 'spitting' of a flue organ. It is a crucial part of the sound, and organ builders try to design it in a way to sound most interesting. It is also helpful in a polyphonic score to identify new tones coming in an already complex texture of pitches. So it sounds good and helps to follow a score.

Classical guitars have a knocking sound at the beginning of each note. The string is driving the soundboard and forces it to vibrate with the frequencies of the string, the self-organization of the instrument, as discussed above. But at the very beginning of the sound, the guitar body is not willing to take it and tries to vibrate with its own frequencies. But it loses the game, and after about more or less 30 ms, it is forced to the strings frequencies. Still, within this sound piece, it vibrates with its own eigenfrequencies, which is the same sound as if one would knock on the guitar body, for example, with a finger. This knocking sound is in the timbre of each note being played on the instrument.

Wind instrument players also need to get into a tone. They might start from scratch or change the note while already playing with normal playing pressure. In both cases, a new tone needs to build up, very much like with the organ pipe, which is not working in the first place and produces a noisy sound at the beginning or when changing pitches. One cannot avoid this, but one can deal with it. Classical musicians often try to sound as harmonious as possible. When doing so, a trick is to start with a low lung pressure, making the noisy income low in volume, and only then raise the pressure. One might also want this noise as an articulation, like with the so-called speaking saxophone in Jazz music. As the saxophone tone only comes in at a certain lung pressure, a player might blow into the horn some seconds with low pressure, only producing noise, and only then go beyond the pitch threshold of this pressure. Such sounds might be found intimate. They also might contribute to harsh noise as wanted in free jazz or experimental music.

Violin players have about the same problems as wind instrument players. They cannot start harmonious and can decide to start with a hard attack, emphasizing the noise, or with a low one, where they use low bowing speed to keep the noise low in volume, only then the rise the bowing speed making the sound loud.

In any case, there is a noise at the beginning of a sound, and people use this when distinguishing musical instrument sounds one from another. Indeed, these are the three most important cues listeners use when judging the similarity of musical timbre, brightness, fluctuations, and inharmonicities at the tone onset.

But this is not the whole story. In investigations where sounds were designed in a certain way, listeners did often use this design to tell timbre one from another. So one might design the amplitudes of the overtones of instruments, lowering all even or all odd harmonics in volume, or changing the steepness of amplitude decay with higher harmonics. Then listeners use these cues in their judgments. Cutting off the initial transients from sounds, their noisy beginning, also leads to very different judgments, with a shift to other aspects not used for a judgment before. It appears that when the presented sounds become more and more the same, according to the parameters discussed above, listeners go deeper into the sounds to look for slighter differences.

So there seems to be a hierarchy in timbre perception, just like the hierarchy of pitch being more prominent than timbre overall. Within this hierarchy, brightness is most important, followed by others like fluctuations, noise at the beginning, the shape of overtones, vocality of sounds sounding like 'a', 'e', 'i', 'o', or 'u', and many others. As with the less salient features, it is hard to say which one is more prominent than the other, building a timbre theory is hard to do. Still, a general tendency has been found using this method of MDS.

But again, this is only one possible method of investigating timbre. Indeed, it asks for one perceptual task, that of judging similarities. Another perceptual task is instrument identification. Here the situation changes, and the perception strategy of listeners changes completely. They no longer take pitch as important, but first of all, listen to the initial transient of a sound. Indeed, when cutting off this transient, one has a hard time identifying which instrument is playing. The first to investigate this was Carl Stumpf at the beginning of the 20th century [242]. He used two rooms, which were connected by a tube that could be closed or opened. In one room was the musician. In the other listeners were asked which instrument was playing. When the tube was closed, the instrument could not be heard in the other room. When asking to identify the instrument in the other rooms while opening the tube only after the tone onset, listeners were very often mismatching the instruments.[92]

In modern tests, the two rooms are replaced by a computer algorithm, cutting-off the initial transient. Then the rate of correct responses about what instrument is playing decreases tremendously. Still, there are also other aspects of a sound that are important. With violins, it is the typical vibrato, which eases identification. With reed instruments, it is a typical sound region around three kilohertz, always present no matter which pitch is played. Such frequency regions with strong energy are called formant regions in human speech and singing. With saxophones and clarinets, they are also present as the reeds of these instruments have a frequency they vibrate on their own. Indeed, basically, they are forced to vibrate with the air tube, as discussed above. Still, they are only valves opening and closing, and such values do vibrate very shortly after they have been moved on their own. This movement with saxophones is about double the speed these reed would vibrate without the wind. Normally their eigenfrequency is about 1.5 kz, but when additionally driven by the wind and pressure, they end up in these three kilohertz, making the frequency region around 3 kHz strong in every pitch played on the instrument. So the ear and nervous system of humans use every cue they can get to identify instruments, which basically is the initial transient.

Different task, different results. Asking for similarities gives a different picture of timbre than when asking for identification. This is one of the reasons which makes a timbre theory so complicated. Finding timbre in the brain has not been tried too often and does not give too many satisfactory results yet. Indeed, yet it mainly has been searched for in the neocortex, and not too much in the auditory pathway. In the neocortex, some kind of identification of brightness and other parameters have been found, but results are not clear yet.

Music as Physical Adaption

The investigations on timbre so far were based on perception tests, combined with Signal Processing analysis, like estimating frequencies, fluctuations, or noise, at tone onsets. Yet another idea is fundamentally different, namely assuming the perception of timbre is based on the same physical principles the production of sound is, on Musical Acoustics, or the sound production processes in nature. This is reasonable because both systems, the physics of musical instruments as well as the neural networks in the brain, are self-organizing systems, working with impulses, sound impulses in instruments, or neural spikes in the brain. These impulses interact in a complex system, leading to a self-organized state, which is either the sound produced by instruments or the perception of this sound in the brain. Yet another reason for estimating this parallelism is that musical instruments have been developed by humans to fit their perception, and therefore it is likely that we built instruments working right like our brains. Going one step further, evolutionary our brains have developed along with our environment, what is around us, what we hear, see, touch, or smell [3]. As we need to get along with our environment to survive, it is essential that we develop a brain suiting this environment.

This is fundamentally different from a view of the brain as a signal processing tool. When we would only perform signal processing, we would not use any information about how the sounds of the environment are produced, nothing about the physics. We would start from scratch and built up some speculations on what might be of interest in the sound for us, and evolutionary neural networks built up, which, by trial-and-error, get along better with our surroundings. Indeed, this would be less efficient compared to using mechanisms of the physical world around us. The idea that we took over the physical principles of this world, and use it for adapting us to it, is, therefore, more likely in terms of efficiency.

It is also more likely in terms of the fact that we are physical objects. We often see ourselves as something special in the world, and indeed no animal has such an elaborated neocortex as we have. But in the end, we are also nothing but bones and flesh, with a large neural network in our heads, and act along the physical principles holding for everything else in the universe. So it not only would be more efficient to adapt to our environment, using the same principles as the world around does, it is also very much likely as we are nothing but environment ourselves.

Adaptation of species is well known, and maybe most prominently promoted by *exkll* (1864–1944) in this biosemiotics and as a founding father of theoretical biology [254]. A nice example in terms of hearing is that of a small fly that needs crickets to survive. It lays its eggs into these crickets, where they eat it up and become the next generation of flies. Still, the main task is to find such cricket. Everyone knows how hard it is to detect a cricket. One sits nicely in the evening on a balcony and listens to these animals making fascinating sounds, but is rarely able to tell where this sound comes from. This is much harder than to say where a dog or a bird is making noise. But this very fly is able to detect crickets with very high precision. It does so with special ears, which are very primitive. They consist of two masses connected by a small rod. The masses and the rod are inside the head of the fly, and when it turns its head around, not in the direction of the fly, this system is not vibrating. Only when the fly has turned its head right in the direction where the cricket is, this system of masses and rod starts vibrating strongly. The fly has adapted its ears right to this sound of the cricket. Many such examples are found in nature, which is all clearly based on evolutionary developments.

This does not affect our ability to create and imagine whatever we want and deal with the world in a subjective way. Our neocortex allows connections that are not hard-wired, and therefore many associations. Associations can be made and broken-off freely, a fact used in commercial advertising, where peanuts, shampoos, or cell phones are associated with beautiful people, beautiful countryside, social media influencers, ect. These things have nothing to do with the product, but humans can associate them, and therefore connect something useless with something nice to create a wish to have that thing. We are able to associate freely whatever we want, and much in music is working with such associations. As they are free to built-up and break-down, we call them fashion. They might enrich our lives, but we will survive without them. Associations follow no rules. They cannot be investigated by science—and need not to be.

Still, pitch, timbre, rhythms, musical instruments, and all that is intrinsically musical, and not only associated with, seem to be hard-wired with very little variations between people. As we have seen, this hard-wiredness is basically present in the subcortical areas, where consciousness is indeed very likely to be present. This does not mean that we do not deal with music in the cortex too, and we will come to some examples below, still, it means that the higher we get in the brain, the more associative and random music perception and production gets. This is a good thing, as it is part of our freedom. Still, we find that the parameters intrinsic to music are all the same for nearly all people on earth and not dependent on culture, religion, the color of the skin, gender, or the like. This is an interesting aspect of music, that within this view it indeed seems to be political by explicitly not been political.

IPF Comparison of Bowed and Wind Instruments

As we have seen so far, the physics behind both, musical instruments and neural networks, is very complicated. To arrive at a simple method that is able to deal with both cases, we need to break down the details and try a method that is fundamental for both. We have discussed such a method before with musical instruments, the Impulse Pattern Formulation (IPF) [178, 179]. These short impulses are traveling through musical instruments, are transferred from one part of the instrument to the other, from a string to a soundboard, from a reed to a tube, and act back in a loop. We have found that this formulation is able even to tell the details of the initial transients of single parts of a classical guitar, explain multiphonics and complex initial transients. It is also able to predict all nonlinear behavior of the instruments, like bifurcations, hysteresis loops, or sudden tone on- and offsets.

It is interesting to use it for musical timbre production and perception. Indeed, when using it for building a violin sound, it ends up in a scratchy sound, just like that of the initial transient of a real violin. Here the impulses are the skeleton. When does which impulse come back and forth the instrument. This skeleton needs to be filled with a time series. With violins, this is the sawtooth motion. The same can be done with wind instruments. Then we also get the skeleton from the IPF, and now fill it with a typical waveform for wind instruments, a rectangle waveform. Indeed, the spitting found is trumpets or trombones appears as a sound very realistically.

An interesting aspect for the sound production of initial transients of a violin and a trumpet is that for the IPF both systems are the same in their most simple form. Both have a nonlinear driving mechanism, once the bow-string system, once the turbulent air jet, and a linear resonator, once the string, once the air tube. So both systems look the same, and therefore the IPF for both instruments is indeed the same. The difference between the scratchy sound of a violin and the spitting of a trumpet is then only the use of the time series, once a sawtooth, once a rectangular waveform. And indeed, both instruments are very similar. They have a tone onset at a certain pressure threshold, they have bifurcations or hysteresis loops. Also, within the steady-state, after the initial transient, both instruments are often mixed up. This is exemplified in a bowed instrument, the trumscheit or *tromba marina*,[93] which sounds like a trumpet. It has the special feature that its bridge on one side is not attached to the body but is free and therefore hits the body each time the bow is detached from the string. So bowing need not necessarily be resulting in a bowed instrument sound.

This does not hold for the guitar or the piano. Both have multiple reflection points and therefore end in very different IPF transients. They are therefore distinguished from instruments that have only one reflection point, and therefore the same skeleton of the IPF. Indeed, it is more seldom that one takes a violin for a guitar than a violin for a wind instrument.

So there seem to be two elements coming together. The IPF distinguishes between musical instruments having only one resonating point, like violins or wind instruments, and those having more than one reflection point, like guitars and pianos.

Furthermore, distinguishing within such an IPF family of instruments, the waveform is getting more important, which again is caused by the different physical properties of the instruments.

There are endless possible combinations one could think of. Comparing a classical guitar tone played very low and a recorder flute, it appears that in a middle pitch range, and when only looking at the steady sound after the transient, both instruments sound nearly only with their fundamental frequency. Both sounds are nearly like a sinusoidal, and when listening to both, it is hard to tell which is which. But when adding the initial transient, the flute has a noisy beginning, which is typical for an IPF of a system with only one reflection point, while the guitar has an IPF, which is typical for an instrument with multiple reflections. And indeed, both instruments can be distinguished clearly.

When it comes to music perception of timbre, using the IPF is straightforward in a neural network. These networks do right this, send out spikes that come back in a loop. Therefore they necessarily need to produce right the same IPF as musical instruments do, again depending on the amount of reflection point and their strength. This is their very nature. So when imagining a musical sound, like when we remember a musical piece or when a composer is creating new pieces, such a system of neurons would produce right the same IPF as musical instruments do. These instruments have different time series, sawtooth or rectangular waveforms, which are what in the brain are the bursts we have already found. When these bursts come regularly, we hear a pitch. The precise shape of these bursts carries timbre, as we have seen with the cochlear. An irregular bursting, therefore, is like the IPF skeleton, while the waveform is the shape of the bursts. Then in the brain, the process of perception, as well that of creation of sounds, is right the same as the physical production of tones in musical instruments.

Although there is strong evidence that this is the way of music perception and production, we cannot prove it today, as we cannot look into the brains of humans that precisely without damaging the auditory system. Still, more and more evidence is pointing in this direction, and it will be interesting to see how our experimental design develops over the next years, maybe making it possible to get deeper into the brain experimentally. Another method will be to develop sophisticated IPF patterns based on what we know about the brain and see how these IPSs are able to tell how we perceive and create music.

Neural Network Models of Timbre

In terms of modeling on a computer, things look much better. Many attempts have been made to model the perception of timbre, also of rhythm, pitch, or tonality, using neural networks.[94] There are two main approaches to this, the so-called Kohonen maps and the connectionist models.

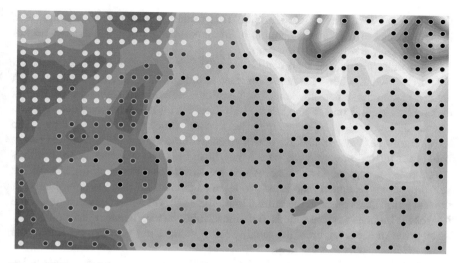

Fig. 1 Self-organizing neural map, as a system of arificial intelligence, sorting three collections of Ethnomusicological Recordings, the Heinitz collection of African music, the 1932 Cairo Congress of Arab Music, and the Collection Bader of recordings from Southeast Asia, China, or India

Kohonen maps are a simplified version of a neural network, which is two-dimensional here. Also, the neurons are not computationally modeled with ions and currents flowing in and out of a nerve cell. Instead, the two-dimensional field is meant to be a grid of neurons, where each one is able to store some features of a sound, like brightness, sharpness, transient time, or the like. As each neuron has different values of these features, one is more bright, the other sharper, etc., they differ one from another.

In Fig. 1 such a map is shown. It is able to distinguish three ethnomusicological collections, two historical, the 1932 Cairo Congress of Arab Music, and the collection Heinitz of African music from the 1910th–1930th, and the Collection Bader of contemporary, traditional music of Southeast Asia, China, India, and related countries.[95] The map consists of a regular grid of neurons, where each neuron is a set of numbers representing a feature set of psychoacoustic timbre features. Each dot on the map is a musical piece, which is most closest to the neuron it is placed at. The three collections, white, red, and black, are clearly separated one from another on the map. They are clustered. This map is like a mental representation of music, it can distinguish which collection a recording belongs to.

The interesting thing about Kohonen maps is that the neurons learn. Say the map shall learn one hundred different sounds or timbres. Then the feature values at each neuron are taken as random at first in each neuron and learn each sound one after the other. Learning here means that each sound is presented to each neuron, and it is calculated which neuron is most close or most similar to this sound. The feature values of this neuron, and some around it, are then changed towards the values of the sound they learn. This is repeated for each sound, all sounds are presented to the map many times, as in real learning. This means that the map is shaping itself towards the set of sounds it is about to learn on its own, it organizes itself to best fit the sounds, it is a self-organizing map.

After this training, the map can be used. This is like we expect trained neural networks to work, when a sound is presented to the network, we can calculate which neuron comes closest to this sound. When doing so with all sounds, we will find that sounds similar one to another will be detected by the map in certain regions. The saxophones might be in the upper left corner, while the guitars might be in the lower-right one. We can also use new sounds with the map, sounds which have not been used for training, and see where they fit in best and will find again that they are detected at regions in the map, which are similar to other such sounds. So the self-organizing Kohonen map is able to learn and detect timbre. We do not know if this detection takes place in the human brain like this, still when comparing the time it takes for humans to learn sounds, the time a Kohonen map needs is just right [219]. Also, other musical features like tonality [172] or rhythm [44] have been found to be sorted reasonably in such maps.

Yet another neural network model for learning sounds is the connectionist model. Again we have single neurons that have values of some musical features. But this time, these neurons are connected one to another. Again the model is trained first and can then be used for the detection of sounds or other musical features. In the training phase, the model takes a sound and compares it with some neurons at the input side of the neural net, called the input layer. According to the similarity of the input sound with the input layer neurons, these neurons become active, and fire to neurons in a middle layer. The strength of this firing is set by the connection strength between the input layer neurons and the middle layer neurons. This strength changes each time the neuron fires. If the firing is strong, the weight is changed so that at the next time of firing, the influence on the middle layer neuron will be stronger. Such changes of weights shall represent the so-called neuroplasticity found in real neural networks. Here, indeed, the strength of one neuron acting on another neuron might be stronger or weaker, and this strength might change slowly over time. Then the neural network changes 'plastically', meaning changing its basic setup, and therefore reacts differently on incoming stimuli, is a different neural network.

The middle layer neurons then act on yet a third layer, the output layer of the connectionist neural model. The output neurons are those to detect which stimulus, what music came in. So at best, only one neuron at the output layer is reacting strongly, while all others stay calm. Then the detection of an input stimulus means the activity of one single neuron. This is like with the Kohonen map, where also one neuron acts most strongly on an input and therefore detects the sound.

So, in this case, the cognitive task is detection. In terms of real neurons, this is a problem. Each neuron is the same, and detection only means that one neuron fires while others do not. Still, this firing neuron is no different from any other neuron. So how would we know that this one neuron detects right this sound?

There are several possibilities to solve this problem, where we are not sure which one is correct. It might be that detection only means a comparison to other cases. If one neuron is detecting the sound, it is another neuron, as the neuron, which detected a previous sound was different. Then identification would be comparison by heart, where we would need to remember previous events. This comes close to an understanding of features as dialectic, where features, characteristics, or ratings come into place due to a comparison between things, which are perfectly the same but are somewhere or sometime else. Two neurons are both the same, still, one is on the left, and one is on the right, which makes a difference and therefore is a feature. This idea was most prominently proposed by Friedrich Hegel in his philosophy of dialectics.[96]

Yet another possibility would be that the neurons are not the same at all. As we have discussed in terms of nanoneuroscience, the cell skeleton of neurons changes when they learn and form a new electrical circuit. This electrical circuit, when active, produces an electromagnetic field, which could be the content of consciousness, the features of the sound we model when using a Kohonen, or a connectionist model in one neuron.

Yet a third possibility would be that such a model of detection is too simple and that every single neuron in the model, in reality, is a neural network of many neurons, a spatio-temporal electromagnetic field, as we have discussed above. Indeed, neurons in a certain area very often fire in bursts, where many neurons fire around the same time. The spatio-temporal pattern of this firing could represent the features or the character of the detected sound. This possibility includes the sensation of pitch, as we have seen. Still, on a more abstract level of music theory, we would assume to happen in the neocortex, where we only deal with music on a theoretical basis, the first option might be correct, the relation of musical pitches, which are all the same, but related. This corresponds to the dialectic foundation of functional Harmonics *Funktionsharmonic* of Hugo Riemann [220], who found the chord progression I–IV–I–V–I to be thetic, antithetic, and synthetic, a dialectical progression.

Next to the neural maps and the connectionist model, there are neural network models coming closer to real neurons and their physical and physiological properties. These models take the ion channels, the cell current flow, cell voltage, and other parameters of neurons into account and solve the mathematical equations of many of such neurons being connected one to another. In recent years many experiments about the brain of mice, rats, or other animals have been made to find connections between neural nuclei.[97] Still, building these models to model the whole brain is only starting, as discussed above.

Rhythm, Musical Form, and Memory

A musical phrase is most often about 3–5 s long. The reason is that it needs to fit into the short-term memory, which is about this size, as we have seen above. Longer melodies need a different form of memory, the long-term memory which seems to work in a different way. Short term memory is most often understood as the brain echoing a sound input. The nervous cells are triggered by the cochlear, which transmits the input to the neocortex via the auditory pathway, as we have seen already. The neurons are heavily connected, therefore, the longer the neurons transmit the first sound impulse in the brain, the more it fades out and vanishes.

So it is reasonable for a musical phrase, a groove, or a sample to last as long as short-term memory lasts. Still, in addition to the echoes from the past, there is an expectation of what is about to come. All longer time spans need long-term memory to work and therefore are considerably different.

Neurophysiology of Time

Neural spiking is faster than musical rhythm, therefore, we do expect rhythm and short melodies to be processed in the subcortical, as well as in the cortical regions. Indeed, rhythm perception and production activate the cerebellum, as well as the forebrain [248]. In the localization view of the brain, different regions process decisive tasks. The cerebellum is known to process temporal events where rhythm is a part of. The forebrain is known for higher thinking and pattern recognition, and therefore it is reasonable to expect rhythm patterns to be processed or identified in the forebrain.

Still, there are other explanations of time perception. Also, those working over longer time spans as short-term memory, but without the use of plasticity. Such a process is the pacemaker-accumulator model [56]. Here, storage in the brain able to collect dopamine, a neurotransmitter, is filled over time. The more dopamine is in the storage, the more we expect a time-event to happen. As the storage might be filed faster or slower, our subjective perception of time is different. Sometimes we find the same physical amount of seconds lasting very long once we find it passing very fast. Indeed, when more dopamine is present in the brain, our subjective time expectation speeds up.

Still, there is another process of time perception, using synchronization of different brain regions. It is about expecting a certain event to come in the future. When we stop our car at a red traffic light, we stand there, expecting that the light will turn green at a time-point in the future. Then some neurons start to trigger a process of synchronizing different brain regions. These regions fire neurons with a certain frequency, most often in the range of 40–50 Hz. These neurons start their 'vibration' at different times and are therefore phase-shifted. So, although all fire at about the same frequency, at the beginning of the process, the neurons of different brain regions fire at different time-points. So within one period of 50 Hz, which is 20 ms, one neuron fires at time-point 0 ms, another might fire at 5 ms, another at 37 ms, etc. They all repeat this firing each 20 ms each on their own, still they all fire at different time-points, they are out-of-phase.

The firing appears within a general loop in the brain, which connects the cortex with the thalamus, so a subcortical region, where, next to many other parts, the medial geniculate nucleus of the auditory cortex lies, and the striatum, a region between cortex and thalamus. As the thalamus projects to many parts of the cortex, this loop is a hub between subcortical and cortical regions.

But this phase difference changes over time, the neurons are getting closer and closer in terms of the time-point they fire until nearly all of them fire at nearly the same time-point. Such a process is synchronization, as all neurons now synchronize to a common phase, fire at right the same time-point each about 20 ms.

This process is increasing the synchronicity of brain regions, which also seems to spread to the motor region, which is right at the top of the head in its middle. The neurons seem to trigger the motor cortex. They can do so, as they all act on it at the same time-point, which is a strong trigger. At the beginning of the process, when there is no synchronization, the neurons act on the motor cortex at different time-points and therefore are not able to trigger it considerably. Still, the more the neurons synchronize, the more the motor cortex starts to 'vibrate'. In terms of the red traffic light we are waiting at, the rhythmic triggering of the motor cortex makes us start moving our fingers, legs, or feet, nervously awaiting the light to turn green.

Musical Form as Tension

The same holds for musical form, which is built by several factors. One important is the change between tension and relaxation. The cadence of a tonic as a base, a subdominant as leaving this base, the dominant building up tension, which is then released in the tonic again, is a classic example. Indeed, Hugo Riemann [220], a highly influential musicologist at his time, who invented the functional harmonic, the terms tonic, subdominant, dominant, etc., thought that all music works with cadences, not only the art music of European Renaissance and Baroque, or Vienna Classical music, but also Flamenco, Indian Raga music, many folk music styles of Europe and so on. Today we find musical styles that are not easily explained by cadence, like Jazz or Rock music. Still, also many musics have their own kind of form. So indeed both, Flamenco and Javanese *gamelan gong gede* have a rhythmic cycle of twelve strikes, where the loudest and most prominent strike is on the last, the twelfth beat, not the first as in Western music, which might cause some confusion in listeners nor familiar with this music.

Still, tension is not only present in the cadence, but also in this twelfth strike, and possibly introduced in music through many other features. Tension can be built up by increasing loudness up to a climax. We automatically expect the music to have such a climax when loudness is continuously increased. Another possibility is density. A musical piece might become denser and denser with more and more notes, percussion sounds, or timbre changes, making the music denser and denser. Yet a third compositional tool for increasing tension is brightness, where music becoming more and more bright is also heard to increase tension. Of course, all three features, loudness, density, and brightness, can be applied in many different ways and over longer and shorter time intervals, producing different kinds of tension.

Fractal Dimensions as Musical Density

Even Parkers solo saxophone piece *Broken Wings* over about seven minutes is a nice example for building up and slowing down tension with a seven-minute arc, by in- and decreasing the density of musical events [21]. Parker is known in the free improvising scene and uses so-called extended techniques with his saxophone, techniques which are normally not used by regular saxophone playing, like using the tongue, splitting, or singing in the instrument, ect.. This example might sound strange—or refreshingly interesting—to some people. Still, it is a good example of a tension built-up using density. The free improvising music scene mixes Free Jazz with Contemporary Classical music and other styles, therefore not using traditional melodies, rhythms, or harmonies in the classical sense, and therefore need other features to create a musical form. Here density is one of them.

This density can be understood as a fractal dimension of a sound. A fractal is a self-similar geometry which repeats itself in smaller and smaller scales [5, 188]. The snowflake is the classic example of such a structure. It was the mathematician Georg Cantor (1845–1918) who was calculating the length of the line of a snowflake. This is not a trivial task at all, as it seems to be endless at first. It is endlessly elaborated. The deeper we look into it, the more detailed it becomes. At the same time, it is finite, as it has a closed geometry quite small. The problem of infinity in finiteness was solved by Cantor by formulating the mathematical rule of how much longer the line becomes within one step of differentiation. Then taking the mathematical limes towards infinity, the actual length of the line can be calculated precisely, assuming the line will continue with the differentiation at smaller and smaller scales endlessly.

This length compares to the length of the very first straight line the snowflake is built of. This line is one-dimensional, while all further differentiations make the snowflake enlarge into two dimensions. Still, all these differentiations will never lead to a complete filling of the two-dimensional space, which would be a rectangle completely filled. The filling of the two-dimensional space is only partly. Therefore the dimension of the snowflake is more than one-dimensional but less than two-dimensional. It has a dimensional number in between, which can be expressed by a fraction like 3/4 or 2/3, so it has a fractal dimension.

To get an idea about the complexity of a sound, in Fig. 1 four sounds are shown as examples, a saxophone, a bass drum, a cello, and a crash cymbal. On the left, the wave-form is shown, as recorded by a microphone. On the right, respectively, the same sounds are embedded in a so-called pseudo-phase-plot. A phase-plot is not moving from left to right, as does the wave-form, but is circling. On both axes, the sound is plotted over time, still with a delay, here of 20 sample points. Such a plot is called *pseudo* phase-plot, as other phase-plots might combine electrical voltage and electrical current, or the like, so two different signals. Here it is the same signal used twice for the x- and y-axis.

Still, this is exactly what we want. If a sound is repeating like with the saxophone, a regular circling structure appears. Such a sound has several harmonic overtones and has a fractal dimension of one. The same holds for the cello sound (third from top). It increases amplitude. Therefore, the phase-plot on the right is more complicated. Still, regular repeatings appear. Such a sound has again a fractal dimension of one, except for the initial transient, where it might be higher when scratchy, inharmonic sounds come in too. Yet the crash cymbal (second from top) and the bass drum (lowest plots) are not repeating. These sounds have inharmonic overtone series. Therefore, their fractal dimensions are high. They might start with about ten during the initial transient phase, then reducing to smaller values as higher, inharmonic components die out time after time.

A musical sound can be viewed in just the same way. A musical timbre consists of musical notes and noise. The notes are harmonic overtone structures, and therefore consistent within themselves, while noise consists of frequencies not harmonic one to another. This builds up a multi-dimensional space, where each harmonic overtone structure, each musical note, opens another dimension. So a chord consisting of three notes, say C-E-G, can be seen as a three-dimensional structure. Each inharmonic component of noise will add another dimension to this, as it is heard separately. The notes might have different tone colors. Still, we fuse all overtones to one sensation of a musical note, and therefore we can consider one note as one dimension on its own. Still, timbre consists not only of steady notes but changes constantly, in- and decreasing loudness, and fading in and out frequencies. These events will not be full dimensions, and therefore will add a fractal part to the timbre.

Then by applying the mathematical rules for fractal dimensions to sound, we can calculate the fractal dimension from short sound portions. The triad example C-E-G will end up in a dimension of three. Adding one additional inharmonic noise

Fig. 1 Four examples of musical instrument pseudo-phase-plots, to get an idea about the relation between fractal dimension and sound. On the left is the recorded sound as a wave-form. On the right is the same sound, still now shown in a two-dimensional way, a phase-plot. The saxophone at the top has a fractal dimension of one. It repeats nicely in a circle. The cello also has a dimension of one. It is more complex but still repeating. The bass drum and the crash cymbal are much more complex, and have a fractal dimension which changes in time, very high at the beginning, maybe around ten, then decaying, as the sound is getting more and more simple

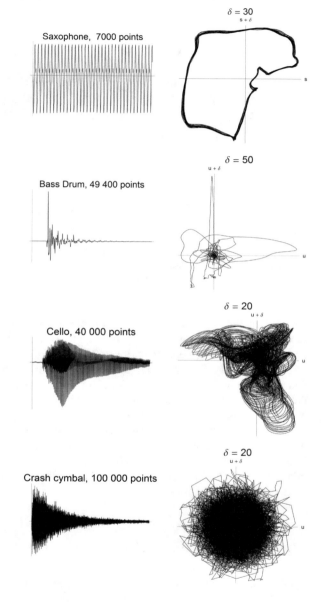

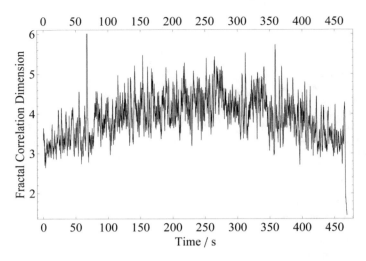

Fig. 2 The tension of a seven-minute saxophone improvisation *Broken Wings* of Even Parker, calculated as fractal dimension. The piece is in- and decreasing tension by in- and decreasing event density, which correlates to a fractal dimension. From: [47]

component will increase it to a dimension of four, and so on. Still, if this noise is low in sound, or when one note is only fading in fast, the dimension will be in between and might end in a dimension of 3.4 or 2.7.

So, in other words, the fractal dimension is counting the number of musical notes and all additional inharmonic components and sum them up to a number. This means the more musical notes and noise present, the higher the fractal dimension. So this is a good measure for musical density, the amount of perceivable events over time.

When applying this to musical pieces, we can analyze the development of density over the whole piece, and therefore the development of density the composer or improviser uses. In the piece of Even Parker, we get a wonderful arc over the seven minutes, displaying the constant in- and decrease of musical events as the overall form of that piece (Fig. 2).

Musical Form as Neural Synchronization

Synchronization in the brain during music perception [120] or music production [177] has been found. The tasks might be different. In terms of musical tension, which correlates with fractal dimension, as well as other features, like changing brightness, or loudness, as mentioned above, a time-point of expectancy in the future leads to an increase of synchronization of different brain regions, up to the time-point of the expected event, where synchronization is strongest. This synchronization is strongest between regions in the forebrain and the temporal lobe, although it is also present to a smaller extent in nearly all brain regions and changing strongly.

When calculating the synchronization, one can use the experimental data recorded by an electroencephalogram (EEG) of a listener listening to a musical piece. The EEG records the electrical field at the scull of the head, using many electrodes attached to it. The electrical field is very small, within a few microvolts. Still, it is measurable. The basic idea is that when the brain is active, when neurons fire, they produce an electrical field, which can be recorded by such electrodes. Therefore the activity of neurons leads to a voltage, while brain regions active not too much lead to very small electrical fields. The advantage of the EEG is that it can record the development of brain activity with a high temporal resolution. We get a slow-motion picture of what is going on. The disadvantage is that the EEG can only record at the scull of a person, and therefore only records the neuronal activity right below the skull. Deeper brain regions are too far away to be measurable. Another feature is that the EEG does not record single neurons, but only bundles of neurons. The electrodes are much larger than the neurons and therefore cover very many of them. If this is a disadvantage or an advantage depends on the question asked. Neurons often fire in bundles, and therefore a higher resolution might lead to more data, but not necessarily to better results. On the other hand, there are cases where single neurons react to an input while neighboring neurons do not, and then a higher resolution of neurons might be necessary. Still, for our case, recording many neurons at once is more an advantage than a disadvantage.

Electronic Dance Music (EDM) is a perfect genre to test synchronization. Each Techno night sees multiple tension buildups. Some ambient sound might be played, which is supplemented by some percussion sounds, other noise-like, floating, or beeping timbres come in, increasing the density of the musical texture and its brightness, a hi-hat might start with a simple rhythm, which is low in volume, still promising something to come, the scene is getting denser and denser, and maybe with a sound of rising pitch or brightness, the piece comes to a climax, maybe one strong boom, from where on the bass drum starts beating, with a four-to-the-floor isochronic rhythm of 120–160 beats-per-minute (BPM). Then the piece most often becomes less dense, and the bass drum dominates. Dancers await this very moment and experience it as a release of the tension, with a strong urge to dance straight to the beat from then on. So while during the tension buildup, the dancing was more free and awaiting, after the expected time-point of the four-to-the-floor income, the dances go into a much more straight dance style.

Using an EEG, and playing such pieces to listeners, parts of their brains are indeed synchronizing more and more towards the point of expectancy, the four-to-the-floor income, from where on the synchronization is gradually decreasing again.

Synchronization is found strongly in the so-called gamma-band of neural activity, around 40–50 Hz. Still, it is also present in all other frequency bands up to at 125 Hz. Synchronization means that the phases of the neurons, the precise time-point they fire with a constant frequency, become synchronized. They all fire at the same time-point. So at the beginning of the tension buildup, the neurons of different brain regions fired more or less independently, many at the same frequency of 40–50 Hz, but at different time-points, only to come closer to fire at the same time-point more and more, when approaching the point of expectation. Synchronization is often calculated numerically to range from zero to one, where zero means no synchronization, and one means a maximum of synchronization. Between the forebrain and the temporal lobe, the synchronization might be around 0 at the beginning and end up at 0.1 at the expectancy point.

In Fig. 3, the brain activity of someone listening to an EDM piece, on the very right top, corresponds to three parameters, calculated from the sound itself. The first is the amplitude, roughly corresponding to the loudness of the piece. The temporal development of brain synchronization and amplitude is clear. The points of maximum tension are the maxima of the amplitudes and the maxima of brain synchronization. Still, the amplitude is much straighter compared to brain synchronization. The fractal correlation, the musical density, and the spectral centroid, the brightness go deeper into details of the development of the piece.

On the left, some examples of brain networks are shown. Different EEG frequency regions have different networks, between which synchronization happens (here the best 5% of neurons qualify for synchronization). This differs for different subjects, and for one subject listening to the piece again.

The strength of synchronization also depends on the attention of listeners. When listening to the same piece three times in a row, the third time synchronization is decreasing, due to a lack of interest in hearing the same piece three times in a row. This is reasonable, as then the tension is no longer felt too much, and therefore we would not expect synchronization to be as strong.

As we have seen in many examples now, the brain is not processing sensory input as a bottom-up/top-down principle, but in many loops and interactions, where the interactions between the cochlear as a mechanical system up to the neocortex, as well as its downward interaction, takes only two stages or neural nuclei in both ways. So a model which takes loops, and therefore their self-organization into consideration is very much plausible.

It is also very much efficient.

Efficient Coding

Self-organization is an efficient way of storing information, as not the whole process needs to be stored by only a parameter telling a production process of what to do. Indeed, the ear is built efficiently, as found with the cat in two ways.

In the world of cats, two sounds are important, ambient sounds, like crying, and crackling sounds of leaves or other things in nature producing a percussive sound. The ear of the cat could have adapted evolutionarily towards the one or the other, as both need another process of transformation of sound into neural spikes. Ambient sounds are long and continuous. Therefore, it would be efficient for the ear to code it

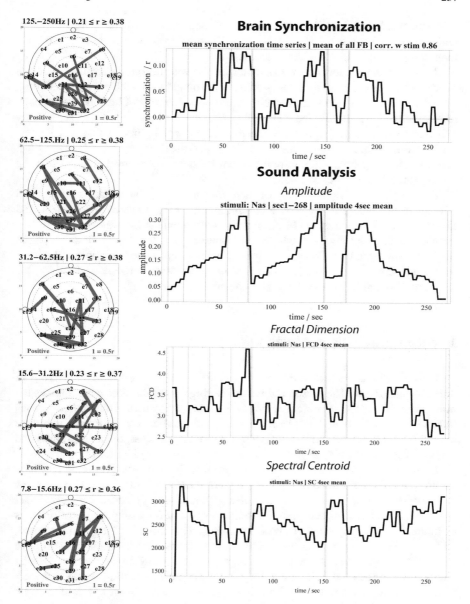

Fig. 3 Brain synchronization correlates with musical tension. On the left top, the synchronization of the brain of someone listening to a piece of Electronic Dance Music (EDM),[99] is shown. The development follows the three parameters this musical piece was analyzed with, the sound amplitude, the fractal dimension, and the spectral centroid. On the left are some examples of brain networks showing the strongest correlations for different EEG frequency bands. From: [121]

with single spikes, which determine when they started, and maybe again when they end. Crackling sounds are short and noisy. Therefore it would be efficient to code them temporally with much higher precision.

Both codings are opposite one to another, but there is only one ear, which has developed over evolutionary time. By analyzing both sound types, ambient and crackling, one can determine which spikes would fit them best in terms of how the spikes need to be produced to represent the input sound with the least amount of spikes possible. When analyzing the real spikes produced by the cat ear, it appears that they lie perfectly in the middle of both cases, the ambient and the crackling sounds [175, 176]. And indeed, when both sounds are important for the cat, it is most efficient to use a coding system that represents both most efficient.

This reminds of Uexküll and his idea that in nature, the animals, plants, and humans have developed in a way to most efficiently adapt to their environment, and therefore are the way they are because their environment is the way it is.

When using the same methodology with musical sounds, e.g., with a xylophone playing, it appears that the ears are indeed able to represent the whole xylophone tone with only a few spikes [26]. The ear is indeed very efficient, and only a few spikes are needed to represent complex music.

Brain Interactions in Music Therapy

Parkinson's disease causes a tremor, the unwilling and uncontrollable shaking of arms, hands, or feet. The reason is an overshooting of a brain loop, which passes the brain stem as well as the motor regions in the brain. Loop means again that neurons interact within a loop, and a spike returns to its origin after one loop cycle, therefore being fed into the cycle again. This is a self-sustaining process, but an unwanted one if it becomes too strong and if it can no longer be suppressed by will.

There is a rude method to stop this process by using neurosurgery. One way is to implant an activator in the brain stem at the place the loop goes through. This activator is connected to a knob the patient can press. Then a strong electrical impulse is set free, which overwhelms this brain region in a way that for a longer time of minutes or hours, it is no longer able to produce spikes. Therefore the loop is interrupted, and the tremor is gone. It is impressive to see this working at patients who have this tremor constantly and cannot stop it, and, after pressing the button, the shaking calms down within seconds, where they can immediately stand up again and walk normally. Still, this is not a healing process, and the tremor comes back after some time. Also, the brain is damaged by it.

Another way to treat the tremor is by music therapy [248]. The idea is that the auditory cortex, when still working properly, may take over the tasks of the motor cortex. This is achieved through training. With a simple task of moving the hand back and forth while listening to a steady beat, the patients learn to connect auditory perception to movement. Often, when training is prolonged over some time, these patients are able to reduce their tremor considerably, and move nearly normal again, a recovery which is permanent. The processes underlying this therapy are now known yet. Still, they again point to the close connection between brain regions and the synchronization between them.

Perception-by-Production as Self-organizing Process

There is another kind of musical form which is present to us every day, and we are seldom aware of that of different auditory sources coming to our ears where we are automatically and easily able to tell what they are and where they come from. Somebody speaks to us, a phone rings, outside a car passes, and some children shout. All these are auditory events, this auditory scene [51], which combine into a sound wave, entering our ears and our brains. We have no problems separating them all one from another, detecting what they are, hearing details of their timbre, and telling us where the sound source comes from. This is so self-evident for us that we do not consider it special. Still, it is astonishing indeed, and decades of research on how to implement this ability in a computer have brought some success, but most of the problems are not solved at all yet.

In music, this task is very useful, as we are able to tell different musical instruments, one from another, the singer from the drums, the guitar from the saxophone. In electronic music, this is more difficult, as we no longer associate musical instruments to the timbres. The sounds are then separated in terms of Gestalt principles, similarity, good continuation, or proximity. A bass sound is in a different register than a high-pitched sound. We align notes close together into a melody, a buzzing sound is heard as a continuous instrument, and split from a more sinusoidal one. Also, stereo effects help. Some sounds always come from the right, some from the left.[100]

How this is achieved in the brain is not yet understood. It might be aligned to the question of how we are able to reproduce music we have heard before in our heads. Most people memorize melodies they like and sing them along in a Karaoke manner. Still, there are cases of people hearing music in their heads, like if they would listen to them on the radio, all details are there, the timbre is very clear, the instruments can clearly be identified. These cases might be pathological [226]. Some appear with composers who are trained to imagine music. This is also known in vision, especially while dreaming, when a dreamer has a vision during his or her dream, which is as precise as everyday vision, a phenomenon called clear vision. Reproducing a whole musical piece in all details means a considerable effort for the brain. How this is achieved by the brain is not clear at all. But basically, there are two possibilities.

Either all details have been stored in the brain and are recalled. The brain is large with billions of neurons, where each neuron has ten thousand connections to other neurons on average. This might be enough to store all input, maybe only heard once in a lifetime right away. Still, the brain is constantly changing, and we perceive information each day, all day, so how can whole musical pieces, whole scenes, or other sensory representations be stored over decades so precisely? Maybe the nanoneuroscience can help there, and find that each neuron is again built of billions of smaller devices of microtubules and actin forming electrical networks to store nearly endless amount of information, as discussed above.

Yet another way to understand this would be the idea of a perception-by-production. As we have seen with the free-energy model of Karl Friston above, the brain only needs to collect a handful of parameters to understand the outside world. If the brain is self-organizing, just as all living beings, and just like music is, then it can reproduce the music, its pitches, timbres, or rhythms using the same mechanisms by only tuning the production parameters accordingly. So the lung pressure and the larynx tension in singing can be the neural connection strength between two neurons in a neural model, working just as the singing voice does. Then the same neural pattern as present in the brain when it is listening to an outside sound would be present in the brain when reproducing the sound in a neural network, using the parameters learned from the outside world before.

This would also be the process of working in the Impulse Pattern Formulation (IPF) discussed above. As musical instruments work with impulses and their interactions and therefore self-organize into harmonic overtone spectra, and as neurons also work with impulses, the spikes interacting between neurons, building a musical instrument in the brain with the same working principle is indeed possible. Then the reproduction of the sound in the brain would lead to the same neural pattern as is present when listening to this sound. Imagining a musical sound or piece would be possible in a very simple manner, and memorizing such pieces would only need very few information, namely only the temporal development of a handful of parameters, those parameters from which the sounds are built of.

This would also align with another principle: nature is generally very effective. As we have seen, the amount of spikes needed to represent a sound is much less than the information of the acoustic sound wave entering the ears. This idea of efficient coding is basic to the brain, and the production-by-perception idea would favor efficiency.

As we have seen already with the cat's ear, it has adapted to its environment as efficient as can be. Indeed, when needing to represent a short percussion sound, like that of a xylophone, the ear needs only a handful of spikes to be able to reproduce the sound nearly perfectly.

So it is reasonable to assume that the brain is indeed working in a production-by-perception manner. Still, we lack evidence yet on this. There is also no other idea persuasively around, and in the light of the brain as a self-organizing system, much is in favor of this idea.

We are now approaching higher temporal and semantic levels of music. In the next stage, we need to consider the nature of meaning and emotion.

Part III

Music, Meaning, and Emotion

Freedom! Jazz! Dance! Yeah!

There is no culture without music. There is also no culture that does not have the octave as a frame interval. No culture exists which does not use musical instruments without a harmonic overtone structure. Some use more, like Western music, some use less, like Indonesian *gamelan*. In all cultures, there is a sense of timbre, rhythm, and form. All perceive loudness, all perceive pitch, all hear different instruments in ensemble music and can identify them.

But then things become more and more culturally related. When Henry Kaiser and David Lindley, two Western guitarists were recording with Madagascan musicians in Antananarivo, the capital of Madagascar, they would come along well one with another, while recording several songs. Still, when it came to the improvising section, keeping a steady beat, the band fell apart rhythmically. At first, they could not figure out why.

Madagascar has much traditional music, often more related to Southeast Asia than to Africa. The *valiha*, a bamboo dulcimer, is a very prominent instrument of the island. Then there was much British ballroom music played at the culture clubs during colonial times, mainly waltzes or counterdances. At least the waltzes are in 3/4 m, a rhythm not known to Madagascans. They interpreted the 3/4 m as triplets, played over a 4/4 m. This indeed worked well until Kayser and Lindley started improvising, where the meter in the musician's heads became audible, and the band could not continue playing. This would not have been noticed without the improvising part, so the musicians did understand the rhythm in two different ways, while the sound was perfectly the same for both.

Indian *sitar* music became very popular in the West after the Beatles went to India, and Ravi Shankar played at the Woodstock festival in 1969. The *sitar* is a long-necked lute which probably derived out of the older Indian and middle-Asian long-necked instruments during the 19th century [115]. During the 70th of the 20th century, there were seven *gharanas*, *sitar* schools in Northern India.

© The Author(s), under exclusive license to Springer Nature Switzerland AG 2021
R. Bader, *How Music Works*, https://doi.org/10.1007/978-3-030-67155-6_16

The Northern Indian art music uses *ragas* [199], modes consisting of certain pitches, like Western major or minor scales, certain musical phrases to be played with it, a musical form, rhythm [64], and a *rasa*, an emotional mood. It most often uses microtones, intervals between the Western semitones, and many melismas, which most probably come from the times where Northern India was ruled by Muslims, bringing their *maqam* music with them. Also, improvisation was introduced in these times to Indian music [249]. There is a strong parallel between the attitude to listen to a *raga* compared to that of a classical *maqam*, which in Egyptian music is called *al-tarab*, [169] an attitude of artistic pleasure while following the melodies, phrases, melismas, rhythms, and forms. The same attitude is present for an Indian listener of a *raga*. He or she follows the melodies concentrated and closely and recognizes all small details of pitch or rhythmic variations.

Western listeners did use *ragas* in sitar music most often for the purpose of being enveloped by the sound. The *sitar* has a curved bridge so that the strings snare a bit at some part of the surface of this bridge. This makes the sound of the instrument unique and very pleasurable. Western listeners most often liked this sound, did not understand the details of pitch of melody, and could very seldom identify a *raga*. For Indian musicians, the sound is of minor importance, and a *raga* played on a keyboard with a not so pleasing sound is still a valid *raga* to them, which will produce the same *rasa*, the same emotion and spirit, when performed properly. So, although both Western and Indian listeners like *sitar* music, they understand two very different things in this music.

Still, this does not mean that one cannot learn the music of other cultures and become bi-musical, a term used in ethnomusicology. But understanding music seems always to involve learning. We are most often not aware of how we learned music from our own culture. We did so by listening a lot, having some kind of music education in school or privately, talking to others about music, dancing to it and seeing others dance, reading about music in books or magazines, listening to radio DJs talking about music, like different style names or musical biographies, etc. Nearly everybody learns about the music of his or her culture every day a bit more. This is how we understand music, our ability to know where a beat is, which musical style a piece is, or even more subtle things, like which synthesizer, guitar, amp, or even microphone was used when we produce music on our own and have experiences here.

The same holds for emotions in music. Most of them are learned. Minor is supposed to be sad, and when a Pop ballad is played in minor, we identify the emotion without the need to really feel the emotion at the time we hear the song. The emotions generated by an Indian *raga* is known to an Indian expert in that music, but not to someone not familiar with the concept of *rasa*. We learn emotions in music from the context first at all. The lyrics of the song might be sad, in live performances, we see the musician suffering. Others tell us that they are so sad when listening to a musical piece. Within this learned framework, there is much in common emotionally within a culture. Still, there are individual associations with certain pieces when a very pleasurable or a very painful experience has happened in our life, and we did listen to a piece of music, which is then related to this very special emotion. Listening again to this piece some years later will most often easily evoke this emotion again.

The same holds for music and color. Aristoteles meant that the fundamental tone of a musical scale is red, a topic used throughout philosophy until now [144]. Still, experimentally there is no evidence of what color is associated with music necessarily. The synaesthetic, the relation between music and color, might be there for individuals, but everyone associates different colors to tones, phrases, tonalities, etc., even when only asking people claiming they are synaesthetes. Indeed, when looking closely, even within such synaesthetes, one tone will not always evoke the same color.

The same holds for the claim that music makes more intelligent. Listening to Mozart will lead to a higher intelligence [3]. The so-called Mozart effect also appears when listening to Heavy Metal or Punk music, and it disappears when using other parameters for the definition of intelligence. Music does also not help when played in the background with tasks like learning vocabulary, increasing sales in warehouses and shops, or motivating in tasks. For some people, music might help in certain cases, indeed. Still, there is no general rule that this works in general, and it is as much likely that music will disturb more than that it helps.

Even if some effect is found, it is often of much doubt if this comes from music or from other tasks. As an example, it was found that children participating in playschools at age 5–6 children show better linguistic skills [181] compared to those not in such playschools. Still, such playschools use singing and rhyming as a major part of playing, which is expected to improve linguistic skills. The additional use of simple musical instruments, like percussion and xylophone, is, therefore, most probably not the main reason for such linguistic improvements.

Still, this does not mean that music is of no use. Music is powerful and can be used in many situations. But on the levels of emotions, mood regulation, increasing sales, synaesthetic, or the like, this is purely individual. Therefore these aspects are not part of scientific investigations, as there is no general rule to find.

But this is indeed good news because it means that music is part of our individual freedom. How would it be if music would work in all these cases discussed above? It would mean that we are slaves to the rhythm, slaves to emotions, and colors. We would be machines that react instantaneously and forced by music played to us! We would not have individual freedom and would start moving in a certain way when a certain rhythm is played, like puppets or slaves. But obviously—and luckily—this is not the case. We are free to decide what to do with music, which associations we want to have with it, for which tasks we could use it, if and how to regulate us emotionally or mentally. As musicians, we are free to build our own world, a whole universe of sounds, melodies, rhythms, noise, or silence, and use it in ways we find appropriate. How wonderful this freedom Jazz Dance is!

Slave to the Rhythm

So are there no restrictions? Yes, there are. During this book, we have found that music is a self-organizing system that most probably has evolutionary evolved to be energy efficient for information exchange with our voice. The self-organization produces a harmonic overtone structure, which puts all energy into very sharp frequencies, rather than distributing it all over the spectrum, as in noise. When listening to the voice, the harmonicity of the sound is enhanced by the ear and the neural nuclei following it, through synchronization of these frequencies, and fuse them into one tone sensation, a single pitch. This pitch can easily and efficiently represent information, and therefore we can communicate very efficiently.

This leads to a strong separation between harmonic and inharmonic sounds, pitches, and noise. Then noise can be used as an additional feature, the other compared to pitch, to again transport information, and enhance information exchange. Additionally, pitch is associated with intelligent beings, as it appears in nature only very rarely. One needs a nonlinear system producing a self-organized output to arrive at harmonic spectra. This produces meaning and associates pitches with 'something to say'. Many people have the idea that music is a universal language. It indeed seems to say something to us, as we know that pitches are produced only by intelligent beings who communicate with us. Still, of course, compared to language, we cannot speak with music the way we speak with language, as there is no general consensus about pitches or melodies, meaning things like table, chair, or tree.

So in this sense, we are true 'slaves to the rhythm'. Everyone hears a pitch immediately when a sound with a harmonic overtone spectrum enters our ears. We cannot decide if we want to hear it as pitch or to hear it differently. The pitch sensation is there. We all hear brightness, roughness, sharpness in musical timbre, we have an integration time of about 50 ms, in which all sound events are fused into one conscious sensation. We all have a short-term memory span of about 3–5 seconds, in which a melody or a groove is most often finished to avoid the need to use our long-term memory, which is more demanding. These, and several other things discussed in the book so far, are fixed in all of us, and we cannot avoid them. They form the basis of what we know as 'music as universal language'. Still, that is it. And other features, like color, emotion, style, etc. are associated and part of our individual freedom.

Politics and Philosophy

As we have seen, epistemology was always strongly related to politics. Systematic Musicology as Comparative Musicology, finding universals about music, and therefore stressing what we all have in common, rather than what makes us different, is therefore political by not discussing politics. Its method is right this: considering the system of music, how music works. So although there are several universals as features, like the octave present in all cultures, or simple rhythms, universals are also systems of impulse patterns, where the output might be most often one feature, still, due to interactions with other impulse patterns in the environment, the output might deviate from time to time. Although most music played in the world today is based on just temperament, many deviations exist, like the Western equidistant tuning, due to constraints from musical instrument geometries, music-theoretical demands, like wanting to go through all keys or the like. Then the universal is the system, and the output has a mean but also deviations.

Still, in the history of philosophy, alternatives to thinking in systems have been proposed. One epistemological alternative is scholastics, the philosophy of the Western Middle Ages, based on Christianity. Ernst Cassirer (1874–1945) was pointing to this difference by examining the history of epistemology and found two notions, that of function and that of substance [60, 61]. He wrote several books to understand the epistemology of Nazi Germany. Cassirer, of German-Jewish origin, needed to emigrate from Germany in 1933. Bahle's considerations on the process of composition just mentioned were also pointing against German Nazi ideology, which was maintaining an esoteric tradition.

With Cassirer, the notion of function is that of a system in its various forms. The notion of substance comes from Aristotelianism. Substance is derived from the Latin word *sub stare*, meaning 'standing below', 'having as foundation'. This again is a translation of the Greek *hypokeimenon*, as used in his book called Metaphysics. In his writings on categories, he finds that many things can be said from a thing, its quality, its quantity, how many it is, where and when it is, etc. And it can be said what its foundation is, its *hypokeinmenon*, its substance. This notion was very dark during the history of philosophy, and many interpretations have been proposed. Indeed, Aristoteles deals with this notion again in his Metaphysics. There he says that a substance is every feature that belongs to a thing, and if one of these features would be split off, the thing would no longer be that thing. This is contrary to the *symbebekon*, the features which occasionally come with a thing and might be there or not. It would not affect the core of the thing. So Sokrates is a human being wearing a white toga. If he would wear a red toga, he would still be Sokrates, but if he would be an animal, he would no longer be Sokrates. So the color of his toga is occasional, while him being a human is part of his substance.

That is all Aristoteles said about this, and it is easy to interpret it in terms of the question of what a notion is. What makes a tree a tree, and when does it become soil again? What makes an ape an ape, and when does it become a human. This question is fundamental in many aspects. One of them is in today's computer science, where a computer shall identify a thing, a face, a situation. Another aspect is an ethical one. In medicine, when is a state of life, like having good or bad times, an illness, like a bipolar disorder, needing drug treatment [150]?

Still, the Middle Ages made this idea the core of their thinking, considered substance by denying a system. A famous example is that air is moving up. In terms of the notion of a substance, warm air has it as its intrinsical attribute, as part of its substance, to move up. Still, today we know this is wrong. Warm air, as well as cold air, have mass, and therefore the mass of the earth is acting on them with force, dragging them towards the earth, downwards. Still, cold air is denser, and therefore heavier, and therefore is dragged down stronger than warm air. So it is the cold air which is pushing the warm air up.[101]

In music it is very common to give musical pieces attributes, call them cool and groovy, sentimental or hard, noisy or harmonic, and millions of other attributes too. This is indeed Middle Age thinking! Musicology made thousands of empirical studies on the attributes the musical pieces have, asking listeners to judge a musical piece along with adjective lists and say if the piece has a certain brightness of three, or five, or a softness of one, or nine. No doubt, people give their judgments, still, the methodological background of asking these questions is Middle Ages. The same holds for music journalism, which still uses these adjectives often. As this is easy to do, and everybody can use his or her adjectives for music, this method is open to manipulations and ideology at random, so politically dangerous.

There are alternatives, indeed. We have discussed the self-organizing maps, which do consider music in terms of the relations of features, and therefore in terms of a system. Neural networks do the same, and many methods have been invented there. Multidimensional Scaling is another such method, which does not ask for adjectives but for similarities, and sorts the results in a multidimensional space. Politically speaking, these are democratic methods.

As we have seen, the Middle Age philosophy of substance opens the way to randomly choosing a reality and therefore opens up a way to ideology or dictatorship. A philosophy influential for many artists in the second half of the 20th century was Martin Heidegger considering the origin of the artwork [126]. Heidegger was a student of Edmund Husserl, and therefore deep into phenomenology, perception, but first wanted to become a priest, and therefore also has roots in theology. In his writing *Being and Time*, he follows the ideas of Aristoteles of *dynamis* and *energeia*, of a possibility becoming a reality in time, and combines it with the notion of Being *Seyn*. He is asking why in the history of philosophy, nobody had ever asked why there is something and why there is not nothing, therefore asking for the very existence itself. This Being is considered as the *substantia* of things, which develop according to their possibilities. He very much criticizes what he calls metaphysics, ideas, methods, and systems, which, according to him, had to be artificially (Gestell) put over the truth, Greek *aleteia*, which he interprets as a-lethe, away from forgetting, recovering. *Hypokeimenon*, the *substantia*, for him shall not be translated as sub-stantia, sub-stare, as laying below a thing, but as laying right in front of perception, the immediate perception, what is right there. Therefore he was in favor of instantaneous perception not failing, an idea withdrawn by his teacher Husserl, with whom he overthrew himself. All systems, all counting and calculating, would never lead to the truth. The truth, the Being, for Heidegger, is the immediate perception, free from misconceptions or theories.

This thinking was also present in Nazi Germany in physics. Philipp Lenard (1862–1947) was proposing what he called a 'German Physics' [173], which was physics nearly without math, physics, mainly, based on perception. The opposite of this German Physics for Lenard was the Jewish Physics, with complex math, where he was pointing to Albrecht Einstein [131]. Lenard may be the first to have produced x-rays in tubes he had constructed but was not able to understand what these x-rays were, as he did not develop a system on it. He delivered his tubes to Whelm Röntgen (1845–1923), who built the theory, and got the Nobel Prize for it, making Lenard quite annoyed. Lenard was also opposing Werner Heisenberg (1901–1976) for his mathematical formulation of quantum mechanics, for the fact that he used such complex math, in the *Stürmer*, the political newspaper of Nazi Germany. Heisenberg was trying to build an atomic bomb in a cave in Southern Germany then and was asking Heinrich Himmler (1900–1945), one of the leading Nazi politicians and head of the SS, to stop the attacks of Lenard, which succeeded.

Heidegger's ideas about arts were based on his later thinkings, where he made a turn to introduce the notion of earth in this thinking. He, therefore, made his ideas about the development of the Being in time more historical and practical. A work of art consists of matter it is built of, and an idea forms its shape, an old Greece idea also found by Aristoteles (*eidos* as idea, and *hyle* as matter). Still, contrary to Aristoteles, he is still opposing the notion of a system and sticks to the pure perception of the artwork during the creation process. These ideas became very influential for artists, as it lacked the notion of a genius and kept the process in the artist, rather than looking at the artist as a pure medium, where the ideas are only flowing through. It made the artist the core, the substance of the creative work, where he could express himself.

Today we take this for granted, that everyone is creative and expresses himself. Philosophically this notion, and therefore the acceptance that this can be and is desirable was developed only recently and is based very much on the work and influence of Heidegger.

Heidegger himself claimed to have problems with Nazi Germany during his professorship through the 40th, and there is evidence also in his lectures in this time, that he was opposing the actual politics of Nazi Germany on a philosophical ground, e.g., in his lecture about Nietzsche. Still, he seems to have stuck to his idea of a Being developing in time, according to its possibilities intrinsically in it, which for him did also hold for the 'German people'. Therefore in his diaries, he tried to interpret many military events during WW II, according to the 'German Being'.

In terms of creativity, it is interesting to see that the age of enlightenment, where people were claiming their right to rule themselves and where the possibilities for modern democracies were formed, did not lead to a notion of creativity as we know it today, it leads to the notion of a genius, very close to shamanistic or animistic traditions around in Europe at that time. Only through the introduction of phenomenology, and the notion of possibility and reality, as introduced by Aristoteles and as shaped by Heidegger, the idea of people expressing themselves could come into place. Then people started having a style, were expressing and developing themselves and their career as a creative artist, developing their own ideas, and wanting to bring them to the world.

Still, as we have seen, there are these two parts of the coin, the physical and physiological laws forming the world and making us humans, where we are slaves to the rhythm, and the side of free will and association, where we are free to combine and 'be creative' as much as we want. Music missing one of the parts is often not too successful. Music not taking into account the physical and physiological constraints cannot be perceived the way the music was intended, and music not taking the creative part in it soon becomes boring.

Logic of Music

In his textbook on logic, Petrus Hispanus (13th century AD), a Middle Age philosopher, derives logic from sound [132]. Several sounds combined become syllables. Several syllables become a word. A word points to a notion, tree, human, music, etc. Combining words leads to a sentence, which is an assertion. When combining sentences, one is able to make connections between notions, and therefore we arrive at logic.

The only logical system existing until the end of the 19th century was syllogistic, invented by Aristoteles. The idea is just this, combining notions using two sentences. Both sentences have two notions, where one of them is the same in both sentences. So if sentence one has notions A and B, and sentence two has notions B and C, then one might be able to see a connection between A and C through B. Sokrates (A) is a human (B), all humans (B) must die (C), so Sokrates (A) must die (C). There are several combinations of the notions, one might say that A is B, or A is not B, or A is sometimes B, or A is sometimes not B. Also, the sentences might be arranged, that A is saying something about B or vice versa, B is saying something about A.

Checking all combinations, one can see that there is only one combination showing that C need necessarily be with A so that Sokrates necessarily must die (which he did). This only case is when both sentences say that A is B and B is C (and not that A is not C or A is sometimes C, etc.). Still, there are a few combinations where one sees that A is sometimes C or that A is not C. But indeed, most combinations do not have any connection between A and C, one cannot say is A is C, or A is not C, or A is sometimes C.

One such case, where one cannot say anything between the two notions, A and C, is with reasonings like: 'Harmony is in music', 'harmony is in nature', therefore 'music is in nature' or 'nature is in music'. Syllogistically speaking, the sentences are harmony (A) is in music (B), and harmony (C) is in nature (B), or in short, A is in B, and C is in B. Only because we find one thing at two other things that we find harmony in music and harmony in nature, does say nothing about the connection between music and nature [25].

Such reasonings are called associations. We often do this in daily life, and it is the main tool in advertisement. Cars are often advertised in TV spots showing it with beautiful people, driving through beautiful landscapes. Hopefully, the next time we see such a car, we will associate beauty with it. Still, only because beauty is with the people driving the car in the TV ad does not at all mean that beauty needs to be with cars. We all know the trick, it is only an ad, wanting us to have this association to buy the car.

So it seems that we can build associations, like our neocortex is able to do, without the need to do so. When speaking logically, Sokrates needs to die because he is a human, no matter what we think about it, which kind of ad we produce about the connection between humans and dying. Therefore many reasonings why music is like nature, like cosmic laws, like string theory, like planetary harmonics, or that music is cool because the musicians in the videos look or act cool, are only associations and not necessarily true. Indeed, there are rules necessarily present in music, pitch perception, rhythm, the octave, etc., still many others are not, which can easily be seen when using such a simple system as syllogistic.

But coming back to the derivation of logic from sound it is interesting to see a similar thing as we have seen already with the self-organizing nature of the human voice and the harmonic overtone spectrum it results in. Sound is the basis for building language. Following the Hispanus reasoning, it could be that combining sounds would lead to notions, words saying something. In language, the words used for things have been agreed upon by speakers, so the same thing has different words in each language. If one agrees upon sounds as words in music, one has a regular language. One example is the talking drum, a drum used in West and Central Africa, which has several sounds, according to how it is played, which drum pattern is used.[102] Special sounds combined with special rhythms form words and combining these words lead to sentences. These sentences are as real sentences, a regular language, but one is able to speak with these drums over much larger distances as one could with the human voice.

Music can also be symbolic, like national anthems, with which one cannot speak as with normal language. Still, they mean something. Everyone taught the melody, and knowing that it is the anthem of a certain country, will recognize its meaning. Other such examples were often used in Baroque music, where musical phrases were denoting meaning everybody would realize. One example is the 'lament second' (*Klagesekunde*), a minor second musical interval going down, representing sadness and lament. As discussed already with emotions, it does not mean that everybody became immediately sad when hearing this phrase, but everybody knew that it means sadness. Using such phrases it was possible to compose so-called program music, overtures, or preludes in operas before a new act, representing the overall emotion of the plot to follow.

Going one step further towards logic, in Western music the functional harmony is giving the opportunity to produce cadences. Although in polyphonic music of Renaissance times, cadenzas were known and used, these are forerunners of the cadence we know today of a tonic, a sub-dominant, dominant, and tonic chord, so, e.g., the chords C-F-G-C, a cadence in C-major tonality. As mentioned above, it was Hugo Riemann who invented this terminology giving the steps of the chords, IV-V-I a semantic meaning [220]. He derived this meaning using the dialectics of Hegel. Then the tonic is the thesis, where the music is still within itself, the subdominant and the dominant are the anti-thesis, where the music is leaving its origin, but where the musical note B in the dominant chord as a minor second interval to the fundamental tone C, is "strongly demanding" to return to the tonic chord, which is the synthesis.

Writing this with syllogistic notation, one finds that this is not the case, and the note B is not necessarily needing the C note in the end. And indeed, the musical syntax also knows the deceptive or interrupted cadence, where the chord G (the V) is followed by the chord A-minor (the VI). The German term for this is *Trugschluss*, false conclusion, or false syllogism, relating to the logical nature of the cadence. So although we do not give our reasoning an extra-musical meaning, the music is still logical, using chord progressions. This is also expressed by what is called 'musical thought', when speaking of Western tonal music, and the chord progressions or the polyphony in it.

But is this still music when we only deal with notes, which are small dots on a piece of paper? Indeed, when only taking the musical score, we could as well analyze a bus schedule, where we can see at which time point which event happens. Only the timbre, the sounding music, makes music, and building dots on a score does not at all mean that someone hears what the composer intends. The dots represent pitches, self-organized harmonic spectra, which are perceived as one pitch event, as the special feature of human auditory perception.

In the so-called mathematical music theory, a reasoning of the tonal system we use in the Western world was found in a vice-versa way [191]. If we use seven tones per octave, one might ask why we often prefer a major or minor scale. There are very many other possibilities, like if we would use only the first seven half-tone steps and leave the rest up to the octave out, or any other combination of tones. Now when assuming constraints to a tonal system, things we expect this tonal system is built from, we can reduce the amount of possible tonal systems meeting these constraints. Using constraints like flexibility, how many different combinations are possible with it, assume symmetry conditions, and other constraints, we arrive just at the major scale. Indeed, this is an experience of composers, that when using some sort of tone material for a composition, it might become boring after some time, or one finds a lack of tension or variability, while other combinations of tones allow much more interesting and differentiating music.

Therefore music can have syntax. Indeed, this syntax is processed in the brain, in the same region as the syntax of language, in the Broca region [157]. Still, the processing is a bit different, and until now, we have no clear picture of how it happens in terms of neural networks within this Broca region.

Physical Culture Theory

Music is part of what we call art and culture. We have seen that music is a self-organizing system by its very nature. Self-organizing systems are more than 'dead matter', they are living. So at least this part of our culture is a life we have invented as living beings. The musical instruments we have invented, the timbre we have created, pitches, chords, as well as musical meaning and emotion, are all part of a whole system, which is alive.

Therefore we consider music as lively, as inspiring, as meaningful, and also as a language. To us, it appears like a living being, although it is not a human, an animal, or a plant. Music is maintaining itself as all self-organized systems do. It has a history, as all living things have. It develops according to its own rules, likes, and its history, again as all beings do.

At the same time music is not only within one human or stored on CDs and MP3s. The musical memories, abilities, and likes, the musical instruments as physical systems with the rules of playing them, the music stored in all formats, the human ear and auditory system with its physiological constraints, music performances, and dance altogether form a self-organizing system which maintains itself, develops, and therefore is a living whole.

Music, culture, nature, humans, we all are electricity in the end. There is no difference between culture and nature. Sound, neural activity, cultural objects, all interaction between bodies is electrical. We have seen that musical instruments, as well as conscious content, works with impulses, short energy bursts. These impulses are spread out, reflected at different points, come back, and form complex, self-organizing patterns. These patterns are highly organized. They maintain form and order, like harmonic overtone spectra, consciousness, or most probably also social communities, economics, or politics. All these are complex spatio-temporal patterns. To maintain this whole system as a living being, time developments include phases of breakdown, chaos, change. Such changes might come slowly, they might be very fast. Still, they are necessary to maintain the system.

R. Bader, *How Music Works*, https://doi.org/10.1007/978-3-030-67155-6_17

The Impulse Pattern Formulation (IPF) is capable of modeling musical instruments to such a precision that we can expect it to be able to model other cultural phenomena too. We might do this computationally, we might make simple calculations in our heads. Just as a simple example, think of the task of writing a letter by hand on a piece of paper lying on a desktop. The desktop is manufactured by us, it is a cultural object, as is the pencil and the piece of paper. The content of the letter we want to write is the conscious content we want to transform into another form, the letter. Now, the desktop is a flat surface to allow us to write the letter. If it had an uneven surface, cracks, ups and downs, the task of writing a letter would be much more difficult. But why is this? When writing the letter on an uneven desktop, our writing hand would constantly experience impulses of mechanical force acting back to the writing hand while it moves over the piece of paper. This is annoying. It complicates things because the mechanical impulses are transferred into neural spikes in the nerves of our hand. These impulses are transmitted to our brains, where they interact with the spatio-temporal electric field of the conscious content of the letter we want to write. Such disturbances will make the task of writing the letter much more difficult. The self-organized system of brain, hand, and desktop, which is supposed to have the conscious content as the one winning the game over the other subsystems, is so much disturbed that the uneven desktop takes over and determines the content of our consciousness: we no longer think about the letter, but about the unevenness of the desktop and stop our task. What to do? We take another desktop, or even out the surface of the present one. All these tasks again work with impulses, carving the desktop, taking a new one. All these are actions we take to maintain a high level of organization, low entropy, which in this case is the successfully written letter. This letter makes communication, it again organizes things, and reduce entropy, it is about to make the system we are in a stable one. As all these tasks necessarily happen deterministic, in the end, we would not have acted differently. There is no nature or culture. There is only a self-organizing process reducing entropy.

Still, this is not a materialistic view. Consciousness is part of all these actions humans or animals are involved in. We experience these actions instantaneously, reflect on them, reconsider, make decisions, and perform all conscious tasks we are able to do. In terms of music, there is no reduction of experience, still, there is also no reduction of reflection and understanding. The joy and pleasure, emotions, and experiences are perfectly part of the whole process, as is music theory and understanding.

Also, this determinism does not dismiss us from considering what to do, using ethical principles as the basis of our actions, or leads to nihilism, considering all actions senseless anyway. We are part of the game, do not know all its elements, and therefore need to consider alternative actions and make decisions. Still, all this is only part of the deterministic organization. The future is determined completely.

Still, as we need to consider ethical principles for our actions, it is a good idea to help this self-organizing system maintain itself, as it is what we call life. We can differentiate it, we can simplify it, sometimes it makes sense to built structures up, sometimes it is necessary to pull them down. But all these actions should lead to a slow-down of the increase of entropy, should resist life-destroying actions, should help maintain humans, animals, plants, culture. The alternative is a desert, a moon, a place where no life exists. Actions destroying our environment need to be counteracted by building structures of low disorder, low entropy, which are able to maintain themselves as part of nature and culture. Music is such a structure. Arts, and all aspect of what politically is meant with the term 'culture' is, as well as all other living Beings.

Still, not all structures are helpful in this sense. Complex bureaucracy, tools with hundreds of parameters, or organizations too large to handle might be very complex. Still, they are often more chaotic than structured, have much higher entropy than more simple organizations, and are often a huge waste of energy. It is not the amount of complexity that makes a system maintain life and culture. It is its low entropy.

Predicting Culture

Building an Impulse Pattern Formulation (IPF) allows to model systems with very high precision, also including chaotic parts. We can understand culture by formulating IPFs. We can also make predictions about what the future of culture might be. We can try different scenarios, which possible actions we take will result in which consequences.

Prediction is what we do every day. Most actions we take are there to achieve an aim. We sit down, composing a song with the aim of having a song in the end. Therefore, we expect it a good idea to use a musical instrument, write a score, make some sounds. When we buy some food, we expect it to feed ourselves. Therefore, we consider it a good idea to go to a food store. Most of our actions are meant to reach a goal.

Now the methods we use to reach that goal are most often not clear to us. Most actions are learned. We have experienced that these actions lead to that goal. Still, maybe this time the situation in our environment has changed, the food store is not open. We do not know this. Therefore, our action fails. But as most of the time we succeed, of course, we use the method of experience.

Another method is that of calculating. Political actions taken are supposed to be based on research of the political, social, or economic situation. Then a certain action is often hotly discussed and decided in the end. Here most people consider the world a complex system, where the complexity is often way too high to really be sure where the action leads to. Prognostics in economy is very often failing, although economy is mainly based on numbers. Still, many find emotions, narratives, and unreasonable behavior more important in economic actions than reason [106, 235]. So our tools to predict culture, politics, or economy, which is all culture in the end, are still very poor.

The Physical Culture Theory, using IPF or related mathematical tools, could improve this considerably in the future. It takes culture as a whole, brains, cultural objects, nature, and nurture in a whole system. It is scalable, one can calculate the whole in a reasonable time of modern ultra-scale computing, it could also look at tiny little aspects. And finally, it is based on physical impulses, electricity, energy, a unit existing without a doubt. It does not need to include any speculations, assumptions, or heuristics.

Physical Modeling or Artificial Intelligence?

The Physical Culture Theory is physically modeling culture. It does so by using the systems, its components, and connections and understand culture or calculate the future. This is fundamentally different from most AI algorithms. They most often use neural networks that learn from the past. As we have just found, this is very human-like, base action on experience.

Still, experience might fail, especially when it comes to sudden changes. Predicting the political situation in Africa, using machine learning algorithms is able to predict long-term developments, but fails when it comes to sudden changes, like revolutions [125]. Predictive policing, as used in Los Angeles and other cities for many years, and only just withdrawn due to political pressure, try to predict where criminal actions will take place based on machine learning algorithms using the past records as a basis. They have the problem of expecting criminal actions in communities, already known to be more critical. This enhances the visibility of even smaller law conflicts, present in other communities too, where they are not recorded that strongly. Such a machine learning algorithm is subject to prejudice, not too helpful.

The Physical Culture Theory, on the other hand, bases its analysis not on experience but on knowledge about physics, using a physical model. As a self-organizing method, it is able to predict long-term, as well as short-ranging, and sudden changes, as we have seen before.

In a world getting more and more complex, we need to improve our methods of prediction. We use predictions every day, hundreds of times, still we do so most often based on experience, poor data, or poor methods. Computational methods able to deal with Big Data will be a tremendous improvement. Methods based on physical reality, and not only on experience, will do even better.

Notions as Spatio-Temporal Patterns

One task of the future, following PCT, is to find the spatio-temporal electric fields in our brains, combined with cultural objects, that are certain conscious contents. Musical pitch seems to be the massive spatial and temporal presence of a certain spike frequency in the auditory pathway. Connecting several areas in the neocortex in the gamma-band frequency range is the experience of rising or falling tension in a musical large-scale form. But what about the millions of other content we might have, identification of a genre, the pleasure when hearing a sound, a perception of fatness of a dance beat? They are all spatio-temporal patterns we will one day be able to visualize as a 3D field.

Examining these fields, finding similarities between similar states, similar conscious content, maybe sorting and ordering them, and finding a system of conscious content is a huge task, waiting for a new generation of scientists to investigate.

Also, notions are part of it, semantics, logic, language, meaning. Notions are often blurred. Attention, expectation, tension, etc. are not perfectly defined notions and not clearly distinguished one from another. This blurring is a fundamental feature of language, it is a feature of modal and fuzzy logic. Such blurred notions could correspond to blurred spatio-temporal fields, which might overlap, show similar patterns in space or time. Replacing notions by such spatio-temporal patterns would abandon endless discussions about the definition of such notions by displaying the spatio-temporal patterns they are. A new kind of logic will emerge from such 3D representations of conscious content.

The Musical System

Music has developed over evolutionary, as well as cultural progress. Both are necessary to develop what we call music today. These developments appear to us to be more than an amount of played pieces, performances, musicians, styles, or storages like CDs, LPs, or downloads. Music is all this as a living system. Therefore it is not reasonable to define music. Jean-Paul Satre once asked what a Beethoven Symphony is, is it the musical score, the dots written on paper, is it the performance during a concert, is it the musical idea Beethoven had in mind while composing? Of course, it is all this, and much more in a whole system of reasonings, a self-organized system.

All these findings are part of a system, where each part can be proven experimentally. This distinguishes the Physical Culture Theory from philosophical speculations of analogies and personal experiences, hermeneutics, where one is reflecting his ideas and thoughts about music, music writing in terms of journalism or feuilleton papers, or similar methods. Contrary, PCT is strict science. Although some of the findings might be similar from time to time between a scientific and a hermeneutic method, with hermeneutics, one can never be sure if these findings are right or wrong, if one found a general law or only reports a personal experience. This is especially problematic with such a thing as music, where everybody has his own, often highly emotional experiences, relate music to social or political convictions, and has his or her personal taste of styles, bands, or instruments. There is nothing wrong with that, of course. When it comes to understanding music and culture, these personal experiences might be a first starting point, but they are also very easily leading us wrong. Only experimental prove can give us confidence and lead to a sound understanding, which can be shared by everybody in the end.

For some, this might be too much thinking and reasoning in terms of music. They even might claim that one is not able to understand music with physics at all. As we have already seen, this leads to a belief in a world outside our world, not related to it, a position which we have seen is not necessary to understand all experiences possible with music, and not favored in this book.

Another reason for rejecting a Physical Culture Theory might be bad experiences with Artificial Intelligence or with technical inventions in the past. Still, this is not contradicting the reasoning of a Physical Culture Theory at all. All that it means is that the development of new self-organizing, and therefore living systems, extending humans, might fail and prove to be more devastating or restricting rather than increasing human freedom and the protection of animals and plants. Adjusting these problems is more intelligent than just skipping the whole process. And indeed, life will find its way. Might this life be what we understand as evolution, or might it be the extensions of humans, what we have built by our physical culture? It is not Artificial Intelligence suppressing people, it is people suppressing people. In former Eastern Germany, this suppression worked with technical primitive methods, like microphones, evaporating letters, and spying.

Is Culture Conscious?

Is music a living being with consciousness? Would it be right to associate a higher being with it, a God of song, which we can communicate with, which has ideas, a self-consciousness, has feelings and wishes?

Following the discussion above in terms of consciousness, we might be skeptical about this idea. Consciousness seems to appear when a neural network is in a medium degree of complexity. As the human mind shows such high complexity in a very small space, the space in our heads, with a complex electric field, such a complex field is not present in the system of music the way we see it now. This does not mean that it is not a living system. Still, we do not have any evidence that it has consciousness on its own.

So, if at least music, as part of culture, is self-organizing, may culture in general be just the same? We cannot tell here for sure because, therefore, we would need to go deep into the physical mechanisms of other art forms and other parts of culture, like economics or politics. Still, if this would be true, then culture is another form of life, a life that follows the same rules as all life does. It is born, lives, and dies. It has a history, tries to maintain itself, reacts to things around it, and appears to us like a living thing, what it indeed is.

The search of an origin of music, evolutionary, cultural, technical, or social is then really a search for the plural, the origins of music, and the question of the chicken and the egg. But if we view music as what it really is, a self-organizing, therefore living system, we do no longer need to search for such an origin, music has gradually developed and will ever do so.

So the question is not too much of what music is, but of how music works. It is the system of this living being music, which is of interest, and the interactions between all its parts. There is no 'must' or 'do not' in music. There are only decisions and consequences, developing the living being music on and on. Everyone does act as part of it and can examine its actions and how it works. We have brought music to quite some height. Let us see where it goes next, and let us act on its development in the future.

Notes

1. Ernst Florens Friedrich Chladni (1756–1827) wrote the first comprehensive book on musical sound [63].
2. Wilhelm Eduard Weber (1804–1891) in 1846 was still promoting such a force from a distance he called *Fernwirkung* within his framework of electrostatics.
3. A wonderful narrative description of this view can be found in the novel *The island of the day before* from Umberto Eco [72].
4. Faraday was a famous man at his times as he performed his experiments publicly. He also had the best-equipped laboratory known in his times and was able to perform experiments no one else could. Therefore his regular publications were read a lot (see, e.g., [83]). Still, when he introduced his concept of force-lines, his major finding, it was not much appreciated in public or by colleagues, as it was too much ahead of his times.
5. Chladni patterns can also be produced by moving water with a loudspeaker, displayed in the wonderful book of Alexander Lauterwasser [170].
6. She suggested the correct differential equation of fourth-order still used today, but did have wrong boundary conditions. Today several plate theories are known and used which differ in one or another way, still, they are all based on the equation of Germain.
7. Inventing new musical instruments or varying existing ones, of course, is as old as mankind. Adding a seventh string to an ancient Greek Kithara [265], or a Renaissance *viola de gamba* has been seen as a symbol of resistance, or a one of completion of the instrument [215]. A journal *Experimental musical instruments* edited by Bart Hopkins was dedicated to such new inventions [134]. See also http://www.oddmusic.com/. There are many attempts to add sensors to electronic instruments to enlarge their articulation [143] and many ideas and thoughts about possible new instruments [48].
8. Of course adding a soundpost to a violin leads to other changes too [66, 137], still, they seem to be of minor importance.
9. The loudness-war is known from Pop music, where pieces are compressed to make them sound louder. This cancels dynamics in music, and often the

R. Bader, *How Music Works*, https://doi.org/10.1007/978-3-030-67155-6

compression is so strong that listening to such pieces is sometimes hard to stand.

10. There are many publications about the history of guitars, especially the American brands. A very good introduction is [77] or the much illustrated [261]. Reference [109] is a guide to all the models and their specifications, including serial numbers, used to identify vintage guitars.

11. A collection of the paintings in their excepts can be seen at http://www.prydein.com/pipes/paintings/brueghel/brueghelspage.html (seen 2018).

12. Temple performances can be heard in the COMSAR archive at http://esra.fbkultur.uni-hamburg.de/. For videos, see https://vimeo.com/albums/5261751.

13. Videos at https://vimeo.com/album/5259277/video/277447431.

14. Videos of this marching band can be found at https://vimeo.com/album/5259277/video/277448638.

15. Young was changing the air pressure at an organ pipe. When the pressure was high, a sound was heard. When the pressure was very low and when adding smoke to the air, the pipe output "...was changed into a succession of distinct puffs, like the sound produced by an explosion of air from the lips; as slow, in some instances, as 4 or 3 in a second. These were undoubtedly the single vibrations, which, when repeated with sufficient frequency, impress on the auditory nerve the sensation of a continuous sound. On forcing a current of smoke through the tube, the vibratory motion of the stream, as it passed out at the lateral orifice, was evident to the eye;..." [270] p. 115.

16. Young used this principle also for combination tones. These are tones that appear when two tones sound at the same time. Then additional tones can be heard below and above the two tones. Beating, a rais and fall of tone volume, was also his theoretical reasoning understanding superposition. In the discussion on combination tones, many prominent researchers took part in the 191th century, among them Weber, Ohm, or Helmholtz. Young also suggested a new tuning system derived from Valotti, the Valotti-Young scale. He also was the first to perform the double-slit experiment of light in 1802, showing for the first time that light is also a wave, an important finding in the development of quantum mechanics about a century later. Young argued between sound and color in close analogy [208].

17. Indeed, beatings were already discussed by Young, Weber, and others in the first half of the 19th century as a possible cause of combination tones. There are new tones not present in the original sounds but appear due to the combination of the sounds. There are additive combination tones that appear at frequencies resulting from adding two frequencies of the original tones and subtractive tones. Combination tones only appear in the ear, and different mechanisms are the cause of them. It was often discussed if these tones are influencing music perception of harmonics, chords, or timbre.

18. http://magneticmusic.ws/telharmonium%20video.htm.

19. The sawtooth motion of the violin bow was first visualized by Helmholtz using a stroboscope.

20. A historical demonstration video of the voder can be found at https://www. youtube.com/watch?v=5hyI_dM5cGo.

21. For textbooks about the mathematical background of self-organization and synergetics see, e.g., [5, 113, 114, 209, 240].

22. For a review on the literature about temporal lobe epilepsy see [192].

23. This leads to women experiencing loud music as even louder than men, at least statistically, as the outer ears of women are built such that the sound coming to the ear is transferred better to the middle ear.

24. Nerves were considered scientifically since Galen (129–216), Avicenna, or Leonardo and already associated with sensory input and activation of limbs.

25. *Magnes sive de arte magnetica*. For a very entertaining summary of this work and the reaction of his contemporaries like René Descartes or Marin Mersenne, see [102].

26. Some of the sounds produced on the terpodion of the Viandrina in Frankfurt a.d. Oder can be heard at http://www.rolfbader.de/terpodion.

27. More about the lounuet can be found here: https://rolfbader.de/lounuet.

28. For an overview on Nepalese musical instruments, see [152].

29. Masking is a fundamental task performed already by the human ear [82]. When two sounds come in at the same time, and both have about the same loudness, both sounds are heard. Still, if one sound is considerably louder, the other sound is completely ignored by the neurons.

30. For those familiar with brain area identifiers, the frontal, associating cortex are found in Brodmann-Areas 9–12, 46–47, the limbic parts are Brodmann-Areas 28, 34, and the parietal areas are Brodmann-Areas 5, 7, 37, 39, 40.

31. Ernst Cassirer (1874–1945) made the first major attempt to write a history of epistemology, discussing many more of these cases [60].

32. As an example, in Germany, the legislation only considers dB(A) and not dB(C). That different noise levels need different weights was found by Fletcher & Manson in 1933 [90]. Still, this is not part of German legislation until today!.

33. People prefer very low-noise indoor devices, like refrigerators, ventilators or computer fans, but want extremely loud devices for outdoor, like reapers, motor saws or cars.

34. Such ideas of a development of culture was also the result of Hegel's philosophy. His dialectics, thesis, antithesis, and synthesis form a development of things, humans, culture, or nature. In his Phänomenologie des Geistes [Phenomenology of Spirit] of 1807, he finds mankind to develop from a primitive state into a state of a naturalistic religion. Hegel's philosophy was very influential in the 19th century.]

35. See Discography: L. H. Corrêa de Azevedo: Music of Ceará and Minas Gerais.

36. For more on Chinese music and politics see: [71]. The history of art music in China in the 20th century is struggling with politics all along, for a detailed view see [180] and a critical review on this book by [161].

37. The recordings of the Cairo congress have been digitalized: https://esra. fbkultur.uni-hamburg.de/.

38. Modern China has officially 56 ethnic groups. Although this was officially declared by the Communist Party after fieldwork in 1954, according to the recommendations of the Ethnic Classification Project [200], the foundation of most of the groups was laid by Rodolph Davies [68], an English colonialist. Traveling through Yunnan, Southwest China, he had a list of common things with him, like tree, river, or man, and asked the people of Yunnan how their words for these things were. He then sorted the tribes according to similar words.

39. The example we saw with the drum-and-sawm bands in Southeast Asia and India is an example of just this, the mixture of a global musical style, the military band of the Janizary with local traditions. So strictly speaking, these fusions are not really new.

40. The signal processing for audio effects, like reverb, chorus, flanger, etc. is also called DAF-X, Digital Audio Effects [272].

41. Sound examples can be found at http://www.extras.springer.com when inserting the ISBN of the book [21], which is ISBN 978-3-642-36098-5.

42. It also estimates the minimum bow force much better than previous methods, as it assumes the process to be dynamic rather than static. It also works for multiphonics of wind instruments, where one instrument plays several pitches at the same time. For a mathematical description see [21, 178].

43. A brilliant textbook on this subject is [5], also going into historical details. See also the memories of Mandelbrot [187].

44. A wonderful book on music and evolution is [3].

45. An excellent example is the CD Michel Doneda: 'Everybody Digs', Relative Pitch Records 2014.

46. Their homepage is http://srl.org A subjective history of noise music is, e.g., [111].

47. E.g. Throbbing Gristle finds: 'We're interested in information, we're not interested in music as such. And we believe that the whole battlefield, if there is one in the human situation, is about information.' https://de.wikipedia.org/wiki/Throbbing_Gristle.

48. To which extend these suggestions follow other rules, and maybe manipulated towards suggestions of music engaging economically towards the streaming platforms is not known.

49. For an overview, see [73].

50. The second solved problem is the Poincaré conjecture, which Grigori Perelman solved and in 2006 got the Fields medal for, the Nobel Price in mathematics. It is concerned with geometry. A torus is a mathematical object like a doughnut. Connecting two points on a doughnut surface is not always possible using straight lines, mathematically speaking, the object is not simply connected. This is obvious in 2D. If it is also the case in 3D is the question asked by Henri Poincaré (1854–1912) in the 19th century. The interesting history of that proof

is starting with Euklid and runs through Gauss, Riemann, and Einstein until the present [205].

51. Leonard Euler (1707–1783) was formulating a fundamental flow equation in 1755, which was developed further by dÁlambert (1717–1783), we already know from the string equation. Only in the first half of the 19th century Claude Louis Marie Henri Navier (1785–1836), Poisson (1781–1840), we also know already, Barré de Saint-Venant (1797–1886), and George Gabriel Stokes (1819–1903) developed the full Navier-Stokes equation which was published by Saint-Venant in 1843 and Stokes in 1845.

52. Ludwig Prandtl (1875–1953) developed solutions for flow at boundaries. His textbook starting in 1931 is still one of the best [214].

53. See [5] p. 643.

54. Many very good textbooks on wind instruments are available for further reading: [4, 79, 80, 88, 201].

55. A wonderful collection is [164]. For measurements see e.g. [97, 98].

56. Some nice examples of multiphonics produced by IPF can be found in [178, 179].

57. For details see [21].

58. The best expert on the *khaen* is Jim Cottingham, see, e.g., [65, 190].

59. Videos of *hulusheng* playing can be found in the appendix, Yunnan—China.

60. Nice videos can be found here: https://www.youtube.com/watch?v=CoxnhjLMVBo.

61. Measuring the radiation of wind instruments is not trivial. Using a microphone array of many microphones, in this case, 128 microphones, one can make photos of sound, just like with a camera taking photos of visible objects. Then the radiation of the instrument from its different sound holes can be viewed, like in [24].

62. For more details see https://rolfbader.de/lounuet and [22].

63. For more details, see https://rolfbader.de/terpodion.

64. For these instruments see [122].

65. The literature on bowing is very large. An excellent introduction textbook is [66]. The whole process is still under debate, see, e.g., [195, 196]. Some illustrations are found here: https://rolfbader.de/violin.

66. For a whole-body guitar model see [28], for reviews see [27, 41].

67. This does hold when inserting into the IPF normal reflection strength from different guitar parts. When using other reflection strength of arbitrary nature, this stability might disappear. So it is not a general rule that many reflection points lead to a stable system. Still, for stringed musical instruments, this stability holds.

68. There is a huge literature about the singing voice, its physics, physiology, and perception [243]. The hysteresis is shown beautifully in [251, 252].

69. For a detailed mathematical review see [21].

70. For a technical overview, see [151].

71. A deep discussion on these and slower rhythms and how such rhythms come into place is found in [57].

72. A wonderful compendium of psychoactive drugs is [217]. Rätsch extensively studied Nepalese shamanism and described many examples and recipes..

73. The philosophical tradition of phenomenology, as founded by Edmund Husserl (1859–1938), formulates a field of consciousness (*Théorie du champ de la conscience*, as called by Aron Gurwitsch (1901–1973) [110]) containing all conscious content we have at a certain moment. Intentionality then is focusing on some content and having the rest in the background, so to speak. Husserl took over this notion from Brentano [54], a major figure in developing psychological tests, the 'inner measurements'.

74. Indeed, phenomenologically, there is no such thing as unconscious because if there would, we would not know about it, it would not be conscious.

75. A comprehensive overview of the different kinds of neural models taking more or less of these processes into consideration is [151], see also [139].

76. The development of synchronization during epilepsy is complex, and measurement techniques are too, see [146].

77. The reason for this kind of epilepsy might be enhanced connectivity between the visual and supplementary motor cortex in patients with this kind of epilepsy, next to other factors. For a review see [67].

78. For a good review see: https://plato.stanford.edu/entries/qt-consciousness/#PenrHameQuanGravMicr. The most elaborated in the OrchOR theory of Hameroff and Penrose [116]. It assumes many microtubules in the brain to act as one coherent quantum state, which breaks down according to an uncertainty relation, causing one conscious moment. The breakdown is supposed to happen about each 20 ms, causing the 50 Hz oscillation in the brain. The reason for consciousness in this theory is found in gravitation, using quantum gravity theory.

79. Inhibitatory interneurons often build circles with pyramidal neurons to built 50 Hz oscillation [57].

80. An interactive tool to examine many experiments showing connections of many brain areas of the mouse brain, including hearing, is the Allen Brain Atlas https://portal.brain-map.org/.

81. There are many textbooks and Wikipedia pages going into details of ear physiology and mechanisms. More detailed descriptions are found in [59, 69], or [140]. Historically interesting is [39].

82. That phases cannot be heard in a sound has been a saying over many decades, claiming that Helmholtz had found this. Still, reading Helmholtz closer [128], he finds that we cannot hear phases in static sounds, the ones he produced with his synthesizer, the first of its kind as discussed below. But he also mentions that with changing sounds, initial transients, vibrato, sound manipulations, the phases might very well be heard. Nowadays, using all kinds of audio effects like phasing, flanging, etc. we use the audibility of phases tremendously.

83. More details are discussed below. Some papers on the subject are [45, 101, 140, 148, 149, 185].

84. For a review see [59].

85. Adriaan Fokker (1887–1972), who studied with Einstein, Rutherford, or Bragg, and coinvented the Fokker-Planck equation of probability physics, was promoting a 31-scale, and invented organs able to play music he composed himself. The idea was to come closer to a pure tuning of simple harmonic ratios for as many keys as possible. The Archifooon, a small version of the Fokker organ, still exists: http://120years.net/the-archifooon-or-archiphone-anton-de-beer-herman-van-der-horst-the-netherlands-1970/.

86. Harry Partch (1901–1974) also wanted to come closer to just intonation, and, inspired by Helmholtz, invented a 43-tone system. He also invented new musical instruments, many made of glass [100].

87. Another approach is to play music with microtones, without the aim to make keys sound like just as possible. This is known as microtonal or enharmonic music.

88. As performed by Izhikevich and Edelman [138] for the thalamocortex [140], which includes cortex and medial geniculate, but does not include the lower parts of the auditory cortex.

89. The Saṅgītopaniṣat-sāroddhāraḥ [194], an Indian music theory probably from the 14th century is a rare source giving a possible meaning to ornaments by associating them with tantric features.

90. Of course ornamentations are found in many other musical styles, like in the Scottish bagpipe music, Baroque cadenzas, Blues music where Gerhard Kubik finds them derived from the Arabian maqam tradition [167], and many other styles.

91. For a review and for literature in what follows in this section, see [21].

92. Carl Stumpf describes this in his book on speech sounds [242], where he was using a complex tube system connecting several rooms (p. 44) to erase or synthesize single overtones of a speech sound. Using a part of this system, he was also investigating musical timbre (p. 374–412).

93. A simple sketch of its sound production is found here: http://www.oriscus.com/mi/tm/index.htm.

94. For an overview on neural models, in general, see [151]. For details on the literature on music neural networks, see [21].

95. The map is online, and can interactively be used, playing the pieces, uploading own recordings, and comparing them with the collections: https://esra.fbkultur.uni-hamburg.de/home. For details see [12].

96. There is also a relation to Martin Heidegger's phenomenological ontology, especially his interpretation of the 'Sentence of Identity', which is A = A. Logically, A = A is a tautology, it does not say anything. Still, phenomenologically, the two A's are different in terms of time, one might come after the other. Then the tautology resolves, and the Sentence of Identity becomes *innig*, a closed-world scenario.

97. Like, e.g., found in the Allen Brain Atlas https://portal.brain-map.org/.

98. Nas: One Mic, from: Nas: CD *Stillmatic*, Columbia (Sony), 2001, track 7.

99. Nas: One Mic, from: Nas: CD *Stillmatic*, Columbia (Sony), 2001, track 7.

100. Literature on Music Information Retrieval (MIR) in this field is huge, some aspects are found in [1, 36, 156].
101. Descartes had a similar problem when co-inventing modern mechanics. Scholastics claimed that all things moving want to go to rest again. This feature was understood to be part of the substance of these things. Only by freeing his thinking from this idea, Descartes could come up with the idea of gyration, that a thing moves on forever, and slowing it down needs another body hitting it.
102. For a discussion on music and language see [86], for a music example see https://www.youtube.com/watch?v=CbMMw-88eT4.

Discography

The musical styles or recordings below are those mentioned in this book.

(1) Sound archives and collections

http://esra.fbkultur.uni-hamburg.de/
Computational, interactive analysis of ethnomusicological recordings from music all over the world.
https://vimeo.com/rolfbader/albums
Videos of local music I collected from different countries in Southeast Asia and China.

(2) Countries

Myanmar

https://vimeo.com/albums/5259277
Videos I have collected in Myanmar over the years of Bama and Kachin music. The Kachin are people in the very north of Myanmar often still living as anarchists [227].

Burma. Classical Theatre Music.
2 CD set recorded by Philip Yampolsky with notes of Ward Keeler on the Bama classical *hwaing waing* orchestra released in 2010.

Mhagitá. Harp and Vocal Music of Burma
CD of the *saun gauk* harp music of Myanmar and its acccompanying singing released by Smithsonian Folkways Recordings in 2003. The 'big music' *maha gita* is a collection of harp tunes. Harps were found all over Southeast Asia in ancient times through the Middle Ages, the Burmese is the last remaining. Once an instrument of the court, it is now used for private entertainment.

© The Editor(s) (if applicable) and The Author(s), under exclusive license
to Springer Nature Switzerland AG 2021
R. Bader, *How Music Works*, https://doi.org/10.1007/978-3-030-67155-6

Wunpawng band

The Kachin ethnic group of Nothern Myanmar have a marching band playing at weddings and in church ceremonies (the Kachin are mainly Christians). They also play at the Manao, a yearly festival of Kachin people, which has not taken place over the last years as Kachin are still in a civil war against the Myanmar government and other ethnic groups. The *dum ba* is a double-reed instrument built from wood. The *sum pyi* flute playing with the band is made of bamboo.

https://vimeo.com/rolfbader/bur-25

Old Southeast Asian 78 rpm recordings

A really wonderful release of old 78 rpm schellack recordings of Southeast Asian music, with extensive booklet is: *Longing for the past. The 78 rpm era in Southeast Asia.* Dust-to-Digital, 2013.

Cambodia

Khmer Rouge Survivors, "They will kill you, if you cry", glitter beat

CD about traditional Cambodian music of musicians having survived the Khmer Rouge Regime 1975–78. A highly interesting, personal recording of the events is that of François Bizot [42].

Khmer music

The 3 CD collection *The music of Cambodia. 9 Gong Gamelan, Royal Court Music, Solo Instrumental Music* by David Parsons, Celestial Harmonies is a wonderful overview on traditional music in Cambodia.

Xinjiang, China, Uyghur ethnic group

Qetiq. Rock from Taklamakan Desert

CD about an Uyghur Rock group who, despite the problems between the Uyghur minority and the Han Chinese, became very popular all over China. The Uyghurs live around the Taklamakan desert in the very west of China, north of Tibet, where the old and the new Silk Road goes through. There is also a film about their music and lives:

https://www.youtube.com/watch?v=VI6y5DNElU8&t=26s

by Mukaddas Mijit.

Sanubar Tursun. Arzu

CD with extensive booklet of an Illi valley based singer and *tanbur* player Sanubar Tursun released by Rachel Harris at Department of Music, SOAS, University of London in 2012.

Uyghur Music. Muqam Nava.

CD and extensive booklet released by Jean During and Radio France, where a whole *muqam nava* is performed. Like all Uyghur *muqam* it is a suite of songs sung in ancient Uyghur language. The booklet gives a very good introduction on Uyghur musical instruments and the structure of the music.

Uyghur twelve *muqams*. Official performance. Twelve videos of the twelve *muqam* of traditional Uyghur music recorded by the official state ensemble. https://vimeo.com/album/5342385

Sri Lanka

Puja Buddhist temple music

A drum-and-shawm band at the tooth temple in Kandy, Sri Lanka during the daily celebration called puja. The shawm is called *horanewa* there and derived from the Turkish military orchestra *zurna*. https://vimeo.com/rolfbader/sl-14

Indonesia

Balinese Gong Kebyar *gamelan*

Gong Kebyar is the most prominent *gamelan* music of Bali. The CD *Music of the Gamelan Gong Kebyar* with works of I Nyoman Windha, or *Music of the Gamelan Gong Kebyar* o the STSI Denparasr National Institute of the Arts in Bali, both Vital Records, 1996, are good good examples with extensive booklets.

Balinese music

The CD series *Anthologie des Musiques de Bali*, Buda Records gives an extensive overview on musical styles and orchestras in Bali.

Historical Balinese *gamelan*

The CD *the roots of gamelan. the first recordings, bali, 1928, new york, 1941*, Arbiter Recordings Company, 1999 is a collection of historical recordings with an informative booklet about the interactions between Bali and Western musicians and scholars.

Experimental

A very interesting fusion of javanese traditional music with contemporary electronic, free improvisation, and experimental music is the duo *Senyawa*. See e.g.: https://sublimefrequencies.bandcamp.com/album/sujud.

Japan

shakuhachi

The homepage of the *International shakuhachi Society* features many recordings and additional material to the instrument: https://www.komuso.com/top/index.pl

gagku

A very nice introduction to *gagaku* music, its instruments, or repertoire can be found at: http://gagaku.stanford.edu/en/

Yunnan, China

Yi ethnic group—*hulusheng* (mouth organ)

Li Quing Kuang playing the *hulusheng*, the traditional mouth organ of Yun-

nan and related regions in Southwest China in the town of Yongren at hist
restaurant during a lunch party.
https://vimeo.com/297466412

Yi ethnic group—*hulusheng, yue xian, jing hu*
Again Li Quing Kuang playing *hulusheng* together with Li Fa Ming, and Li
Ze Wei playing the 'moon string' *yue xian* (the guitar) and *jing hu*, the bowed
instrument, similar to the better-known *er hu* of China mainland.
https://vimeo.com/297466527

Musou ethnic group—*hulusheng*
The Musou musician Niu Ge (his Chinese name, lit. 'cow song') playing the
hulusheng at Luguhu, a mountain lake at the border between Yunnan and
Sichuan, China.
https://vimeo.com/297466823

Brazil

L. H. Corrêa de Azevedo: Music of Ceará and Minas Gerais Corrêa de
Azevedo was a Brazilian composer and ethnomusicologist who recorded
music from remote regions of Brazil in the 1940th. Among them are contradanzas and waltzes from Minas Gerais. João Gilberto invented the Bossa
Nova during a trip there, where he smoothened the Samba into this new Bossa
Nova style, maybe inspired by the smoothness of the waltzes and contradanzas still alive in *Minas Gerais* at his time.
https://folkways.si.edu/l-h-correa-de-azevedo-music-of-ceara-and-minas-
gerais/latin-world/music/album/smithsonian.

Madagascar

A world out of time Henry Kayser & David Lindley, two guitarist, were
performing with Madagascan musicians, and, in the CD booklet, they tell
an interesting story about meter in Madagascan and their music. Shanachie
Records, 1992.

valiha The *valiha* is a bamboo tube dulcimer, played in Madagascar. On a
bambo tube, sometimes more than 18 strings are attached at both tube ends.
Putting a bridge between tube and strings makes them playable and freely
vibrating. The instrument sound is very low, still the rhythmic patterns are
highly complex. There are many recordings of the *valiha* out, an informative
one is from the Radio France collection: Madagascar. Pays Bara.

Madagascar Music Collection The 3 CD set *Malagasy Music Box*, Feuer
und Eis, Moers, 1990, gives a wonderful overview on the great variety of
Madagascar music.

India

sitar The *sitar* is a long-neck lute, only existing since the mid of 19th century,
and played in Northern India [115], South India has its own music tradition.

The amount of *gharana*, *sitar* schools in Northern India has decreased from about seven in the 70th up to two or three nowadays. There are endless recordings released, one example of the fusion of *sitar* and Western Classical music is: Ravi Shankar & Andre Previn: Shankar Concerto for Sitar & Orchestra, London Symphony Orchestra, EMI 1971.

sarod The *sarod* is again a stringed instrument with resonator strings, like the *sitar*, but with a steel soundboard and a membrane, on which the bridge is placed. A prominent musician is Amjad Ali Khan, with many recordings available.

The Indian violin The Indian violin is played more downwards, not straight up like in the West. It is again used in *raga* music. A recording in the South Indian style is: Shankar with Zakir Hussain & Vikku Vinayakram, Raga Aberi. World of Music, 1995.

mohan veena The *mohan veena* is the Indian slide guitar . Vishwa Mohan Bhatt, together with Ry Cooder, in 1994 won a Grammy Award for best world music album for their CD *A meeting by the river*. Another virtuoso is Debashish Bhattacharya releasing several albums, among them *Slide guitar ragas from dusk till dawn,* Riverboard Records, 2015.

Learning Ragas The *Raga Guide* [46] is a collection of Ragas, their scales, typical melodies, rhythms, etc. with CDs for those wanting to get into the music deeper.

Egypt

History of Egyptian music An excellent introduction to the history of Classical Egyptian music is *al-tarab* [169], roughly translated as attitude of pleasure and educated interest when listen to music, with a CD. The book is only published in French and German.

(3) Musical Styles

Noise

Survival Research Laboratories Using machines and explosives constructed for Hollywood blockbuster movies in the 80th they performed concerts in the Mojave desert, see movie: . The band is still active today: http://srl.org.

Michel Doneda The CD Michel Doneda: 'Everybody Digs', Relative Pitch Records 2014 is an excellent example of what sounds can be produced on a soprano saxophone, in terms of multiphonics and all kinds of extended techniques.

Throbbing Gistle Starting end of the 60th Throbbing Gristle combines performances and philosophical considerations on noise and music. Since they called their label 'Industrial' this music is also known as 'Industrial'.

Einstürzende Neubauten Raised in the Berlin music scene in the 80th before the fall of the Berlin Wall, the band experimented with all kinds of things procing noise. CD: Eunstürzende Neubauten: Haus der Lüge, Rough Trade 1989.

Merzbow Merzbow is a legendary 'Japanoise' band started by Masami Akita in 1979. The name was derived from Kurt Schwitter's 'Merzbau', a rearrangement of his house. Akita also wrote several books and articles on music, animal rights, or environmental issues http://merzbow.net/.

Even Parker, Broken Wings This piece is played using extended techniques for wind instruments, in- and decreasing tension continuously over about seven minutes by gradual event density in- and decrease.

The Wire The main magazine discussing noise music. Started in 1982 as a jazz magazine it gradually shifted into experimental, improvisational, or noise music https://www.thewire.co.uk/.

(4) Musical Instruments

Friction instruments

Terpodion A friction keyboard instrument of the 19th century. This one is part of the collection of the Städtische Junge Kunst und Viadrina in Frankfurt a.d. Oder. Some videos can be found here: https://vimeo.com/329772238. Some details can be found here: https://rolfbader.de/terpodion.

Lounuet The *lounuet* is a friction instrument of New Ireland, extinguished yet. Still some examples remain in museums. The one found here is from the Überseemuseum in Bremen, Germany https://rolfbader.de/lounuet.

Lounuet Recordings are very rare. Some can be found in: A. L. Kaeppler & J. W. Love (eds.): *Garland Encyclopedia of World Music*. Vol. 9 on the supplemental CD.

Glassharmonica Invented by Benjamin Franklin in 1766 is is a friction instrument, where glass cylinders rotates through water, and when a player presses his fingers on the glass a sound is produced, just like with singing glasses. Wonderful recordings are from the Wiener Glasharmonica Duo http://www.glasharmonika.at/.

Experimental musical instruments

Daxophone The daxophone is a bowed instrument, consisting of several wooden plates driven by a bow. It sounds like a mixture of saxophone and human voice: Hans Reichel: Yuxo—A new daxophonoe operetta. all, www.free-music-production.de, 2002.

Novachord The novachord is a polyphonic synthesizers, working with vaccum tubes, built by the Hammond Instrument Comparny from 1938–1942.

In a project one instrument was reconstructed, sounding very spherical: Phil Cirocco: Music of the Electron. The Novachord Restoration Project, CMS music USA, 2007.

Ice Percussion Instruments built out of ice often sound very similar to wooden instruments, ice is a solid too. Terje Isungset, Sidsel Endresen: Igloo, All Ice Records, Norway, 2006.

Experimental voice Extreme extensions of the human voice was performed by: Alfred Wolfson's Experiments in Extension of Human Vocal Range, Smithsonian Folkways Recordings, 1956 (2007).

Experimental Musical Instrument Sampler A wonderful collection of experimental musical instruments is Bart Hopkins: Gravikords, Whirlies & Pyrophones. Experimental Musical Instruments (with extensive booklet), sllipsis arts, 1998.

References

1. Alexandraki, Ch., Bader, R.: Anticipatory networked communications for live musical interactions of acoustic instruments. J. New Music Res. **45**(1), 68–85 (2016)
2. Almeida, A., Vergez, Ch., Caussé, R.: Quasistatic nonlinear characteristics of double-reed instruments. J. Acoust. Soc. Am. **121**, 536–546 (2007)
3. Altenmüller, E.: Vom Neandertal in die Philharmonie: Warum der Mensch ohne Musik nicht leben kann [From Neandertal into Concert Hall: Why man cannot live without music,] Springer (2018)
4. Angerster, J., Miklós, A.: Properties of the Sound of Flue Organ Pipes. In: Bader, R. (ed.) Springer Handbook of Systematic Musicology, pp. 141–155. Springer, Heidelberg (2018)
5. Argyris, J.H., Faust, G., Haase, M., Friedrich, R.: An Exploration of Dynamical Systems and Chaos: Completely Revised and Enlarged, 2nd edn. Springer, Berlin, Heidelberg (2015)
6. Arndt, M., Nairz, O., Vos-Andreae, J., Keller, C., van der Zouw, G., Zeilinger, A.: Wave-particle duality of C_{60} molecules. Nature **401**, 680–682 (1999)
7. Aschhoff, V.: Experimentelle Untersuchungen an einer Klarinette. [Experimental investigations of a clarinet,] Akustische Zeitschrift **1**, 77–93 (1936)
8. Attali, J.: Noise. The Political Economy of Music, Translated from French edition 1977, Theory and History of Literature, Vol. 16 (1985)
9. Baars, B.J., Franklin, S., Ramsoy, Th. Z.: Global workspace dynamics: cortical binding and propagation enables conscious contents https://pubmed.ncbi.nlm.nih.gov/23974723/ (2013)
10. Bader, R.: Neural coincidence detection strategies during perception of multi-pitch musical tones. arXiv:2001.06212 [q-bio.NC] (2020)
11. Bader, R., Plath, N.: Impact of damping on oscillation patterns on the plain piano soundboard, https://zenodo.org/record/4133250 (2020)
12. Bader, R. (ed.): Compuational Phonogram Archiving. Springer Series: Current Research in Systematic Musicology, vol. 5 (2019)
13. Bader, R., Linke, S., Mores, R.: Measurements and impulse pattern formulation (IPF) model of phase transitions in free-reed wind instruments. J. Acoust. Soc. Am. **146**(4), 2779 (2019). https://doi.org/10.1121/1.5136626
14. Bader, R.: Temperament in tuning systems of southeast asia and ancient India. In: R. Bader (ed.): Computational Phonogram Archiving, Springer Series: Current Research in Systematic Musicology, vol. 5, pp. 75–108 (2019)
15. Bader, R. (ed.): Springer Handbook of Systematic Musicology. Springer, Heidelberg (2018)
16. Bader, R.: Cochlear spike synchronization and coincidence detection model. Chaos **023105**, 1–10 (2018)

17. Bader, R., Mores, R.: Cochlear detection of double-slip motion in cello bowing, arXiv:1804.05695v1 [q-bio.NC] 16 Apr 2018

18. Bader, R.: Pitch and timbre discrimination at wave-to-spike transition in the cochlea, arXiv:submit/2066467 [q-bio.NC] 12 Nov 2017

19. Bader, R.: Finite-Difference model of mode shape changes of the Myanmar pat wain drum circle using tuning paste. Proc. Mtgt. Acoust. 29(035004), 1–14 (2016)

20. Bader, R.: Phase synchronization in the cochlea at transition from mechanical waves to electrical spikes. Chaos 25(103124), 1–9 (2015)

21. Bader, R.: Nonlinearities and Synchronization in Musical Acoustics and Music Psychology, Springer Series: Current Research in Systematic Musicology, vol. 2. Springer, Heidelberg, Berlin (2013)

22. Bader, R.: Outside-instrument coupling of resonance chambers in the New-Ireland friction instrument lounuet. Proc. Mtgs. Acoust. 15, (2012). https://doi.org/10.1121/2.0000167

23. Bader, R.: Buddhism, Animism, and entertainment in Cambodian Melismatic chanting smot. In: Schneider, A., von Ruschkowski, A. (eds.), Hamburg Yearbook of Musicology vol. 28, pp. 283–305 (2011)

24. Bader, R., Münster, M., Richter, J. & Timm, H.: Microphone array measurements of drums and flutes. In: Bader, R. (eds.), Musical Acoustics, Neurocognition and Psychology of Music/Musikalische Akustik, Neurokognition und Musikpsychologie, Hamburger Jahrbuch für Musikwissenschaft 25, 13–54 (2009)

25. Bader, R., Diezt, M.-K., Elvers, P, Elias, M., Tolkien, L.: Foundation of a syllogistic music theory. In: Bader, R. (ed.), Musical Acoustics, Neurocognition and Psychology of Music/Musikalische Akustik, Neurokognition und Musikpsychologie, Hamburger Jahrbuch für Musikwissenschaft, vol. 25, pp. 177–196 (2009)

26. Bader, R.: Efficient Auditory Coding of a Xylophone Bar. In: Schneider, A. (ed.) Concepts, Methods, Findings: Studies in Systematic Musicology, Hamburger Jahrbuch für Musikwissenschaft 24, pp. 197–212. Peter Lang Verlag, Hamburg (2008)

27. Bader, R., Hansen, U.: Acoustical analysis and modeling of musical instruments using modern signal processing methods. In: Havelock, D., Vorländer, M., Kuwano, S. (eds.), Handbook of Signal Processing in Acoustics, Springer, pp. 219–247 (2008)

28. Bader, R.: Computational Mechanics of the Classical Guitar. Springer (2005)

29. Bader, R.: Turbulent k-ε model of flute-like musical instrument sound production New trends in theory and applications. In: Lutton, E., Lévy-Véhel, J. (eds.) Fractals in Engineering, pp. 109–122. Springer, NY (2005)

30. Bader, R.: Berechnung fraktaler Strukturen in den Études für Klavier von György Ligeti. [Calculating fractal structures in the Études for piano of György Ligeti.]. In: Stahnke, M. (ed.) Mikrotöne und mehr. Bockel-Verlag Hamburg, Auf György Ligetis Hamburger Pfaden (2005)

31. Bahle, J.: Der musikalische Schaffensprozess. Psychologie der schöpferischen Erlebnis- und Antriebsformen, [The musical creative process. Psychology of creative experiences and motivations,] Hirzel, Leipzig (1936)

32. Baily, J.: Can You Stop the Birds Singing? The Censorship of Music in Afghanistan, Freemuse Report, 1 (2001)

33. Bandyopadhyay, A., Pati, R., Sahu, S., Peper, F., Fujita, D.: Massively parallel computing on an organic molecular layer. Nature Phys. 6, 369–375 (2010)

34. Barbieri, P.: Violicembalos and Other Italian Sostenente Pianos 1785–1900. Galpin Soc. J. 62, 117–139 (2009)

35. Barker, A.: Pythagorean harmonics. In: Huffman, C.A. (ed.), A History of Pythagoreanism, pp. 185–203. Cambridge University Press (2014)

36. Benetos, E., Dixon, S.: Multiple-instrument polyphonic music transcription using a temporally constrained shift-invariant model. J. Acoust. Soc. Am. 133(3), 1727–1741 (2013)

37. Berlioz, H.: Instrumentationslehre. Peters, Leipzig (1904)

38. Battacharya, J., Petsche, H., Pereda, E.: Long-range synchrony in the γ Band: role in music perception. J. Neurosci. 21(16), 6329–6337 (2001)

39. von Bekesy, G.: Experiments in Hearing. McGraw-Hill, New York, McGraw-Hill (1960)
40. Belafonte, H., Shnayerson, M.: My Song: A Memoir. Alfred A. Knopf, NY (2011)
41. Bilbao, S., Hamilton, B., Harrison, R., Torin, A.: Finite-difference schemes in musical acoustics: a tutorial. In: Bader, R. (ed.) Springer Handbook of Systematic Musicology. Springer, Heidelberg (2018)
42. Bizot, F.: The Gate. Harvill (2003)
43. Blaß, M., Fischer, J., Plath, N.: Computational Phonogram Archiving: a generic framework for knowledge discovery in music archives. Physics Today **73**, 12 (2020)
44. Blaß, M., Bader, R.: Content Based music retrieval and visualization system for Ethnomusicological music archives. In: Bader, R. (ed.), Computational Phonogram Archiving, Springer Series: Current Research in Systematic Musicology, Vol. 5, 145–174. Springer, Heidelberg (2019)
45. Bones, O., Hopkins, K., Krishnan, A., Plack, C.J.: Phase locked neural activity in the human brainstem predicts preference for musical consonance. Neuropsychologia **58**, 23–32 (2014)
46. Bor, J.: The Raga Guide. A Survey of 74 Hindustani Ragas. Boydell and Brewer (1992)
47. Borgo, D.: Sync or Swarm: Improvising Music in a Complex Age. Continuum, NY (2005)
48. Bovermann, T., de Campo, A., Egermann, H., Hardjowirogo, S.I., Weinzierl, S.: Musical Instruments in the 21th Century. Springer (2017)
49. Brattico, E., Pearce, M.: The Neuroaesthetics of music. Psychol. Aesthetics Creativity Arts **7**(1), 48–61 (2013)
50. Braun, J.: Geschichte der Kunst: in ihrem Entwicklungsgang durch alle Völker der alten Welt hindurch auf dem Boden der Ortskunde nachgewiesen, [History of art: in its development through all people of the old world, substantiated by geographical places,] Kreidel und Niedner, Wiesbaden, pp. 1856–1858
51. Bregman, A.S.: Auditory Scene Analysis: the Perceptual Organization of Sound. MIT Press (1994)
52. Brémaud, I., Gril, J., Thibaut, B.: Anisotropy of wood vibrational properties: dependence on grain angle and review of literature data. Wood Sci. Technol. **45**, 735–754 (2011)
53. Brentano, F.: Das Genie. Vortrag gehalten im Saale des Ingenieurs- und Architektenvereins in Wien, [The genius. Talk given at the venue of the Engineer- and Architecture Society in Vienna], Verlag von Dunder & Humbolt, Leipzig (1892)
54. Brentano, F.: Psychologie vom empirischen Standpunkt, [Psychology from an empirical standpoint], Leipzig 1874, Routledge Classics (2015)
55. Brown, S., Savage, P.E., Ko, A.M.-S., Stoneking, M., Ko, Y.-C., Loo, J.H., Trejaut, J.A.: Correlations in the population structure of music, genes and language. Proc Biol Sci. **281**(1774), 1–7 (2014)
56. Buhusi, C.V., Meck, W.H.: What makes us tick? nature reviews neuroscience **6**, 755–765 (2005)
57. Buzsáki, G.: Rhythms of the Brain. Oxford University Press (2006)
58. Caclin, A., Smith, B.K., Giard, M.-H.: Interactive processing of timbre space dimensions: an exploration with event-related potentials. J. Cognit. Neurosci. **20**, 49–64 (2008)
59. Cariani, P.: Temporal codes, timing nets, and music perception. J. New Music Res. **30**(2), 107–135 (2001)
60. Cassirer, E.: Das Erkenntnisproblem in der Philosophie und Wissenschaft der neueren Zeit, 4 Bd. [The problem of epistemoloty in the philosophy and science in recent times, 4 Vol.], pp. 1906–1957
61. Cassirer, E.: Substanzbegriff und Funktionsbegriff. Untersuchungen über die Grundfragen der Erkenntniskritik, [Substance and Function, and Einstein's theory of relativity, (Chicago 1923)], Verlag von Bruno Cassirer (1910)
62. Chladni, E.F.F.: Beyträge zur praktischen Akustik und zur Lehre vom Instrumentbau, enthaltend die Theorie und Anleitung zum Bau des Clavicylinders und damit verwandter Instrumente, [Contributions to the practical acoustics and principles of instrument building, containng the theory and the manual for building the Clavicylinder and related instruments]. Breitkopf & Härtel, Leipzig (1821)

63. Chladni, E.F.F.: Die Akustik [The Acoustics]. Breitkopf & Härtel, Leipzig (1802)
64. Clayton, M.: Time in Indian Music. Rhythm, Metre, and Form in North Indian Rag Performance. Oxford Monographs on Music, Oxford University Press (2000)
65. Cottingham, J.: Acoustics of free-reed instruments. Phys. Today **64**(3), 44 (2011), https://physicstoday.scitation.org/doi/full/10.1063/1.3563819
66. Cremer, L.: The Physics of the Violin. MIT Press, Cambridge (1985)
67. da Silva, A.M., Leal, B.: Photosensitivity and epilepsy: current concepts and perspectives—a narrative review. Seizure **50**, 209–218 (2017)
68. Davies, H.R.: Yün-nan: The Link Between India and the Yangtze. Cambridge University Press (1909)
69. de Boer, E.: Auditory physics. Physical principles in hearing theory. Phys. Rep. **203**, 127–229 (1991)
70. Desmore, F.: Teton Sioux Music and Culture. University of Nebraska Press, (Reprint from 1918), (2001)
71. Dewoskin, K.J.: A Song for One or Two: Music and the Concept of Art in Early China. The University of Michigan, Michigan, Ann Arbor (1982)
72. Eco, U.: The Island of the Day Before, Secker & Warburg (1994)
73. Eerola, T.: Music and emotions. In: Bader, R. (ed.), Springer Handbook of Systematic Musicology, pp. 539–554. Springer (2018)
74. Elejabarrieta, M.J., Ezcurra, A., Santamaria, C.: Coupled modes of the resonance box of the guitar. J. Acoust. Soc. Am. **111**, 2283–2292 (2002)
75. Elsner, J.: Maqām–Raga–Zeilenmelodik: Konzeptionen und Prinzipien der Musikproduktion; Materialien der 1. Arbeitstagung der Study Group "*Maqām*" beim International Council for Traditional Music vom 28. Juni bis 2. Juli 1988 in Berlin. [*Maqām* - Raga -line melodicity: Conceptions and principles of music production; Materials of the 1st symposium of the "*Maqām*" study group of the International Council for Traditional Music from 28. June - 2. July 1988 in Berlin,] Nationalkomitee DDR, Berlin (1989)
76. Elwood, R.W., Adams, L.: Electric shock causes physiological stress responses in shore crabs, consistent with prediction of pain. Biol. Lett. **11**, 20150800 (2015). https://doi.org/10.1098/rsbl.2015.0800
77. Evans, T., Evans, M.A.: Guitars—Music, History, Construction and Players—From the Renaissance to Rock. Paddington Press, NY (1977)
78. Eybl, M., Fink-Mennel, E. (eds.): Schenker-Traditionen : eine Wiener Schule der Musiktheorie und ihre internationale Verbreitung / A Viennese school of music theory and its international dissemination, Wiener Veröffentlichungen zur Musikgeschichte 6. Böhlau, Wien (2006)
79. Fabre, B., Gilbert, J., Hirschberg, A.: Modeling of wind instruments. In: Bader, R. (ed.) Springer Handbook of Systematic Musicology, pp. 121–139. Springer, Heidelberg (2018)
80. Fabre, B., Gilbert, J., Hirschberg, A., Pelorson, X.: Aeroacoustics of Musical Instruments. Ann. Rev. Fluid Mech. **44**, 1–25 (2012)
81. Farhat, H.: The Dastgāh Concept in Persian Music. Cambridge University Press (1990)
82. Fastl, H.: Psychoacoustics. Springer, Facts and Models (2006)
83. Faraday, M.: Experimental Researches in Chemistry and Physics. Taylor and Francis, London (1859)
84. Fell, J., Axmacher, N.: The role of phase synchronization in memory processes. Nature Neurosci. Rev. **12**, 105–118 (2011)
85. Fingelkurts, A.A., Fingelkurts, A.A.: Altered structure of dynamic 'Electroencephalogram oscillatory pattern' in major depression. Biol. Psychiatry **77**(12), 1050–1060 (2015)
86. Finnegan, R.: Oral Literature in Africa. Open Book Publishers (2012)
87. Fischer, J.L., Bader, R., Abel, M.: Aeroacoustical coupling and synchronization of organ pipes. J. Acoust. Soc. Am. **140**(4), 2344–2351 (2016)
88. Fletcher, N.H., Rossing, ThD: The Physics of Musical Instruments. Springer (2000)
89. Fletcher, N.H.: Mode locking in nonlinearly excited inharmonic musical oscillators. J. Acoust. Soc. Am. **64**, 1566–1569 (1978)

90. Fletcher, H., Manson, W.A.: Loudness, its definition, measurement and calculation. J. Acoust. Soc. Am. **5**, 82–108 (1933)
91. Freely, J.: Light from the East: How the Science of Medieval Islam helped to shape the Western World. I.B, Tauris (2011)
92. Freeman, W.: A Socitey of Brains. Psychology Press (2014)
93. French, A.D., Johnson, G.P.: Advanced conformational energy surfaces for cellobiose. Cellulose **11**, 449–462 (2004)
94. Fries, P., Neuenschwander, S., Engel, A.K., Goebel, R., Singer, W.: Rapid feature selective neuronal synchronization through correlated latency shifting. Nature Neurosci. **4**(2), 194–200 (2001)
95. Friston, K.J., Friston, D.A.: A free energy formulation of music performance and perception. helmholtz revisited. In: Bader, R. (ed.), Sound–Perception–Performance, Springer Series: Current Research in Systematic Musicology, vol. 1, pp. 43–70. Springer, Heidelberg (2013)
96. Gallagher, Th.F.: Rydberg Atoms. Cambridge University Press (1994)
97. Gibiat, V.: Phase space representations of acoustical musical signals. J. Sound Vibr. **123**(3), 529–536 (1988)
98. Gibiat, V., Castellengo, M.: Period doubling occurences in wind instruments musical performance. Acustica **86**, 746–754 (2000)
99. Gilbert, J.: Stepwise regime in elephant trumpet calls: similarities with brass instrument behaviour. J. Acoust. Soc. Am. **141**, 3724 (2017)
100. Gilmore, B.: Harry Partch: A Biography. Yale University Press (1998)
101. Glackin, C., Maguire, L., McDaid, L., Wade, J.: Synchrony: a spiking-based mechanism for processing sensory stimuli. Neural Netw. **32**, 26–34 (2012)
102. Glassie, J.: A Man of Misconceptions: The Life of an Eccentric in an Age of Change. Riverhead Books (2012)
103. Godøy, R.-I.: Quantal elements in musical experience. In: Bader, R. (ed.) Sound–Perception–Performance, Springer Series: Current Research in Systematic Musicology, vol. 1. Springer, Heidelberg (2013)
104. Goodman, S.: Sonic Warfare. Affect, and the Ecology of Fear, MIT Press, Sound (2010)
105. Goodwin, J.: Lords of the horizons. Vintage, A history of the Ottoman Empire (1999)
106. Graeber, D.: Bullshit Jobs: A Theory. Simon & Schuster (2018)
107. Grekow, J.: Music Emotion Maps in Arousal-Valence Space. (2016). https://doi.org/10.1007/978-3-319-45378-1_60
108. Grüsser, O.: Justinus Kerner 1786–1862. Springer (1987)
109. Gruhn, G., Carter, W.: Gruhn's Guide to Vintage Guitars. An Identification Guide for American Fretted Instruments, 3rd ed., Backbeat Books (2010)
110. Gurwitsch, A.: Théorie du champ de la conscience, [Field of Consciousness,] Dusquesne University Press, Pittsburgh (1964)
111. Hacke, A.: Krach, [Noise,] Metrolit Verlag, Berlin (2015)
112. Haken, H.: Brain Dynamics. An introduction to models and simulation, 2. ed., Springer (2000)
113. Haken, H.: Synergetics, 3rd edn. Springer (1990)
114. Haken, H.: Advanced Synergetics. Springer (1983)
115. Hamilton, J.S.: Sitar Music in Calcutta. An Ethnomusicological Study, Performing Arts Series Vol. III, Motilal Banarsidass Publishers, Selhi (1994)
116. Hameroff, S., Penrose, R.: Consciousness in the universe. Rev. 'Orch OR' Theory Phys. Life Rev. **11**, 39–78 (2014)
117. Hansing, S.: Das Pianoforte in seinen akustischen Anlagen, [The pianoforte in its acoustical foundations,] Schwerin 1909, second ed., published by friends of Siegfried Hansings (1950)
118. Harris, R.: The making of a musical canon in Chinese Central Asia: the Uyghur Twelve Muqam. Ashgate (2008)
119. Harris, I.: Cambodian Buddhism. University of Hawai'i Press, Honolulu, History and Practice (2005)
120. Hartmann, L., Bader, R.: Neural Synchronization of Music Large-Scale Form, arXiv:2005.069.38 [q-bio.NC] (2020)

121. Hartmann, L., Bader, R.: From musical form to neural network synchronization: neural correlates of musical form perception (in preparation for 2021)
122. Heise, B.: Some key friction instruments: the Clavicylinder, the Melodion and the Terpodion. Society J. **58**, 160–167 (2005)
123. Haynes, B: A History of Performing Pitch: The Story of 'A'. Scarecrow Press (2002)
124. Hegarty, P.: Noise/Music. A History, The Continuum International Group, NY (2007)
125. Hegre, H., Allansson, M., Basedau, M., Colaresi, M., Croicu, M., Fjelde, H., Hoyles, F., Hultman, L., Högbladh, S., Mouhleb, N., Muhammad, S.A., Nilsson, D., Nygård, H.M., Olafsdottir, G., Petrova, K., Randahl, D., Rød, E.G., von Uexkull, N., Vestby, J.: ViEWS: a political violence early-warning system. J. Peace Res. **56**(2), 155–174 (2019)
126. Heidegger, M.: Der Ursprung des Kunstwerks, [The Origin of the Work of Art, In: Off the Beatne Track, 1950.], Reclam, Ditzingen (1986), 1935/36
127. Heins, E., den Otter, E., van Lamsweerde, F.: Jaap Kunst: traditional music and its interaction with the West. KIT Publishers (1994)
128. von Helmholtz, H.: Die Lehre von den Tonempfindungen als physiologische Grundlage für die Theorie der Musik [On the Sensations of tone as a physiological basis for the theory of music]. Vieweg, Braunschweig (1863)
129. von Helmholtz, H.: Über die akademische Freiheit der deutschen Universitäten. [On the academic freedom of German universities,] Berlin (1878)
130. Hendy, D.: Noise. Profile Books, London, A Human History of Sound and Listening (2013)
131. Hillman, B., Ertl-Wagner, B., Wagner, B.C. (eds.): The man who stalked Einstein: how Nazi scientist Philipp Lenard changed the course of history. Connecticut, Lyons Press, Guilford (2015)
132. Hispanus, P.: Tractatus/Summulae Logicales. Philosophia Verlag, Transl. to German by B. Pabst & W. Degen (2006)
133. Hoffmann, D.W.: Die Gödel'schen Unvollständigkeitssåtze : eine geführte Reise durch Kurt Gödels historischen Beweis, [The Gödel incompleteness sentences: a guided tour through kurt Gödels historical proof,] Springer, Heidelberg (2013)
134. http://barthopkin.com/books-cds/experimental-musical-instruments-complete-back-issues-on-cd-rom/ (1999)
135. von Hornbostel, E.M., Sachs, C.: Systematik der Musikinstrumente. Ein Versuch, [Systematic of musical instruments. A try,] Zeitschrift für Ethnologie **46**, 553–590 (1914)
136. Huffman, C.A.: A History of Pythagoreanism. Cambridge University Press (2014)
137. Hutchings, C. (ed.): Research Papers in Violin Acoustics, Vol. I & II, Publication by the Acoustical Society of America (1997)
138. Izhikevich, E.M., Edelman, G.M.: Large-scale model of mammalian thalamocortical systems. PNAS **105**(9), 3593–3598 (2008)
139. Izhikevich, E.M.: Dynamical Systems in Neuroscience. MIT Press, The Geometry of Excitability and Bursting (2007)
140. Izhikevich, E.M., Gally, J.A., Edelman, G.M.: Spike-timing dynamics of neuronal groups. Cerebral Cortex V, (14) N 8, 933–994 (2004)
141. Jackson, M.W.: Harmonious Triads. Musicians, and Instrument Makers in Nineteenth-Century Germany, MIR Press, Physicists (2006)
142. Jansson, E.V.: A study of acoustical and hologram interferometric measurements of the top plate vibrations of a guitar. Acustica **25**, 95–100 (1971)
143. Jensenius, A.R., Lyons, M.J.: The NIME Reader. Fifteen Years of New Interfaces for Musical Expression, Springer (2017)
144. Jewanski, J.: Ist C = Rot?: eine Kultur- und wissenschaftsgeschichte zum Problem der welchselseitigen Beziehung zwischen Ton und Farbe, vor Aristoteles bis Goethe, [Is C = red?: a culture and science history about the problem of the mutual relation between tone and color, from Aristoteles zu Goethe,] Sinzig, Studio, (1999)
145. Jeyapalina, S.: Studies of the hydro-thermal and viscoelastic properties of leather. Univ. of Leichester, Ph.D. (2004)

146. Jiruska, P., de Curtis, M., Jefferys, J.G.R., Schevon, C.A., Schiff, St. J., Schindler, K.: Synchronization and desynchronization in epilepsy: controversies and hypotheses. J. Physiol. **591**(4), 787–797 (2013)

147. Jones, G.S.: Karl Marx: Greatness and Illusion. Belknap Press (2016)

148. Joris, P.X., Carney, L.H., Smith, P.H., Yin, T.C.T.: Enhancement of neural synchronization in the anteroventral cochlear nucleus. I. Responses to tones at the characteristic frequency. J. Neurophysiol. **71**(3), 1022–1036 (1994)

149. Joris, P.X., Carney, L.H., Smith, P.H., Yin, T.C.T.: Enhancement of neural synchronization in the anteroventral cochlear nucleus. II. Responses in the tuning curve tail. J. Neurophysiol. **71**(3), 1037–1051 (1994)

150. Jureidini, J., McHenry, L.B.: The Illusion of Evidence-Based Medicine. Wakefield Press (2020)

151. Kacprzyk, J., Pedrycz, W. (eds.): Springer Handbook of Computational Intelligence. Springer, Berlin, Heidelberg (2015)

152. Kadel, R.P.: Musical Instruments of Nepal. Nepali Folk Musical Instrument Museum, Kathmandu, Nepal (2007)

153. Kandel, E.: In Search of Memory: The Emergence of a New Science of Mind. Norton, NY (2006)

154. Kerner, J.: Die Seherin von Prevorst: Eröffnungen über das innere Leben des Menschen und über das Hineinragen einer Geisterwelt in die unsere, [The seeress of Prevorst: Disclosures about the inner life of man and über the extension of the world of the spirits into ours,]. 1829. https://archive.org/details/seeresssprevorst00crowgoog/page/n5/mode/2up

155. Kausel, W., Zietlow, D.W., Moore, T.R.: Influence of wall vibrations on the sound of brass wind instruments. J. Acoust. Soc. Am. **128**(5), 3161–3174 (2010)

156. Klapuri, A., Davy, M.: Signal Processing Methods for Music Transcription. Springer, New York (2006)

157. Koelsch, S.: Brain and Music. Wiley-Blackwell (2012)

158. Kolmogorov, A.N.: A new metric invariant of transient dynamical systems and automorphisms of Lesbesgue spaces. Dokl. Akad. Nauk SSSR **119**, 861–864 (1958)

159. Kornman, R. (transl.), Bstan-vdzin-rgya-mtsho, Dalai Lama XIV.(foreword): The epic of Gesar of Ling : Gesar's magical birth, early years, and coronation as king, Shambala (2015)

160. Korteweg, D.J., de Vries, G.: On the change of form of long waves advancing in a rectangular canal, and on the new type of long stationary waves. Philos. Mag. **39**(240), 422–443 (1895)

161. Kouwenhoven, F.: Liu Ching-chin—a Critical History of New Music in China, Chime 18–19, Leiden, http://home.wxs.nl/~chime, 216–221 (2013)

162. Kozma, R., Freeman, W.J.: Cognitive Phase Transitions in the Cerebral Cortex–Enhancing the Neuron Doctrine by Modeling Neural Fields. Studies in System, Decision and Control, vol. 39. Springer (2016)

163. Kozma, R., Freeman, W.J.: Intermittent spatio-temporal desynchronization and sequenced synchrony in ECoG signals. Chaos **18**(037131), 1–8 (2008)

164. Krassnitzer, G.: Multiphonics für Klarinette mit deutschem System und andere zeitgenössische Spielarten. [Multiphonics for clarinet with German system and other contemporary performance techniques,] edition ebenos Verlag, Aachen (2002)

165. Kratzky, K., Wallner, F.: Grundprinzipien der Selbstorganisation. [Foundations of self-organization,] Wissenschaftliche Buchgesellschaft (1990)

166. Krause, C.M.: Event-related Desynchronization (ERD) and synchronization (ERS) during auditory information processing. J. New Music Res. **28**(3), 257–265 (2010)

167. Kubik, G.: Africa and the Blues. University of Mississippi Press (1999)

168. Kubik, G.: Die Amadinda Musik in Buganda. In: Musik in Afrika, pp. 139–156. Berlin (1983)

169. Lagrange, F., Abou-Khalil, R.: Al-Tarab. Die Musik Ägyptens, Palmyra (2010)

170. Lauterwasser, A.: Water Sound Images: The Creative Music of the Universe. MACROmedia Publishing (2007)

171. Legge, K.A., Fletcher, N.H.: Nonlinear generation of missing modes on a vibrating string. J. Acoust. Soc. Am. **76**(1), 5–12 (1984)

172. Leman, M.: Music and Schema Theory. Springer, Berlin (1995)
173. Lenard, Ph.: Deutsche Physik in vier Bänden, [German Physics in four volumes,] Lehmanns Verlag, München, 1936–1937
174. Lerch, Th.: Vergleichende Untersuchung von Bohrungsprofilen historischer Blockflöten des Barock, [Comparison of bore-profiles of historical recorder flutes of Baroque age,] Staatliches Institut für Musikforschung. Preussischer Kulturbesitz Musikinstrumentenmuseum (1996)
175. Lewicki, M.S.: Efficient auditory coding. Nature **439**(23):978–82 (2006)
176. Lewicki, M.S.: Efficient coding of time-varying signals using a spiking population code. In: Rajesh, R., Olshausen, B.A., Lewicki, M.S. (eds.), Probabilistic Models of the Brain, pp. 243–56. MIT Press (2002)
177. Lindenberger, U., Li, S-Ch., Gruber, W., Müller, V.: Brains swinging in concert: cortical phase synchronization while playing guitar. BMC Neurosc. **10**(22) (2009)
178. Linke, S., Bader, R., Mores, R.: The impulse pattern formulation (IPF) as a model of musical instruments-Investigation of stability and limits. Chaos **29**, 103109–1 (2019)
179. Linke, S., Bader, R., Mores, R.: The Impulse Pattern Formulation (IPF) as a nonlinear model of musical instruments. Proceedings of International Symposium on Musical Acoustics, 13–17. Sep. Detmold, 2019
180. Liu, C.-C.: A Critical History of New Music in China. The Chinese University Press, Hong Kong (2010)
181. Linnavalli, T., Putkinen, V., Lipsanen, J., Huotilainen, M., Tervaniemi, M.: Music playschool enhances children's linguistic skills. Nature Scientific Reports **8**, 8767. https://doi.org/10.1038/s41598-018-27126-5 (2018)
182. Lomax, A.: The land where the Blues began. American Patchwork DVD Series (1979)
183. Lomax, A.: A worldwide evolutionary classification of cultures by subsistence systems, Current anthropology Chicago. Univ. Chicago Press **18**(4), 659–708 (1977)
184. Loos, S., Claussen, J.C., Schöll, E., Zakharova, A.: Chimera patterns under the impact of noise. Phys. Rev. E **93** (2016)
185. Louage, D.H., van der Heijden, M., Joris, P.X.: Fibers in the trapezoid body show enhanced synchronization to broadband noise when compared to auditory nerve fibers. In: Pressnitzer, D., de Cheveigne, A., McAdams, St., Collet, L. (eds.), Auditory Signal Processing: Physiology, Psychoacoustics, and Models, pp. 100–106. Springer, New York (2004)
186. Myers, A., Pyle Jr., R.W., Gilbert, J., Campbell, D.M., Chick, J.P., Logie, S.: Effects of nonlinear sound propagation on the characteristic timbres of brass instruments. J. Acoust. Soc. Am. **131**, 678–688 (2012)
187. Mandelbrot, B.: The Fractalist: Memoir of a Scientific Maverick. Pantheon (2012)
188. Mandelbrot, B., Wallis, J.R.: Some long-run properties of geophysical records. Water Resour. Res. **5**, 321–340 (1969)
189. Manley, G.A., Fay, R.R., Popper, A.N.: Active Processes and Otoacoustic Emissions. Springer(2008)
190. Martin, W.C., Nash, K., Cottingham, J.P.: Finite element simulation and experiment in khaen acoustics. Proc. Mtgs. Acoust. **39**(035003), 2019 (2019). https://doi.org/10.1121/2.0001178
191. Mazzola, G., Göller, St., Müller, St.: The topos of music. Geometric Logic of Concepts, Theory, and Performance. Birkhäuser (2002)
192. McCrae, N., Elliott, S.: Spiritual experiences in temporal lobe epilepsy: a literature review. British J. Neurosci. Nursing **8**(2) (2012/13)
193. Messner, F.: Friction blocks of New Ireland. In: Kaeppler, A.L., Love, J.W. (eds.), Garland Encyclopedia of World Music, vol. 9: Australia and the Pacific Islands. pp. 380–382. Routledge, London (1998)
194. Minder, A. (ed.): The Sa?gītopaniṣat-sāroddhāraḥ A Fourteenth-century Text on Music from Western India. Composed by Vācanācārya-Śrī Sudhākalaśa. New Delhi, Indira Gandhi National Center for the Arts (1998)
195. Mores, R.: Further empirical data for torsion on bowed strings. Plos One **14**(2) (2019)
196. Mores, R.: Complementary empirical data on minimum bow force. J. Acoust. Soc. Am. **142**(2), 728–736 (2017)

197. Moon, F.C.: Chaotic and Fractal dynamics. NY (1992)
198. Münxelhaus, B.: Pythagoras musicus : zur Rezeption der pythagoreischen Musiktheorie als quadrivialer Wissenschaft im lateinischen Mittelalter [Pythagoras musicus: reception of pythagorean music theory as quadrivial science in the latin middle ages,] Verlag für Systematische Musikwissenschaft, Bad Godesberg (1976)
199. Mukherjee, P.: The Scales of Indian Music. A cognitive approach to *Thāt / Melakartā*, Indira Gandhi National Center for the Arts, Aryan Books International (2004)
200. Mullaney, Th.S.: Coming to Terms with the Nation. Ethnic Classification in Modern China. University of California Press (2011)
201. Nederveen C.J.: Acoustical Aspects of Musical Instruments. Rev. Ed. Northern Illinois University Press (1998)
202. Nickerson, L.M., Rossing, T.D.: Acoustics of the Karen bronze drums. J. Acoust. Soc. Am. **106**, 2254 (1999)
203. Ohl, F.W., Scheich, H., Freeman, W.J.: Change in pattern of ongoing cortical activity with auditory category learning. Nature **412**, 733–736 (2001)
204. Oliver, D.L., Cant, N.B., Fay, R.R., Popper, A.N. (eds.), The Mammalian Auditory Pathways. Synaptic Organization and Microcircuits. Springer Handbook of Auditory Research, Vol. 65, Springer (2018)
205. O'Shea, D.: The Poincaré Conjecture. In Search of the Shape of the Universe, Penguin, UK (2007)
206. Pelat, A., Gautier, F., Colon, St.C., Semperlotti, F.: The acoustic black hole: a review of theory and applications. J. Sound Vibr **476**(115316), 1–23 (2020)
207. Peretz, I., Zatorre, R.J.: The Cognitive Neuroscience of Music. Oxford University Press (2003)
208. Pesic, P.: Thomas Youngs's musical optics, In: Hui, A., Kursell, J., Jackson, M.W. (eds.), Music, sound, and the laboratory from 1750–1980, Osiris 28, Rhode Island (2013)
209. Pikovsky, A.: Synchronization : a Universal Concept in Nonlinear Sciences. University of Cambridge Press (2001)
210. Pressnitzer, D., de Cheveigne, A., McAdams, St., Collet, L. (eds.): Auditory Signal Processing: Physiology, Psychoacoustics, and Models. Springer, New York (2004)
211. Pickering, N.C.: Noninear behavior in overwound violin strings. J. Catugut Acoust. Soc. **1**(3), 46–50 (1989)
212. Pierce, A.: Intrinsic damping, relaxation processes, and internal friction in vibrating systems. POMA **9**, 1–16 (2010)
213. Pinto, T.de.O.: Capoeira, Samba, Candomble. Afro-brasilianische Musik im Reconcavo, Bahia, [Capoeira, Samba, Candomble. Afro-brasilian music in Reconcavo, Bahia,] Museums für Völkerkunde, Reimer, Berlin (1991)
214. Prandtl, L.: Führer durch die Strömungslehre, [Guide through flow dynamics]. Springer Vieweg, 13. ed. (2012)
215. Quignard, P.: Tous Les Matins Du Monde. Paris (1991)
216. Müller-Ebeling, C., Rätsch, Ch., Shahi, S.B.: Shamanism and Tantra in the Himalayas, Thames & Hudson (2002)
217. Rätsch, Ch. & Müller-Ebeling, C.: The Encyclopedia of Aphrodisiacs: Psychoactive Substances for Use in Sexual Practices. Park Street Press (2013)
218. Rao, R.P.N., Olshausen, B.A., Lewicki, M.S.: Probabilistic Models of the Brain. MIT Press (2002)
219. Rohrmeier, M., Pearce, M.: Musical syntax I: theoretical perspectives. In: Bader, R. (ed.) Springer Handbook of Systematic Musicology, pp. 473–486. Springer, Berlin, Heidelberg (2018)
220. Riemann, H.: Präludien und Studien: gesammelte Aufsätze zur Aesthetik, Theorie und Geschichte der Musik [Preludes and Studies: collected writings on aesthetics, theory, and history of music], (1895&. 1900), Nendeln, Kraus (1976)
221. Romanillos, J.L.: Antonio De Torres: Guitar Maker. His Life and Work, Harper Collins Distribution Service (1987)
222. Rossing, T.D.: Science of Percussion Instruments. World Scientific, Singapore (2008)

223. Ruschkowski, A.: Elektronische Klänge und musikalische Entdeckungen. [Electronical sounds and musical discoveries], Reclam (2010)
224. Russell, J.S.: Report on Waves. Report of the fourteenth meeting of the British Asociation for the Advancement of Science, York, September 1844, 311–390, London (1845)
225. Ryugo, D.K., Fay, R.R., Popper, A.N. (eds.): Auditory and Vestibular Efferents. Springer (2011)
226. Sacks, O.: The Man Who Mistook His Wife For A Hat And Other Clinical Tales. Summit Books (1985)
227. Sadan, M.: Being and Becoming Kachin. Histories Beyond the State in the Borderworlds of Burma. Oxford University Press (2013)
228. Sahu, S., Ghosh, S., Fujita, D., Bandyopadhyay, A.: Live visulizatins of single isolated tubulin protein self-assembly via tunneling current: effect of electromagnetic pumping during spontaneous growth of microtubule. Scientific Reports **4**, 7303 (2014). https://doi.org/10.1038/srep07303
229. Sahu, S., Gosh, S., Hirata, K., Jujita, D., Bandyopadyay, A.: Multi-level memory-switching properties of a single brain microtubule. Appl. Phys. Lett. **102**, 123701-1-4 (2013)
230. Sathej, G., Adhikari, R.: The eigenspectra of Indian musical drums. J. Acoust. Soc. Am. **126**(2), 831–838 (2009)
231. Schneider, A.: Systematic musicology. A historical interdisciplinary perspective. In: Bader, R. (ed.): Springer Handbook of Systematic Musicology, pp. 1–24. Springer, Heidelberg (2018)
232. Schneider, A.: Tonhöhe, Skala, Klang. Akustische, tonometrische und psychoakustische Studien auf vergleichender Grundlage, [Pitch, Scale, Timbre. Acoustic, tonometric, and psychoacoustic studies on comparing foundations,]. Verlag für Systematische Musikwissenschaft, Bonn (1997)
233. Schofield, B.R.: Central descending auditory pathways. In: Kyugo, D.K., Fay, R.R., Popper, A.N. (eds.) Auditory and Vestibular Efferents. Springer, Springer Handbook of Auditory Research (2011)
234. Sethares, W.A.: Tuning, Timbre, Spectrum, Scale, Berlin. Springer (2004)
235. Shiller, R.J.: Narrative Economics. Princeton University Press (2019)
236. Simon, A.: Southeast Asia: musical syncretism and cultural identity. Fontes Artis Musicae **57**(1), 23–34 (2010)
237. Simon, A. (ed.): Das Berliner Phonogramm-Archiv 1900-2000. Sammlungen der traditionellen Musik der Welt, [The Berlin Phonogram Archive 1900-2000. Collections of traditional music of the world,] VWB, Berlin (2000)
238. Snyder, R.: Music and Memory. Bradford (2001)
239. Steptoe, A.: Mesmer, Mozart and Cosi Fan Tutte. Music Lett. **67**(3), 248–255 (1986)
240. Strogatz, S.H.: Nonlinear dynamics and chaos. Perseus Books Publications (1994)
241. Stumpf, C.: Tonpsychologie. [Tonepsychology]. Hirzel Verlag, Leipzig, 2 Bd., 1883–1890
242. Stumpf, C.: Die Sprachlaute. Experimentell-Phonetische Untersuchungen. Nebst einem Anhang über Instrumentenklänge, [The Speech Sounds. Experimental-Phonetic Investigations. With supplement on instrument sounds.], Verlag von Julius Springer, Berlin (1926)
243. Sunderberg, J.: The Science of the Singing Voice. Northern Illinois University Press (1989)
244. Suzuki, Miamoto, Y.: Resonance frequency changes of Japanese drum (nagado daiko) diaphragms due to temperature, humidity, and aging. Acoust. Sci. Tech. **33**(4), 277–278 (2012)
245. Szwed, J.: So What. The Life of Miles Davis, Cornerstone (2003)
246. Tan, Y.T., McPherson, G.E., Peretz, I., Berkovic, S.F., Wilson, S.J.: The genetic basis of music ability. Frontiers Psychol. **5**(658), 1–19 (2014)
247. Tenzer, M.: Gamelan Gong Kebyar. University of Chicago Press, The Art of Twentieth-Century Balinese Music (2000)
248. Thaut, M.: Rhythm, Music, and the Brain : Scientific Foundations and Clinical Applications. Routledge, NY (2005)
249. Thielemann, S.: The Darbhangā Tradition. Dhrupada in the school of Pandit Vidur Mallik. Indica Books (1997)

250. Tingey, L.: Auspicious music in a changing society: the Dam-ai musicians of Nepal. SOAS musicology series, School of Oriental and African Studies, London (1994)
251. Titze, I.R., Schmidt, S.S., Titze, M.R.: Phonation threshold pressure in a physical model of the vocal fold mucosa. J. Acoust. Soc. Am. **97**(5), 3080–3084 (1995)
252. Tokuda, I.T., Zemke, M., Kob, M., Herzel, H.: Biomechanical modeling of register transitions and the role of vocal tract resonators. J. Acoust. Soc. Am. **127**(3), 1528–1536 (2010)
253. Tompkin, D.: How to Wreck a Nice Beach. Stop Smiling Books (2011)
254. von Uexküll, J.J.B.: Theoretical Biology, Harcourt. Brace & Co, NY (1926)
255. Vogel, M.: Die Enharmonik der Griechen I. Tonsystem und Notation, [Enharmonic of the Greek I. Tonal System and Notation,] Verlag für Musikwissenschaft, Bad Godesberg (1963)
256. Vogel, M.: Die Enharmonik der Griechen II. Der Ursprung der Enharmonik, [Enharmonic of the Greek II. The Origin of Enharmonic,] Verlag für Musikwissenschaft, Bad Godesberg (1963)
257. Vogel, M.: Chiron, der Kentaur mit der Kithara, [Chiron, the centaur with the kithara,] Verlag für Musikwissenschaft, Bad Godesberg (1978)
258. Voss, R.F., Clarke, J.: 1/fnoise in music: music from 1/fnoise. J. Acoust. Soc. Am. **63**, 258–263 (1978)
259. Warneken, B.J.: Schubart. Ein unbürgerliches Leben, [Schubart. An uncivil life,] Eichborn (2009)
260. Wendt, G., Bader, R.: Analysis and perception of javanese gamelan tunings. In: Bader, R. (ed.): Computational Phonogram Archiving, Springer Series 'Current Research in Systematic Musicology', Vol. 5, pp. 129–144 (2019)
261. Wheeler, T.: American Guitars. An Illustrated History, Harper Resource (1992)
262. World Health Organization Regional Office For Europe (2018) Environmental Noise Guidelines for the European Region. http://www.euro.who.int/en/health-topics/environment-and-health/noise/environmental-noise-guidelines-for-the-european-region (2018)
263. Williamson, M.: Burmese Harp. Illinois, Nothern Illinois Monograph Series of Southeast Asia. Vol. 1 (2000)
264. Winer, J.A., Schreiner, Ch.E. (eds.): The Auditory Cortex. Springer (2011)
265. Winnington-Ingram, R.P.: The pentatonic tuning of the Greek Lyra. Theory Exami. Class. Quart. **6**(3/4), 169–186 (1956)
266. Wolpert, R.F.: Táng music theory of ritual calendrical transposition applied. Chime J. **18–19**, 67–82 (2013)
267. Woolf, N.J., Priel, A., Tuszynski, J.A.: Nanoneuroscience. Structural and Functional Roes of the Neuronal Cytoskeleton in Health and Disease. Springer (2010)
268. Woolf, N.J.: Cholinoceptive cells in rat cerebral cortex: somatodendritic immunoreactivity for muscarinic receptor and cytoskeletal proteins. J. Chem. Neuroanat. **6**(6), 375–90 (1993)
269. Wriggers, P.: Nichtlineare Finite-Element Methoden, [Nonlinear Finite-Element Methods]. Springer (2001)
270. Young, T.: Outlines of experiments and inquiries respecting sound and light. Philos. Trans. Royal Soc. Lond. **90**, 150–160 (1800)
271. Yang, M., De Coensel, B., Kang, J.: Presence of 1/f noise in the temporal structure of psychoacoustic parameters of natural and urban sounds. J. Acoust. Soc. Am. **138**(2), 916–927 (2015)
272. Zölzer, U.: DAFX—Digital Audio Effects. Wiley (2011)

Printed in the United States
by Baker & Taylor Publisher Services